Chemical Engineering License Review

Chemical Engineering License Review

Dilip K. Das
Rajaram K. Prabhudesai

2nd Edition

Engineering Press San Jose, CA

ISBN 1-57645-000-7
Library of Congress Cataloging in Publication Data
Das, Dilip K
Chemical Engineering License Review

Includes index
1. Chemical Engineering 1. Prabhudesai, Rajaram K.
II Title

TP 155.P75 1996 660.2 82-22890
ISBN 1-57645-000-7

Contents

Preface

With the current emphasis on professional licensing and registration of engineers, an increasing number of chemical engineers are taking the P.E. examinations. This book is written primarily for chemical engineers who want to prepare for the professional part of the P.E. examination. It is intended to fill their need for a concise book which will help them to review quickly the fundamentals of chemical engineering and to regain their confidence and facility in the application of those fundamentals in solving practical problems within the environment and time constraint of the P.E. examination. The book may also serve as a useful reference companion during the P.E. examination. In addition, it may be used as a valuable reference source by practicing engineers and as an excellent supplemental study aid by senior level chemical engineering students as well.

This book includes the important subjects of fluid mechanics, heat transfer, and unit operations such as distillation, absorption, extraction, humidification. Chapters on chemical engineering thermodynamics and chemical kinetics complete the coverage of the basic principles of chemical engineering. A chapter on engineering economics is included because of the importance given this subject in the P.E. examinations. Thus, in this single book, the chemical engineering candidate will have a ready access to all the essential material needed to prepare for the professional part of the P.E. examination.

The emphasis in this book is on recalling the chemical engineering concepts and methods and their application to the solution of a variety of problems. Therefore, the coverage of the fundamentals is brief, though the essential principles are not sacrificed. Numerous examples are given to illustrate the application of the basic principles to solving practical problems. Another feature of this book is that the problems solved in it are very general and are not restricted to the type of problems set in the past P.E. examinations. This is especially important because the pattern of the problems given in chemical engineering professional examinations undergoes considerable change every year. Therefore, the candidates' review for the P.E. examination should necessarily emphasize basic principles and their application to solve different types of problems.

The International System of Units (SI units) has now been adopted as the standard by most of the nations, including the United States and Great

Britain. This system will eventually replace the existing systems of units in common usage. Meanwhile, the use of the SI units is rightly receiving a progressively greater emphasis in schools and the P.E. Fundamentals of Engineering Examinations. However, in the professional examination, the emphasis is still on the fps units. The SI and the more familiar fps, British, and cgs systems of units are reviewed in the introductory chapter of this book. In addition, a few problems to illustrate the use of SI units are included in some chapters. However, a greater weight is given to fps and cgs units throughout the book because of their current extensive use in industrial practice and in the professional part of the P.E. examinations in the United States.

We have not added a separate nomenclature in each chapter. Instead, we have explained the notation in the text.

We wish good luck to our readers in their P.E. examinations and every success in their professional careers.

Dilip K. Das
Rajaram K. Prabhudesai

Acknowledgments

It is our privilege to acknowledge those people who assisted in the preparation of this book. We sincerely thank the engineering management of Stauffer Chemical Company for allowing us to use word processing and reproduction facilities in the preparation of the manuscript. We also are thankful to the management of Rhone-Poulenc Inc. for assistance in the revisions of the manuscript.

We thank all the readers, specially Mr. Bryon Brandt of UOP Technical Service for the suggestions to improve the book.

Our special thanks go to Mrs. Barbara Veals, who willingly and expertly typed the final manuscript. We also thank Jerri Volpe, Sue Garcia, and several other secretaries at Stauffer and Rhone-Poulenc who participated in the preparation of the preliminary draft.

Finally and most importantly, our sincere appreciation and thanks are due to Mrs. Malancha Das and Mrs. Vimal Prabhudesai for their patience, forbearance, and encouragement during the completion of this book.

Becoming A Professional Engineer

To achieve registration as a Professional Engineer there are four distinct steps: education, fundamentals of engineering (engineer-in-training) exam, professional experience, and finally, the professional engineer exam. These steps are described in the following sections.

Education

The obvious appropriate education is a B.S. degree in chemical engineering from an accredited college or university. This is not an absolute requirement. Alternative, but less acceptable, education is a B.S. degree in something other than chemical engineering, or from a non-accredited institution, or four years of education but no degree.

Fundamentals of Engineering (FE/EIT) Exam

Most people are required to take and pass this eight-hour multiple-choice examination. Different states call it by different names (Fundamentals of Engineering, E.I.T., or Intern Engineer) but the exam is the same in all states. It is prepared and graded by the National Council of Examiners for Engineering and Surveying (NCEES). Review materials for this exam are found in other books like Newnan: Engineer-In-Training License Review and Newnan and Larock: Engineer-In-Training Examination Review.

Experience

Typically one must have four years of acceptable experience before being permitted to take the Professional Engineer exam, but this requirement may vary from state to state. Both the length and character of the experience will be examined. It may, of course, take more than four years to acquire four years of acceptable experience.

Professional Engineer Exam

The second national exam is called Principles and Practice of Engineering by NCEES, but probably everyone else calls it the Professional Engineer or P.E. exam. All states, plus Guam, the District of Columbia, and Puerto Rico use the same NCEES exam.

Chemical Engineering Professional Engineer Exam Background

The reason for passing laws regulating the practice of chemical engineering is to protect the public from incompetent practioners. Beginning about 1907 the individual states began passing title acts regulating who could call themselves a chemical engineer. As the laws were strengthened, the practice of certain aspects of chemical engineering was limited to those who were registered chemical engineers, or working under the supervision of a registered chemical engineer. There is no national registration law; registration is based on individual state laws and is administered by boards of registration in each of the states.

Examination Development

Initially the states wrote their own examinations, but beginning in 1966 the NCEES took over the task for some of the states. Now the NCEES exams are used by all states. This greatly eases the ability of a chemical engineer to move from one state to another and achieve registration in the new state. About 1600 chemical engineers take the exam each year. As a result about 23% of all chemical engineers are registered professional engineers.

The development of the chemical engineering exam is the responsibility of the NCEES Committee on Examinations for Professional Engineers. The committee is composed of people from industry, consulting, and education, plus consultants and subject matter experts. The starting point for the exam is a chemical engineering task analysis survey that NCEES does at roughly five to ten year intervals. People in industry, consulting and education are surveyed to determine what chemical engineers do and what knowledge is needed. From this NCEES develops what they call a "matrix of knowledge" that form the basis for the chemical engineering exam structure described in the next section.

The actual exam questions are prepared by the NCEES committee members, subject matter experts, and other volunteers. All people participating must hold professional registration. Using workshop meetings and correspondence by mail, the questions are written and circulated for review. The problems relate to current professional situations. They are structured to quickly orient one to the requirements, so the examinee can judge whether he or she can successfully solve it. While based on an understanding of engineering fundamentals, the problems require the application of practical professional judgement and insight. While four hours are allowed for four problems, probably any problem can be solved in 20 minutes by a specialist in the field. A professionally competent applicant can solve the problem in no more than 45 minutes. Multi-part questions are arranged so the solution of each succeeding part does not depend on the correct solution of a prior part. Each part will have a single answer that is reasonable.

Examination Structure

The ten problems in the morning four-hour session are regular computation "essay" problems. In the afternoon four-hour session all ten problems are multiple choice. In each category (Thermodynamics, Process design, Mass transfer, and so on) about half of the problems will be in the morning session and half in the afternoon. Engineering economics may appear as a component within one or two of the problems.

Note: The examination is developed with problems that will require a variety of approaches and methodologies including design, analysis, application, economic aspects, and operations.

Taking The Exam

Exam Dates

The National Council of Examiners for Engineering and Surveying (NCEES) prepares Chemical Engineering Professional Engineer exams for use on a Friday in April and October each year. Some state boards administer the exam twice a year in their state, while others offer the exam once a year. The scheduled exam dates are:

	April	October
1996	19	25
1997	18	31
1998	24	30
1999	23	29
2000	14	27

EXAMINATION SPECIFICATIONS
NCEES PRINCIPLES AND PRACTICE EXAMINATION
IN THE DISCIPLINE OF
CHEMICAL ENGINEERING

EFFECTIVE BEGINNING WITH APRIL 1995 EXAMINATION

		Number of Problems
A.	**FLUID MECHANICS** piping network problems; pump sizing or pump performance; compressor sizing or compressor performance; control valve selection problems; fluid flow through beds; two-phase flow; Bernoulli equation applications.	3
B.	**HEAT TRANSFER** industrial heat transfer including but not limited to the following: heat exchanger design and performance; energy conservation; conduction, especially insulation problems; convection; radiation, especially furnace design; evaporation.	3
C.	**CHEMICAL KINETICS** interpretation of experimental data and reaction rate modeling; commercial reactor design from rate model and/or product distribution; comparison of reactor types; reaction control.	2
D.	**MASS AND ENERGY BALANCES** process stoichiometry and material balances; process energy balances; conservation laws.	4
E.	**MASS TRANSFER** typical applications including but not limited to the following: gas absorption and stripping; distillation; liquid-liquid extraction and leaching; humidification and dehumidification; drying.	3
F.	**PLANT DESIGN** process and equipment design including but not limited to the following: optimization of design; general safety considerations; environmental and waste treating; solids separation; vapor-liquid separations; flow sheets; HAZOP (hazard and operational) analysis; fault tree analysis; scheduling techniques; sizing and fabrication of equipment; material selection; life cycle cost; physical and chemical properties of matter; strength of materials; crystallographic structure; phase diagrams (metallurgical); latent heat; PVT data and relationships; molecular structure; sensors; transmitters and controllers; control loops; simulation.	3
G.	**THERMODYNAMICS** estimation and correlation of physical properties; chemical equilibrium; heats of reaction; application of first and second laws; vapor-liquid equilibrium; combustion; refrigeration.	2
	Total number of problems =	20

Note: The examination is developed with problems that will require a variety of approaches and methodologies including design, analysis, application, and operations. Some problems may include engineering economics analyses.

People seeking to take a particular exam must apply to their state board several months in advance.

Exam Procedure

Before the morning four-hour session begins, the proctors will pass out an exam booklet and solutions pamphlet to each examinee. There are likely to be civil, electrical, and mechanical engineers taking their own exams at the same time. You must solve four of the ten chemical engineering problems.

The solution pamphlet contains grid sheets on right-hand pages. Only work on these grid sheets will be graded. The left-hand pages are blank and are for scratch paper. The scratchwork will <u>not</u> be considered in the scoring.

If you finish more than 30 minutes early, you may turn in the booklets and leave. In the last 30 minutes, however, you must remain to the end to insure a quiet environment for all those still working, and to insure an orderly collection of materials.

The afternoon session will begin following a one-hour lunch break. The afternoon exam booklet will be distributed along with an answer sheet. The booklet will have ten 10-part multiple choice questions. You must select and solve four of them. An HB or #2 pencil is to be used to record your answers on the scoring sheet.

Exam-Taking Suggestions

People familiar with the psychology of exam-taking have several suggestions for people as they prepare to take an exam.

1. Exam taking is really two skills. One is the skill of illustrating knowledge that you know. The other is the skill of exam-taking. The first may be enhanced by a systematic review of the technical material. Exam-taking skills, on the other hand, may be improved by practice with similar problems presented in the exam format.

2. Since there is no deduction for guessing on the multiple choice problems, an answer should be given for all ten parts of the four selected problems. Even when one is going to guess, a logical approach is to attempt to first eliminate one or two of the five alternatives. If this can be done, the chance of selecting a correct answer obviously improves from 1 in 5 to, say, 1 in 3.

3. Plan ahead with a strategy. Which is your strongest area? Can you expect to see one or two problems in this area? What about your second strongest area? What will you do if you still must find problems in other areas?

4. Have a time plan. How much time are you going to allow yourself to initially go through the entire twelve problems and grade them in difficulty for you to solve them? Consider assigning a letter, like A, B, C and D, to each problem. If you allow 15 minutes for grading the problems, you might divide the remaining time into five parts of 45 minutes each. Thus 45 minutes would be scheduled for the first - and easiest - problem to be solved. Three additional 45 minute periods could be planned for the remaining three problems. Finally, the last 45 minutes would be in reserve. It could be used to switch to a substitute problem in case one of the selected problems proves too difficult. If that is unnecessary, the time can be used to check over the solutions of the four selected problems. A time plan is very important. It gives you the confidence of being in control, and at the same time keeps you from making the serious mistake of misallocation of time in the exam.

5. Read all five multiple choice answers before making a selection. The first answer in a multiple choice question is sometimes a plausible decoy - not the best answer.

6. Do not change an answer unless you are absolutely certain you have made a mistake. Your first reaction is likely to be correct.

7. Do not sit next to a friend, a window, or other potential distractions.

Exam Day Preparations

There is no doubt that the exam will be a stressful and tiring day. This will be no day to have unpleasant surprises. For this reason we suggest that an advance visit be made to the examination site. Try to determine such items as:

1. How much time should I allow for travel to the exam on that day? Plan to arrive about 15 minutes early. That way you will have ample time, but

not too much time. Arriving too early, and mingling with others who also are anxious, will increase your anxiety and nervousness.

2. Where will I park?

3. How does the exam site look? Will I have ample workspace? Where will I stack my reference materials? Will it be overly bright (sunglasses) or cold (sweater), or noisy (earplugs)? Would a cushion make the chair more comfortable?

4. Where is the drinking fountain, lavoratory facilities, payphone?

5. What about food? Should I take something along for energy in the exam? A bag lunch during the break probably makes sense.

What To Take To The Exam

The NCEES guidelines say you may bring the following reference materials and aids into the examination room for your personal use only:
1. Handbooks and textbooks
2. Bound reference materials, provided the materials are and remain bound during the entire examination. The NCEES defines "bound" as books or materials fastened securely in its cover by fasteners which penetrate all papers. Examples are ring binders, spiral binders and notebooks, plastic snap binders, brads, screw posts, and so on.
3. Battery operated, silent non-printing calculators.

At one time NCEES had a rule that did not permit "review publications directed principally toward sample questions and their solutions" in the exam room. This set the stage for restricting some kinds of publications from the exam. State boards may adopt the NCEES guidelines, or adopt either more or less restrictive rules. Thus an important step in preparing for the exam is to know what will - and will not - be permitted. We suggest that if possible you obtain a written copy of your state's policy for the specific exam you will be taking. Recently there has been considerable confusion at individual examination sites, so a copy of the exact applicable policy will not only allow you to carefully and correctly prepare your materials, but also will insure that the exam proctors will allow all proper materials.

As a general rule we recommend that you plan well in advance what books and materials you want to take to the exam. Then they should be obtained promptly so you use the same materials in your review that you will have in the exam.

License Review Books

There are two rules that we suggest you follow in selecting license review books to insure that you obtain up-to-date materials:

1. Consider the purchase only of materials that have a 1993 or more recent copyright. The exam used to be 25 essay questions, including a full scale engineering economics problem. The engineering economics problem is gone (at least as a separate question) and half the remaining 24 problems are now multiple choice.

2. Even if a license review book has a recent copyright date, is the content up to date? Books with older content probably will also lack the afternoon multiple choice problems.

Textbooks

If you still have your university textbooks, we think they are the ones you should use in the exam, unless they are too out of date. To a great extent the books will be like old friends with familiar notation.

Bound Reference Materials

The NCEES guidelines suggest that you can take any reference materials you wish, so long as you prepare them properly. You could, for example, prepare several volumes of bound reference materials with each volume intended to cover a particular category of problem. Use tabs so specific material can be located quickly. If you do a careful and systematic review of chemical engineering, and prepare a lot of well organized materials, you just may find that you are so well prepared that you will not have left anything of value at home.

Other Items

Calculator - NCEES says you may bring a battery operated, silent, non-printing calculator. You need to determine whether or not your state permits pre-programmed calculators. Extra batteries for your calculator are essential, and many people feel that a second calculator is also a very good idea.

Clock - You must have a time plan and a clock or wristwatch.

Pencils - You should consider mechanical pencils that you twist to advance the lead. This is no place to go running around to sharpen a pencil, and you surely do not want to drag along a pencil sharpener.

Eraser - Try a couple to decide what to bring along. You must be able to change answers on the multiple choice answer sheet, and that means a good eraser. Similarly you will want to make corrections in the essay problem calculations.

Exam Assignment Paperwork - Take along the letter assigning you to the exam at the specified location. To prove you are the correct person, also bring something with your name and picture.

Items Suggested By Advance Visit - If you visit the exam site you probably will discover an item or two that you need to add to your list.

Clothes - Plan to wear comfortable clothes. You probably will do better if you are slightly cool.

Box For Everything - You need to be able to carry all your materials to the exam and have them conveniently organized at your side. Probably a cardboard box is the answer.

Exam Scoring

Essay Questions

The exam booklets are returned to Clemson, SC. There the four essay question solutions are removed from the morning workbook. Each problem is sent to one of many scorers throughout the country.

For each question an item specific scoring plan is created with six possible scores: 0, 2, 4, 6, 8, and 10 points. For each score the scoring plan defines the level of knowledge exhibited by the applicant. An applicant who is minimally qualified in the topic is assigned a score of 6 points. The scoring plan shows exactly what is required to achieve the 6 point score. Similar detailed scoring criteria are developed for the two levels of superior performance (8 and 10 points) and the three levels of inferior performance (0, 2, and 4 points). Every essay problem submitted for grading receives one of these six scores. The scoring criteria may be based on positive factors, like identifying the correct computation approach, or negative factors, like improper assumptions or calculation errors, or a mixture of both positive and negative factors. After scoring, the graded materials are returned to NCEES, which reassembles the applicants work and tabulates the scores.

Multiple Choice Questions

Each of the four multiple choice problems is 10 points, with each of the ten questions of the problem worth one point. The questions are machine scored by scanning. The input data are evaluated by computer programs to do error checking. Marking two answers to a question, for example, would be detected and no credit given. In addition, the programs identify those questions with statistically unlikely results. There is, of course, a possibility that one or more of the questions is in some way faulty. In that case a decision will be made by subject matter experts on how the situation should be handled.

Passing The Exam

In the exam you must answer eight problems, each worth 10 points, for a total raw score of 80 points. Since the minimally qualified applicant is assumed to average six points per problem, a raw score of 48 points is set equal to a converted passing score of 70. Stated bluntly, you must get 48 of the 80 possible points to pass. The converted scores are reported to the individual state boards in about two months, along with the recommended pass or fail status of each applicant. The state board is the final authority of whether an applicant has passed or failed the exam.

Although there is some variation from exam to exam, the following gives the approximate passing rates:

Applicant's Degree	Percent Passing Exam
Engineering from accredited school	62%
Engineering from non-accredited school	50
Engineering Technology from accredited school	42
Engineering Technology from non-accredited school	33
Non-Graduates	36
All Applicants	56

Although you want to pass the exam on your first attempt, you should recognize that if necessary you can always apply and take it again.

This Book

This book is organized to cover the chemical engineering professional engineer (principles and practice) exam. The books contains:

* A review of each topic on the exam.
* Example problems to illustrate the topic discussion along with solutions.

Each chapter begins with a review of the particular topic including example problems where appropriate. NCEES does not allow their problems to be reproduced, so none of the problems in this book came from them. Each one is structured to approximate the scope and difficulty of the actual exam problems you will encounter. The National Council of Examiners for Engineering and Surveying (NCEES), which prepares the chemical engineering examination, calls it an open book examination. Most states accept this and allow applicants to bring textbooks, handbooks and any bound reference materials to the exam. A few states, however, do not permit review publications directed principally toward sample problems and their solutions.

Chapter

1

Units and Dimensions

With the adoption of the International System of Units (SI units) as the standard by most nations, including the United States and United Kingdom, there has been a progressively greater emphasis on the use of SI units. Besides knowing the customary fps (engineering) and cgs units, candidates for the P.E. examination should familiarize themselves with the SI units. The following discussion provides a short review of the different systems of units and their conversions into one another.

Dimensions

A physical quantity consists of two parts: (1) a unit which indicates the dimension and gives its standard of measurement and (2) a quantity which gives the numerical value of the units. For example, the distance between two points has the dimension of length and can be expressed as 3 feet. Here the foot is the unit of the length dimension, and 3 is the numerical value of the corresponding number of length units.

Physical properties of a system are interrelated by mechanical and physical laws. Such relationships allow certain dimensions to be regarded as basic and others as derived. The dimensions chosen as basic vary from one system to another, but it is usual to treat length L, time T, and mass M as basic. In some systems, force F or both mass M and force F are used as the basic dimensions.

Systems of Units

The systems of units which chemical engineers are usually required to use in their work are described next.

Centimeter-Gram-Second (cgs) System. The basic dimensions in this system are length L, mass M, and time T. The units and nomenclature are as follows:

Dimension	Dimension Symbol	Unit
Length	L	Centimeter (cm)
Mass	M	Gram (g)
Time	T	Second (s)

The unit of force in this system is the dyne, defined as the force which imparts an acceleration of 1 cm/s^2 to a mass of 1 g. Thus

$$\text{Force } F = MLT^{-2} \qquad \text{unit one dyne (1 dyn)}$$

British Engineering System. In this system, the foot and second are the units of length and time, but the third fundamental unit is the pound-force. The pound-force is defined as that force which gives an acceleration of 32.17 ft/s^2 to a mass of one pound. This is a fixed quantity and is not the same as the pound-weight which is the force exerted by the gravitational field on a mass of one pound. The pound weight is a variable quantity because of the variation of the gravitational field strength from place to place relative to the center of the earth.

The unit of mass in the British system is the slug. This is the mass to which an acceleration of 1 ft/s^2 is given by 1 pound-force.

$$1 \text{ pound-force} = (1 \text{ slug})\,(1 \text{ ft/s}^2)$$

or

$$1 \text{ slug} = \frac{1 \text{ pound-force}}{1 \text{ ft/s}^2}$$

Foot-Pound-Second (fps) System. The basic dimensions and units of this system are

Dimension	Dimension Symbol	Unit
Length	L	Foot (ft)
Mass	M	Pound (lb)
Time	T	Second (s)

The poundal, which is the unit of force in this system, is that force which gives an acceleration of 1 ft/s^2 to a mass of 1 lb. Thus

$$1 \text{ poundal} = (1 \text{ lb-mass})\,(1 \text{ ft/s}^2)$$

A comparison of the British and fps systems shows that 1 slug = 32.17 lb-mass

and

$$1 \text{ lb-force} = 32.17 \text{ poundals}$$

Confusion between 1 lb-mass and 1 lb-force can be avoided by writing 1 lb-mass as 1 lbm or simply 1 lb* and 1 pound force as 1 lbf. In the United States, the pound-mass and the pound-force are both used as the basic units.

In this system a proportionality factor g_c between the force and the mass is defined:

$$\text{Force (lbf)} = \frac{[\text{mass (lb)}][\text{acceleration (ft/s}^2)]}{g_c}$$

*Hereafter 1 lb-mass will be denoted by 1 lb and 1 lb-force by 1 lbf in this book.

For the purposes of the gravitational force unit Newton's law proportionality factor g_c is taken as*

$$g_c = 32.17 \text{ lb} \cdot \text{ft/s}^2 \cdot \text{lbf}$$

SI Units (Système International). The basic dimensions and units of this system are as follows.

Dimension	Dimension Symbol	Unit
Length	L	Meter (m)
Mass	M	Kilogram (kg)
Time	T	Second (s)

The unit of force in this system is the newton, defined as that force which gives an acceleration of 1 m/s^2 to a mass of 1 kg. Thus

$$1 \text{ N} = (1 \text{ kg})\ (1 \text{ m/s}^2) = 10^5 \text{ dyn}$$

For practical use, the multiples or submultiples[1,3] of the selected SI unit are convenient. These are given in Table 1-1.

Thermal Units. For convenience, thermal energy is expressed in terms of the mass, temperature (introduced as basic unit), and a proportionality constant, which is the specific heat of the material. Since the specific heat varies from material to material, the heat quantities are expressed in terms of the specific heat of water at 15°C (288 K). In the cgs system, one calorie is the amount of heat which will raise the temperature of one gram of water through one Celsius degree at atmospheric pressure.

In the British and fps systems, the British thermal unit (Btu) is the quantity of heat required to raise the temperature of one

*In solving problems in this book, a rounded value of $g_c = 32.2$ will be used frequently.

TABLE 1-1 Prefixes and Symbols for the Multiples and Submultiples of SI Units

Multiple	SI Prefix	Symbol	Submultiple	SI Prefix	Symbol
10^{18}	exa	E	10^{-1}	deci	d
10^{15}	peta	P	10^{-2}	centi	c
10^{12}	tera	T	10^{-3}	milli	m
10^{9}	giga	G	10^{-6}	micro	μ
10^{6}	mega	M	10^{-9}	nano	n
10^{3}	kilo	k	10^{-12}	pico	p
10^{2}	hecto	h	10^{-15}	femto	f
10	deka	da	10^{-18}	atto	a

pound of water through one Fahrenheit degree (from 60 to 61°F) and the pound-calorie or Celsius heat unit is the heat quantity required to raise one pound of water (at 60°F) through one Celsius degree.

The calorie and kilocalorie are redefined in terms of the joule as

$$1 \text{ calorie} = 4.1868 \text{ J} \qquad 1 \text{ kilocalorie} = 4186.8 \text{ J}$$

The specific-heat values for water in various units are

cgs: $1 \text{ cal/g} \cdot {}^\circ\text{C}$

fps: $1 \text{ Btu/lb} \cdot {}^\circ\text{F}$

SI: $4186.8 \text{ J/kg} \cdot \text{K}$

A thermochemical calorie, as sometimes used in chemistry, is given by

$$1 \text{ cal (thermochemical)} = 4.184 \times 10^7 \text{ erg} = 4.1840 \text{ J}$$

Pressure Units. The unit of pressure in the SI system is the newton per square meter and is termed *pascal* (Pa). The pascal is a very small unit and is related to the *bar* as follows:

$$1 \text{ bar} = 1 \times 10^5 \text{ Pa} = 1 \times 10^5 \text{ N/m}^2$$

A convenient unit for pressure that is used in all the systems is the atmosphere; this is defined by

$$1 \text{ atm} = 1.01325 \times 10^5 \text{ Pa} = 1.01325 \times 10^5 \text{ N/m}^2$$

Work, Energy, and Power. In SI units, work and energy are measured in joules (J),

$$1 \text{ J} = 1 \text{ N} \cdot \text{m} = 1 \text{ kg} \cdot \text{m}^2/\text{s}^2$$

Power is measured in watts; watts are defined as joules per second (J/s).

Molar Units. In problems involving chemical reactions, it is often convenient to work in molar units. A gram-mole of a substance is its molecular weight expressed in grams. The kilogram-mole is the molecular weight in kilograms. In a similar manner, the pound-mole (lb · mol) is the molecular weight expressed in pounds.

Conversion of Units. The conversion of units from one system to another is made by expressing the quantities in terms of the fundamental units (length, mass, time, temperature) and using the appropriate basic conversion factors as follows:

Mass: $$1 \text{ kg} = 2.2046 \text{ lb}$$

$$1 \text{ lb} = \frac{1}{32.17} \text{ slug} = 453.6 \text{ g} = 0.4536 \text{ kg}$$

Length: $$1 \text{ ft} = 30.48 \text{ cm} = 0.3048 \text{ m}$$

$$1 \text{ m} = 3.2808 \text{ ft} \doteq 3.281 \text{ ft}$$

Time: $$1 \text{ s} = \tfrac{1}{3600} \text{ h}$$

Temperature difference: $$1°\text{F} = \frac{1}{1.8}°\text{C} = \frac{1}{1.8} \text{ K}$$

Force: $1 \text{ lbf} = 32.17 \text{ poundal} = 4.44 \times 10^5 \text{ dyn} = 4.44 \text{ N}$

The conversion of units from one system to another is illustrated by the following example.

Example 1-1. By using the basic defined dimensions and constants, calculate the conversion factors to convert **(a)** newtons to pound-force, and **(b)** atmospheres to pounds per square inch (psi).

Solution.

a. By definition, $1\ \text{N} = 1\ \text{kg}\cdot\text{m/s}^2$; so

$$1\ \text{N} = \frac{(1\ \text{kg})(2.2046\ \text{lb/kg})(1\ \text{m})(3.281\ \text{ft/m})}{\text{s}^2}$$

$$= 7.233\ \text{lb}\cdot\text{ft/s}^2$$

Now, by definition, $1\ \text{lb}\cdot\text{ft/s}^2 = (1/32.17)\ \text{lbf}$. Therefore,

$$1\ \text{N} = 7.233\ \text{lb}\cdot\text{ft/s}^2$$

$$= 7.233\ \frac{1}{32.17}\ \text{lbf}$$

$$= 0.2248\ \text{lbf}$$

b. A standard atmosphere is defined by

$$1\ \text{atm} = 1.01325 \times 10^5(\text{kg}\cdot\text{m/s}^2)/\text{m}^2$$

$$= \frac{1.01325 \times 10^5[(2.2046\ \text{lb})(3.281\ \text{ft})/\text{s}^2]}{(3.281\ \text{ft})^2}$$

$$= 6.8083 \times 10^4(\text{lb}\cdot\text{ft/s}^2)/\text{ft}^2$$

$$= \frac{6.8083 \times 10^4(\text{lb}\cdot\text{ft/s}^2)/\text{ft}^2}{g_c(\text{lb}\cdot\text{ft/s}^2\cdot\text{lbf})}$$

$$= \frac{6.8083 \times 10^4(\text{lbf}/32.17)}{144\ \text{in}^2}$$

$$= 14.696\ \text{lbf/in}^2 \doteq 14.7\ \text{psi}$$

Conversion factors can be derived in the above manner for the basic units and dimensions. It is, however, more convenient to use a table of already derived conversions. Some of the commonly required conversion factors are listed in Tables 1-2 and 1-3. For a more detailed discussion of units and dimensions, see Coulson and Richardson[1] or McCabe and Smith.[2] Additional conversion factors are given by Perry.[3]

TABLE 1-2 Conversion Factors for SI Units

To convert	To	Multiply by
Length		
in	mm	25.4
ft	m	0.3048
m	ft	3.2808
mi	km	1.609
Mass		
lb	kg	0.4536
kg	lb	2.2046
ton (British)	kg	1016
ton (metric)	kg	1000
ton (U.S.)	kg	907.2
Area		
ft^2	m^2	0.093
m^2	ft^2	10.765
Volume		
ft^3	m^3	0.0283
m^3	ft^3	35.3147
m^3	U.S. gal	264.17
Force		
poundal	N	0.138
lbf	N	4.45
dyn	N	1×10^{-5}
Pressure		
lbf/in^2 (psi)	kN/m^2	6.895
atm	kN/m^2	101.3
bar	kN/m^2	100
ftH_2O	kN/m^2	2.99
inH_2O	N/m^2	249
inHg	kN/m^2	3.39
mmHg	N/m^2	133

To convert	To	Multiply by
Energy		
Btu	kJ	1.055
cal	J	4.1868
erg	J	10^{-7}
ft · lbf	J	1.36
kWh	MJ	3.6
Therm	MJ	105.5
Power		
Btu/h	W	0.293
Btu/s	W	1054.4
hp	W	746
Thermal Energy		
Btu	cal	252
	ft · lbf	778.2
	J	1055.1
Btu/lb	cal/g	0.55556
	kJ/kg	2.326
Btu/lb · °F	cal/g · °C	1.0
	J/kg · K	4186.8
Btu/h · ft^2	W/m^2	3.1546
Btu/h · ft^2 · °F	W/m^2 · K	5.674
Btu/(h · ft^2 · °F/ft)	W/(m^2 · K/m)	1.7307
Btu/ft^3	kJ/m^3	37.26
Density		
lb/ft^3	kg/m^3	16.02
Other Factors		
kWh	Btu	3412.1
hp	Btu/h	2545.1
	kW	0.746
ft^3	U.S. gal	7.48
ft^3/s	U.S. gal/min	448.8
Celsius heat unit or pound-calorie	Btu	1.8

TABLE 1-3 Units of Absolute Viscosity*

Centipoise or 0.01 g/cm · s or 0.01 poise	Slug/ft · s or lbf · s/ft²	lb/ft · s or poundal · s/ft²	Pa · s or kg/m · s
1	2.09×10^{-5}	0.000672†	0.001
47,900	1	32.2 or g	47.9
1,487	$\frac{1}{g}$ or 0.0311	1	1.487
1,000	0.02088	0.672	1

Units of Kinematic Viscosity‡

Centistokes	ft^2/s	m^2/s
1	1.075×10^{-5}	1×10^{-6}
92,900	1.0	0.0929
1×10^{6}	10.7643	1

†Another conversion factor: 1 centipoise = 2.42 lb/ft · h.

‡ν = kinematic viscosity = absolute viscosity/density = μ/ρ.

*Example: To convert 2.5 cP into pascal-seconds, we note from the table that 1 cP = 0.001 Pa · s. Hence, 2.5 cP = 0.0025 Pa · s = 2.5 mPa · s.

References

1. J. M. Coulson and J. F. Richardson, *SI units*, vol. 1 of *Chemical Engineering*, 3d ed., Pergamon, New York, 1977, pp. 1–13.
2. W. L. McCabe and J. C. Smith, *Unit Operations of Chemical Engineering*, 3d ed., McGraw-Hill, New York, 1976, pp. 1–13.
3. *Chemical Engineers' Handbook*, 5th ed., R. H. Perry (ed.), McGraw-Hill, New York, 1973, pp. 1-24 to 1-31, and inside of back cover.

Chapter

2

Fluid Mechanics

The following topics are reviewed in this chapter: fluid statics and pressure measurement, flow measurement, Bernoulli's theorem, and fluid flow and transportation.

Viscosity

Viscosity is a measure of the resistance of a fluid to shear or angular deformation. The shear stress between two adjacent layers of a newtonian fluid varies as the rate of shear at constant temperature and pressure:

$$\tau \propto \frac{du}{dy} \qquad \text{hence} \qquad \tau = \frac{\mu}{g_c}\frac{du}{dy} \tag{2-1}$$

where τ = shear stress, lbf/ft^2
g_c = Newton's law proportionality factor, $\text{lb}\cdot\text{ft/s}^2\cdot\text{lbf}$
u = linear velocity, ft/s
y = distance perpendicular to the direction of flow, ft
μ = viscosity, $\text{lb/ft}\cdot\text{s}$

The conversion factors for viscosity from one system of units into the others are given in Table 1-3.

Fluid Statics

Static Head. The static head of a fluid is the pressure at a given point in a fluid that is exerted by a vertical column of the fluid above that point. Since liquids are more or less incompressible,

$$P_h = \frac{h\rho g}{g_c} \tag{2-2}$$

where P_h = static pressure, lbf/ft^2
h = head of liquid above the point, ft = static head
ρ = liquid density, lb/ft^3
g = local acceleration due to gravity, ft/s^2
g_c = 32.17 $\text{lb} \cdot \text{ft/lbf} \cdot \text{s}^2$, dimensional constant

When the value of $g \doteq g_c = 32.2\ \text{ft/s}^2$, substituting $\rho = s(62.4)$ in Eq. (2-2) gives

$$P_h = h(62.4s)\ \text{lbf/ft}^2 = \frac{h(62.4s)}{144}\ \text{psi} = \frac{hs}{2.31}\ \text{psi} = 0.433hs\ \text{psi}$$

where s is the specific gravity of the fluid with respect to water.

Liquid Manometers. The height of a liquid in an open tube connected to a pressure source (e.g., a liquid in a vessel) is a direct measure of the static pressure at the point of connection. The tubes may be straight or U-type. The pressure-measuring fluid may be different from the fluid whose pressure is to be measured. Both the open and differential tube gauges are used. The height-measuring fluid must be immiscible with the test fluid.

In an open manometer of the type shown in Fig. 2-1*a*, the pressure at point *A* (connecting point) is

$$P_A = (h_m\rho_m - h_A\rho_\ell)\frac{g}{g_c} = h_m\rho_m - h_A\rho_\ell \qquad \text{since } g \doteq g_c \tag{2-3}$$

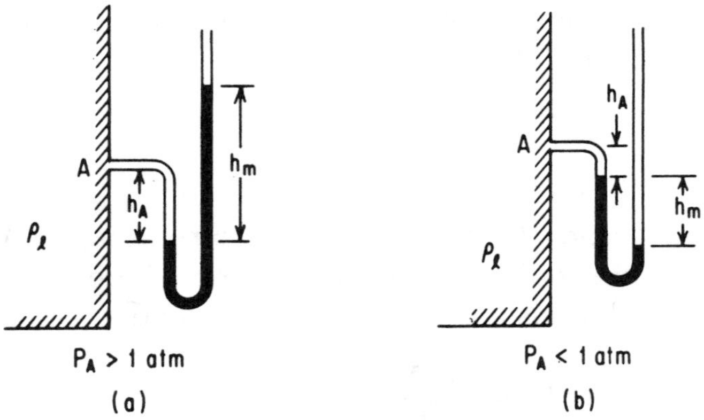

Figure 2-1. Open manometers.

where h_m = differential height of manometric fluid, ft
h_A = distance between point of attachment and the interface between the test fluid and the manometric fluid, ft
ρ_m = density of manometric fluid, lb/ft^3
ρ_ℓ = density of test fluid, lb/ft^3

Therefore the head at point A is

$$h_\ell = h_m\left(\frac{\rho_m}{\rho_\ell}\right) - h_A \qquad \text{ft} \tag{2-4}$$

In the case of gases, ρ_ℓ is so small that the terms containing h_A may be neglected in the above equations. In the case of the differential U-tube manometer shown in Fig. 2-2*b*,

$$P_A - P_B = [h_m(\rho_m - \rho_A) + h_A\rho_A - h_B\rho_B]\frac{g}{g_c} \tag{2-5}$$

where ρ_A and ρ_B are the densities at points A and B.

As a special case when $\rho_A = \rho_B = \rho$, $h_A = h_B$, and s_m and s are the specific gravities of the manometric fluid and process fluid, respectively, Eq. (2-5) reduces to

$$\frac{P_A - P_B}{\rho} = \Delta H = h_m\left(\frac{s_m}{s} - 1\right) \tag{2-6}$$

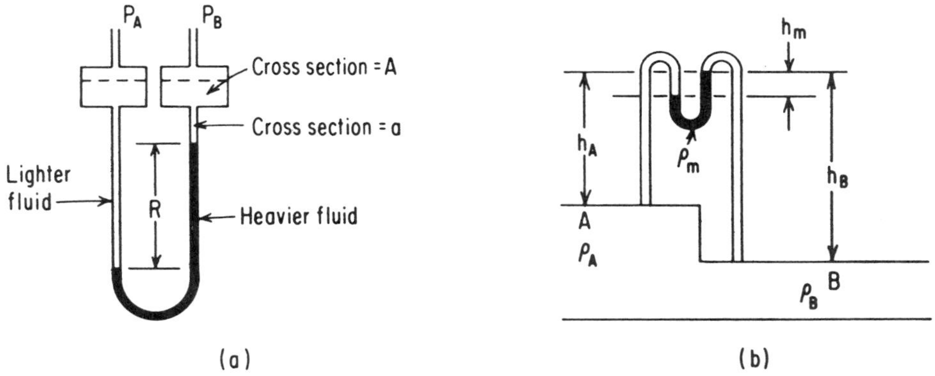

Figure 2-2. U-tube manometers: (*a*) two-fluid; (*b*) differential.

The diameter of a manometer tube should be at least 0.5 in to avoid capillary error. For U tubes, small diameters are permissible because the capillary rises in the two legs tend to compensate for each other.

For a simple manometer (Fig. 2-3*a*), the following relation can be established:

$$\Delta P = P_A - P_B = h_m(\rho_A - \rho_B)\frac{g}{g_c} \tag{2-7}$$

To obtain a better accuracy in measurements, multiplying manometers are used. To multiply the reading of height, a fluid of lower density is used in an open tube. In differential manometers, the difference between the density of the manometric fluid and that of the test fluid should be as small as possible. In an inclined manometer (Fig. 2-3*b*), the fluid head is given by

$$h_m = R \sin \theta \tag{2-8}$$

where R is the distance between the two meniscuses along the tube and θ is the angle of inclination of the manometer in radians.

A draft gauge (Fig. 2-3*c*) consists of an inclined tube connected to a reservoir. If the gauge liquid is other than water, the reading must be multiplied by ρ/ρ_w, where ρ_w is the density of water and ρ is the density of gauge fluid.

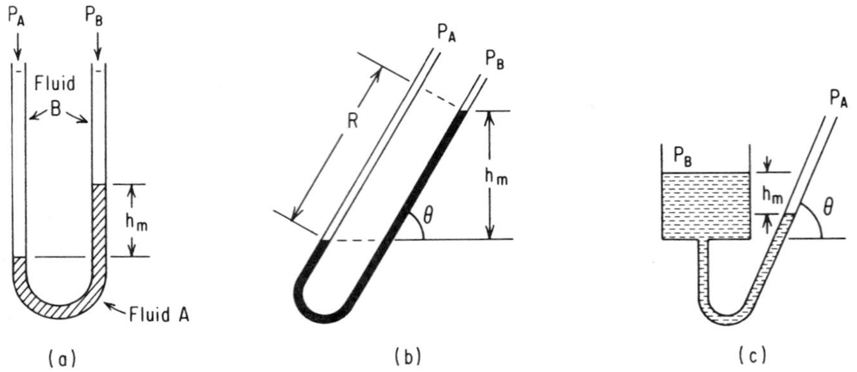

Figure 2-3. (*a*) Simple manometer, (*b*) inclined manometer, (*c*) draft gauge.

The two-fluid U-tube manometer (Fig. 2-2*a*) is very useful in measuring small heads of gases. In this case, the differential pressure is given by

$$\Delta P = P_A - P_B = (R - R_0)\left(\rho_2 - \rho_1 + \frac{a}{A}\rho_1\right)\left(\frac{g}{g_c}\right) \qquad \text{lbf/ft}^2 \qquad (2\text{-}9)$$

where R = reading of heavier manometric fluid, ft
R_0 = reading of heavier manometric fluid, ft, when $P_A = P_B$
ρ_1 = density of lighter fluid, lb/ft^3
ρ_2 = density of heavier fluid, lb/ft^3

If A/a is very large, the term $(a/A)\rho_1$ is negligible, but it may not be omitted without a check.

Bernoulli's Theorem

Applying the principle of conservation of energy (first law of thermodynamics) to a flowing fluid (Fig. 2-4), the following Bernoulli equation is obtained for the steady flow of a fluid:

$$Z_B\left(\frac{g}{g_c}\right) + \frac{u_B^2}{2g_c} + \frac{P_B}{\rho_B} = Z_A\left(\frac{g}{g_c}\right) + \frac{u_A^2}{2g_c} + \frac{P_A}{\rho_A} - F + W \qquad (2\text{-}10)$$

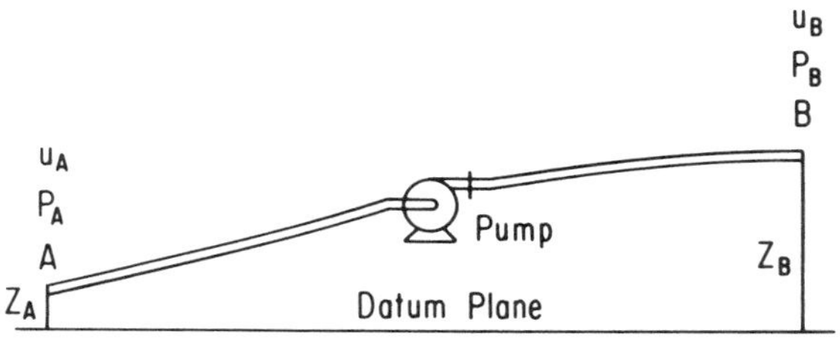

Figure 2-4. Illustration for the Bernoulli equation.

where u_A, u_B = velocities of fluid, ft/s, at A and B, respectively
Z_A, Z_B = liquid heights at A and B with respect to a datum plane, ft
P_A, P_B = pressures at A and B, lbf/ft^2
ρ_A, ρ_B = densities of fluids at A and B, lb/ft^3
$u^2/2g_c$ = kinetic head, ft · lbf/lb
F = energy loss due to friction, ft · lbf/lb
W = work done by the pump on the fluid ft · lbf/lb

The various terms in the Bernoulli equation must be expressed in the same units as energy (ft · lbf/lb). Normally $g/g_c \doteq 1$, and therefore g/g_c can be dropped from the terms containing Z.

Example 2-1. As shown in Fig. 2-5, water flows through the pipe. If the contraction loss is half a velocity head based on the velocity at B, calculate the velocity in feet per second and diameter in inches at B. Neglect the pressure drop due to friction.

Solution. Apply the Bernoulli equation between the points A and B. There is no pump in the line; therefore $W = 0$. The frictional drop is given to be negligible. Hence

$$Z_A + \frac{u_A^2}{2g_c} + \frac{P_A}{\rho} = Z_B + \frac{u_B^2}{2g_c} + \frac{P_B}{\rho} + \frac{0.5u_B^2}{2g_c}$$

$$Z_A = 0 \qquad \text{(taken as datum plane)}$$

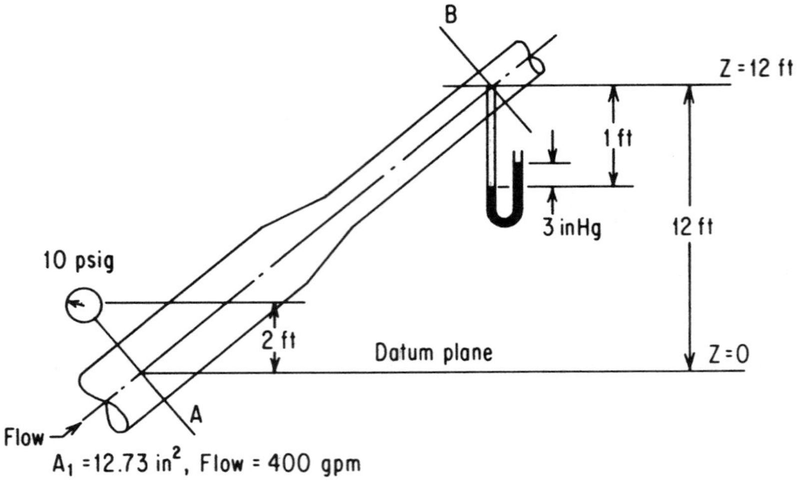

Figure 2-5. Data illustration for Example 2-1.

The cross-sectional area at point A is computed to be

$$\frac{12.73 \text{ in}^2}{144 \text{ in}^2/\text{ft}^2} = 0.0884 \text{ ft}^2$$

$$\text{Flow} = 400 \text{ gpm}$$

$$u_A = \frac{400 \text{ gpm}}{(7.48 \text{ gal/ft}^3)(60 \text{ s/min})} \frac{1}{0.0884 \text{ ft}^2} = 10.08 \text{ ft/s}$$

$$P_A = 10 \text{ psig} + 2 \text{ ft}$$

$$= 24.7(2.31) + 2 \text{ ft} = 59.06 \text{ ft}$$

$$Z_B = 12 \text{ ft} \qquad u_B = \text{required}$$

$$3 \text{ inHg} = 13.6(\tfrac{3}{12}) = 3.4 \text{ ftH}_2\text{O}$$

From fluid statics, the pressure head at point B is equal to

$$P_B = 3 \text{ inHg} + 14.7 \text{ psi} - 1 \text{ ftH}_2\text{O}$$

$$= 3.4 + 33.96 - 1 = 36.36 \text{ ft}$$

Take $g_c = 32.17 \doteq 32.2$ for calculation. Substitution in the Bernoulli equation gives

$$0 + \frac{10.08^2}{64.4} + 59.06 = 12 + \frac{1.5u_B^2}{64.4} + 36.36$$

$$\frac{1.5u_B^2}{64.4} = 12.28 \qquad u_B = 22.96 \text{ ft/s}$$

$$\frac{\text{Volume}}{\text{Second}} = \frac{400}{60(7.48)} = 0.8913 \text{ ft}^3\text{/s}$$

$$\text{Cross section at } B = \frac{0.8913}{22.96} = 0.0388 \text{ ft}^2 = 5.59 \text{ in}^2$$

$$\text{Diameter at } B = \left(\frac{5.59}{0.785}\right)^{0.5} = 2.67 \text{ in}$$

Fluid Measurements

Static Pressure. The static pressure is measured by a piezometer opening or a pressure tap. The piezometer opening in the side of the conduit should be normal to and flush with the surface. A *piezometer ring* is a manifold into which are connected several sidewall static taps. Its advantages are: (1) it gives average pressure; (2) it reduces the possibility of completely plugging all the static openings. The specifications for pressure tap holes are given by Perry.[1a]

Local Velocities, Pitot Tubes. Consider a Pitot tube of the type shown in Fig. 2-6*a*. Apply the Bernoulli equation to the points *A* and *B*:

$$\frac{u_A^2}{2g_c} + \frac{P_A}{\rho_A} + Z_A + W - F = \frac{u_B^2}{2g_c} + \frac{P_B}{\rho_B} + Z_B$$

In the above equation, $Z_A \doteq Z_B$, $u_A = 0$, $W = 0$, $F = 0$. Therefore

$$u_B^2 = 2g_c\left(\frac{P_A}{\rho_A} - \frac{P_B}{\rho_B}\right) \qquad (2\text{-}11)$$

or

$$u_B = C\sqrt{2g_c\left(\frac{P_A}{\rho_A} - \frac{P_B}{\rho_B}\right)}$$

The constant *C* is introduced in the above equation to account

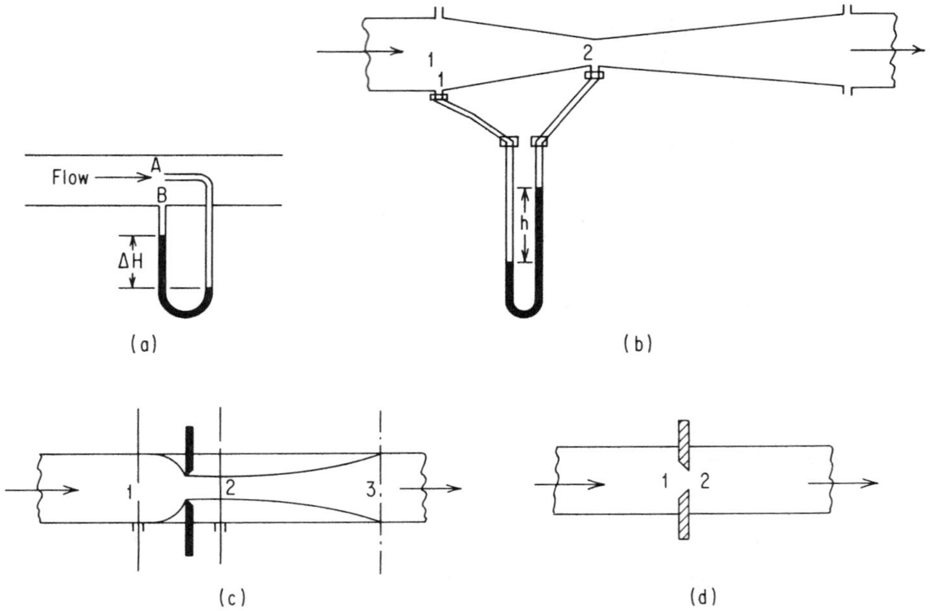

Figure 2-6. (*a*) Pitot tube, (*b*) venturimeter, (*c*) sharp-edged orifice, (*d*) round-edged orifice.

for the fluctuations in the velocity which cause errors in the measurement. For the incompressible fluids, $\rho_A = \rho_B = \rho$, and the above equation becomes

$$u_B = C\sqrt{2g\ \Delta H} = C\sqrt{2g\ \frac{P_i - P_0}{\rho}} \qquad (2\text{-}11a)$$

where u_B = local velocity at the point where tip is located, ft/s
C = correction coefficient, dimensionless
ΔH = differential pressure, ft of liquid
$P_A = P_i$ = impact pressure, lbf/ft^2
$P_B = P_0$ = local static pressure, lbf/ft^2
ρ = density of fluid, lb/ft^3

The coefficient $C = 1 \pm 0.01$ for simple Pitot tubes and $C = 0.98$ to 1 for Pitot static tubes.

For gases at velocities > 200 ft/s, the compressibility is important and the following equation is to be used:

$$u_0 = C\left\{\frac{2g_c k}{k-1}\frac{P_0}{\rho}\left[\left(\frac{P_i}{P_0}\right)^{(k-1)/k} - 1\right]\right\}^{1/2} \tag{2-12}$$

where k is the ratio of specific heats of the gases.

Venturimeters. A venturimeter (Fig. 2-6*b*) consists of a tube with a constricted throat which increases the velocity of the fluid at the expense of pressure. The constriction is followed by a gradually diverging portion where the velocity is decreased with an increase in the pressure accompanied by slight friction losses.

Writing the Bernoulli equation for an ideal case (no friction losses) gives

$$\frac{P_1}{\rho_1} + Z_1 + \frac{u_1^2}{2g_c} = \frac{P_2}{\rho_2} + \frac{u_2^2}{2g_c} + Z_2 \qquad \text{and} \qquad Z_1 = Z_2$$

By the continuity equation, $u_1 = (A_2/A_1)u_2$. Then ideal u_2 becomes

$$\left[\frac{2g_c(P_1/\rho_1 - P_2/\rho_2)}{1-(A_2/A_1)^2}\right]^{1/2} \tag{2-13}$$

where A_1 and A_2 are the cross sections at points 1 and 2, respectively. Because of the frictional losses, the actual velocity will be smaller, and therefore a discharge coefficient is introduced in the above equation. In addition, if ρ is constant, the above equation reduces to

$$u_2 = C\left[\frac{2g\,(P_1 - P_2)/\rho}{1-(A_2/A_1)^2}\right]^{1/2}$$

$$= C\sqrt{\frac{2g\ \Delta H}{1-\beta^4}} \tag{2-14}$$

where $\beta^2 = A_2/A_1 = D_2^2/D_1^2$

D_1, D_2 = diameters at points 1 and 2, respectively

ΔH = differential head, ft

The volumetric flow rate through a venturi throat is given by

$$Q = CYA_T\sqrt{\frac{2g\ \Delta H}{1-\beta^4}} \qquad \text{ft}^3/\text{s} \tag{2-15}$$

where Y is an expansion factor and A_T is the throat area, ft^2. For the flow of liquids, $Y = 1.0$.

The change in the potential energy for the inclined pipes needs to be allowed for. The flow equation is then to be modified to

$$Q = CA_T\sqrt{\frac{2g\ \Delta H + 2g\,(Z_1 - Z_2)}{1-\beta^4}} \qquad \text{ft}^3/\text{s} \tag{2-16}$$

Weight rates of flow can be obtained by using the equation

$$W = Q\rho \qquad \text{lb/s} \tag{2-16a}$$

For gases, the expansion factor Y is not unity and the values of Y are best obtained from the plots[1b] of Y versus $(1 - r)/k$ with β^2 as a parameter. For the venturi tubes, $C = 0.98$ in most cases if $N_{\text{Re}} > 10{,}000$. [For the definition of N_{Re}, refer to Eq. (2-21).]

Orificemeters. An orifice (Fig. 2-6*c* and *d*) is a simple, flat plate with a central opening. The contraction of a stream flowing through an orifice is quite large. The point of the minimum cross section, *vena contracta*, is one or two diameters downstream from the orifice plate. For an orifice plate, the velocity through the orifice is given by

$$u_0 = C_0Y\sqrt{\frac{2g\ \Delta H}{1-\beta^4}} \qquad \text{ft/s} \tag{2-17}$$

and the weight rate of flow can be calculated by

$$W = u_0\rho A_0 = \rho A_0C_0Y\sqrt{\frac{2g\ \Delta H}{1-\beta^4}} \tag{2-18}$$

where β is the usual diameter ratio and Y is the expansion factor. The factor Y is unity for liquids. For gases, Y can be obtained from a plot.[1b] Alternatively, Y may be approximately calculated from[1c]

$$Y = 1 - \frac{0.41 + 0.35\beta^4}{k}\left(1 - \frac{P_2}{P_1}\right) \tag{2-18a}$$

The coefficient C_0 is a function of N_{Re} and β. These values can be obtained from a plot[1c] of C_0 versus N_{Re} with β as a parameter.

Permanent Pressure Loss.

Venturi. Permanent pressure loss through venturis depends upon the diameter ratio β and the discharge cone angle α. The pressure loss for smaller angles (5 to 7°) is 10 to 15 percent of the total pressure differential $(P_1 - P_2)$. It is 10 to 30 percent for large angles (> 15°).

Subsonic Flow Nozzles. For subsonic flow nozzles, the pressure loss is given by

$$\text{Pressure loss} = \frac{1 - \beta^2}{1 + \beta^2}(P_1 - P_2) \tag{2-19}$$

where β is the nozzle throat diameter divided by the pipe diameter.

Concentric Circular Nozzles. In the case of the concentric circular nozzle,

$$\text{Permanent pressure loss} = (1 - \beta^2)(P_1 - P_2) \tag{2-20}$$

Example 2-2. A venturimeter is to be installed in a schedule 40, 6-in line to measure the flow of water. The maximum rate is expected to be 800 gpm at 86°F. A 50-inHg manometer is to be used. Specify the throat diameter of the venturi and calculate the power required to operate it. The discharge cone angle is 5°.

Solution. From Eq. (2-14),

$$u_T = C\sqrt{\frac{2g\ \Delta H}{1-\beta^4}}$$

$C = 0.98$ for venturis if $N_{\text{Re}} > 10{,}000$ (i.e., fully turbulent flow). Assume fully turbulent flow. Then

$$\text{Flow} = \frac{800}{60(7.48)} = 1.7825\ \text{ft}^3/\text{s}$$

$$\text{ID} = 6.6025\ \text{in} = 0.5052\ \text{ft}$$

$$\Delta H = h_m\left(\frac{s_m}{s} - 1\right) \qquad \text{from Eq. (2-6)}$$

$$= \tfrac{50}{12}(13.6 - 1) = 52.5\ \text{ft}$$

$$u_T = 0.98\sqrt{\frac{64.4(52.5)}{1-\beta^4}}$$

$$\text{Flow } u_T A_T = 0.98\ A_T\sqrt{\frac{64.4(52.5)}{1-\beta^4}} = 1.7825$$

$$\frac{A_T}{\sqrt{1-\beta^4}} = \frac{1.7825}{0.98\sqrt{64.4(52.5)}} = 0.03128$$

$$A_T = \tfrac{1}{4}\pi D_T^2 = 0.7854 D_T^2$$

$$\frac{0.7854\ D_T^2}{\sqrt{1-(D_T/D_P)^4}} = 0.03128$$

where D_P is the inside diameter of pipe, ft. Squaring both sides, one obtains

$$\frac{0.617 D_T^4}{1-(D_T/D_P)^4} = 0.0009784$$

Simplifying and solving for D_T gives

$$D_T = 0.198\ \text{ft} = 2.38\ \text{in}$$

Use 2.375 in = 0.1979 ft (a standard size). Check the Reynolds number through the throat of the venturi.

$$u_T = \frac{1.7825}{0.7854(0.1979)^2} = 57.95\ \text{ft/s}$$

$$N_{Re} = \frac{D_T u_T \rho}{\mu} = \frac{0.1979(57.95)(62.4)}{0.85(0.000672)} = 1.253 \times 10^6$$

This N_{Re} is greater than 10,000 and therefore the flow is in the fully turbulent region. Therefore, the size of the venturi is adequate.

Next calculate the permanent pressure loss and the power for operation:

$$\beta = \frac{D_T}{D_P} = \frac{2.375}{6.0625} = 0.3918 \qquad \beta^4 = 0.02355$$

Substitution in the equation for u_T gives

$$57.95 = 0.98\sqrt{\frac{64.4\ \Delta H}{1 - 0.02355}}$$

$$\Delta H = \frac{(57.95)^2(1 - 0.02355)}{(0.98)^2(64.4)} = 53.02\ \text{ftH}_2\text{O}$$

$$= 53.02\frac{144}{2.31} = 3305\ \text{lbf/ft}^2$$

The discharge cone angle is 5°, and the permanent pressure loss for the venturi may be taken as 10 percent of the ΔP:

$$\text{Pressure loss} = 0.1(3305) \doteq 331\ \text{lbf/ ft}^2$$

and

$$\text{Flow} = 800/7.48 = 106.95\ \text{ft}^3/\text{min}$$

$$\text{Power required to operate venturi} = \frac{106.95(331)}{33{,}000}$$

$$= 1.07\ \text{hp} = 0.8\ \text{kW}$$

Example 2-3. Natural gas (viscosity = 0.011 cP) is flowing through a 6-in schedule 40 pipe equipped with a 2-in orifice with flanged taps. The gas is at 90°F and 20 psia at the upstream tap. The manometer reading is 50 inH_2O at 60°F. k for natural gas is 1.3. Calculate the rate of flow of the gas through the line in pounds per hour. Assume that the molecular weight of the gas is 16.

Solution. As defined in Eq. (2-17), the velocity of gas through the orifice is

$$u_0 = C_0 Y \sqrt{\frac{2g_c\,\Delta H}{1-\beta^4}}$$

Since C_0 and Y are not known, a trial-and-error solution is required. Assume $C_0 = 0.61$ (based on the assumption of fully turbulent flow). Also

$$\beta = \frac{2}{6.065} = 0.3299 \qquad \beta^4 = 0.01184 \qquad \beta^2 = 0.109 \doteq 0.11$$

The average density of the gas is

$$\rho_1 = \frac{16}{359}\left(\frac{20}{14.7}\right)\left(\frac{492}{460+90}\right) = 0.0543 \text{ lb/ft}^3$$

From Eq. (2-6),

$$\Delta H = h_m\left(\frac{s_m}{s} - 1\right) = \frac{50}{12}\left(\frac{62.4}{0.0543} - 1\right) = 4784 \text{ ft}$$

$$\Delta P = \rho_1\,\Delta H \tfrac{1}{144} = 1.8 \text{ psi}$$

$$P_2 = 20 - 1.8 = 18.2 \text{ psia}$$

$$\frac{P_2}{P_1} = \frac{18.2}{20} = 0.91$$

Y can be calculated approximately by Eq. (2-18a) as follows:

$$Y = 1 - \frac{0.41 + 0.35\beta^4}{k}\left(1 - \frac{P_2}{P_1}\right)$$

$$= 1 - \frac{0.41 + 0.35(0.01184)}{1.3}(1 - 0.91) = 0.9713$$

$$u_0 = 0.61(0.9713)\sqrt{\frac{64.4(4784)}{1 - 0.01184}} = 330.9 \text{ ft/s}$$

$$N_{Re} = \frac{Du\rho}{\mu} = \frac{\frac{2}{12}(330.8)(0.0543)}{0.011(0.000672)} = 404{,}997$$

Since the Reynolds number is very high and in the fully turbulent region,[1b]

$$C_0 = 0.61$$

and

$$T_2 = T_1\left(\frac{P_2}{P_1}\right)^{(k-1)/k} = 550(0.91)^{0.3/1.3} = 538°\text{R}$$

$$\rho_2 = 0.0543\left(\frac{550}{538}\right)\left(\frac{18.20}{20}\right) = 0.0505 \text{ lb/ft}^3$$

$$\text{Average } \rho = \tfrac{1}{2}(0.0543 + 0.0505) = 0.0524 \text{ lb/ft}^3$$

$$\text{Flow of gas} = 330.8(0.7854)(\tfrac{2}{12})^2(0.0524)(3600) = 1361 \text{ lb/h}$$

Example 2-4. A pitot tube is inserted in a 30-in-ID duct carrying air so that the tip is located at the center of the duct and is aimed in the direction of the flow. The manometer reading is 1.1 inH_2O, and the static pressure at the point of measurement is 33 inH_2O. Calculate the flow of the air in cubic feet per minute if the temperature of the air is 100°F and the viscosity of air is 0.02 cP.

Solution. For a pitot tube, velocity in the pipe is given by Eq. (2-11*a*):

$$u = \sqrt{2g\ \Delta H}$$

$$\text{Air pressure} = 14.7 + \tfrac{33}{407}(14.7) = 15.9 \text{ psia}$$

$$1 \text{ atm} = 14.7 \text{ psia} \doteq 407 \text{ inH}_2\text{O}$$

$$\text{Temperature of air} = 460 + 110 = 570°\text{R}$$

$$\text{Density* of air } \rho = \frac{29}{359}\left(\frac{15.9}{14.7}\right)\left(\frac{492}{570}\right) = 0.0754 \text{ lb/ft}^3$$

From the manometer reading, $\Delta P = (1.1/12)(62.4) = 5.72$ lbf/ ft^2 where the height of the air column is ignored. Since the tip of the pitot tube is at the center,

$$u_{\max} = 0.98\sqrt{2g_c\frac{\Delta P}{\rho}} = 0.98\sqrt{\frac{(64.4)(5.72)}{0.0754}} = 68.5 \text{ ft/s}$$

Neglecting the compressibility correction, the Reynolds number is given by

$$N_{\text{Re}} = \frac{\frac{30}{12}(68.5)(0.0754)}{0.02(0.000672)} = 960{,}733$$

*The density can also be calculated as $\rho = PM_w/RT = 15.9(29)/10.73(570) = 0.0754$ lb/ ft^3.

At this value of the Reynolds number, * see next Page

$$\frac{u}{u_{\max}} = 0.86$$

Therefore $\qquad u_{av} = 0.86(68.5) = 58.91 \text{ ft/s}$

Then $\qquad \text{Airflow} = 0.7854\,(2.5)^2\,(58.91)(60) = 17{,}350 \text{ ft}^3/\text{min}$

Flow of Fluids in Pipes

The nature of the flow of a fluid in a pipe depends upon the Reynolds number which is defined as

$$N_{\text{Re}} = \frac{Du\rho}{\mu} = \frac{DG}{\mu} \qquad (2\text{-}21)$$

where D = inside diameter of pipe, ft
u = velocity of fluid, ft/s
ρ = density of fluid, lb/ft^3
G = mass velocity, lb/h · ft^2
μ = viscosity, lb/ft · s for $Du\rho/\mu$ and lb/ft · h for DG/μ

Flow regimes are defined as follows:

$$N_{\text{Re}} \begin{cases} < 2100 & \text{viscous flow} \\ > 4000 & \text{turbulent flow} \\ = 2100 - 4000 & \text{transition region} \end{cases}$$

Distribution of Velocities. For fluids flowing through a pipe, the velocity distribution will depend upon the type of flow. For the laminar or viscous flow, the velocity distribution is truly parabolic.

Laminar Flow of Newtonian Fluids in Cylindrical Pipes. Starting from the definition of viscosity and by considering the shear stress of an incompressible fluid through the tube, it can be shown that

$$\frac{u}{u_{\max}} = \left(\frac{r}{r_w}\right)^2 \qquad (2\text{-}22)$$

where u is the local velocity at r and u_{max} is the maximum velocity at $r = 0$ (center of tube). For the flow of a fluid in a tube, the average velocity is given by[1d]

Laminar Flow

$$\frac{u_{av}}{u_{max}} = 0.5 \tag{2-23}$$

Turbulent Flow. For turbulent flow,[1d]

$$\frac{u_{av}}{u_{max}} \doteq 0.82 \tag{2-24}$$

The relationship between u_{av}/u_{max} versus Reynolds number $Du\rho/\mu$ is available in a graphical form.[1d]

Frictional Losses in Circular Pipes. The frictional losses for flowing fluids are a function of Reynolds number. ΔP and the head loss from friction are given by

$$\begin{aligned} \Delta P_f &= \frac{4u^2L}{g_cD}\,\phi\left(\frac{Du\rho}{\mu}\right) \\ \Delta H_f &= \frac{4u^2L}{g_cD}\,\phi\left(\frac{Du\rho}{\mu}\right) \end{aligned} \tag{2-25}$$

A more common method is to use the Fanning equation which expresses the pressure drop in terms of a friction factor:

$$\begin{aligned} \Delta P_f &= \frac{4fu^2L\rho}{2g_cD} = \frac{2fu^2L\rho}{g_cD} \qquad \text{lbf/ft}^2 \\ \Delta H_f &= \frac{2fu^2L}{g_cD} = h_L \qquad \text{ft} \end{aligned} \tag{2-26}$$

where f is Fanning friction factor to be obtained from Fig. 2-7 and L is the length of the pipe in feet.

For the laminar flow, if the Fanning equation is combined with the Hagen-Poiseuille equation, the following relation for the friction factor results:

$$f = \frac{16}{N_{Re}} \tag{2-27}$$

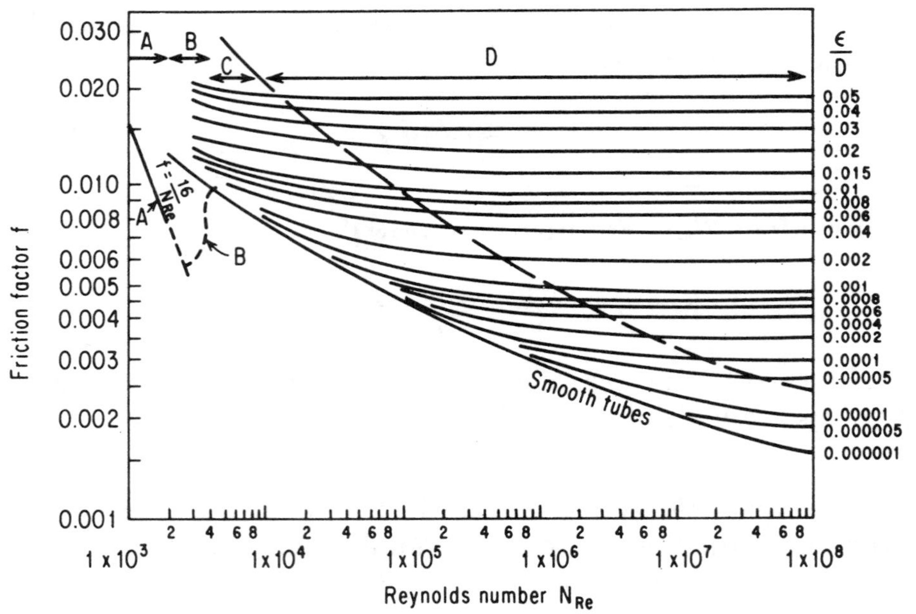

Figure 2-7. Fanning friction factor vs. Reynolds number. (*Reprinted from Chemical Engineers' Handbook, 5th ed., McGraw-Hill Book Company, New York, 1973, with permission.*) (*A*) Laminar region; (*B*) critical region; (*C*) transition region; (*D*) region of complete turbulence.

For both the turbulent flow and the laminar flow, the friction factor f is to be obtained from the Fanning friction factor chart given in Fig. 2-7.

If the upper boundary of the streamline flow is given by $Du\rho/\mu = 2100$, the critical velocity is given by

$$u_{cr} = 2100 \frac{\mu}{\rho D} \tag{2-28}$$

The Fanning friction factor f depends also upon the roughness of pipe ϵ. The effect of the roughness of the pipe on the friction factor is shown in the Fanning friction chart by giving the values of f with ϵ/D as parameter.

Equivalent Diameters for Noncircular Conduits. For noncircular conduits, an equivalent diameter for fluid flow is defined as

$$D_e = \frac{4(\text{cross-sectional area of flow})}{\text{wetted perimeter of channel}} = 4r_h \qquad (2\text{-}29)$$

where the hydraulic radius is

$$r_h = \frac{\text{cross-sectional area of flow}}{\text{wetted perimeter of channel}} \qquad (2\text{-}30)$$

For a square channel,

$$r_h = \frac{b}{4} \qquad \text{and} \qquad D_e = b \qquad (2\text{-}30a)$$

where b is the side of the square. For a rectangular channel of sides a and b,

$$r_h = \frac{ab}{2(a+b)} \qquad \text{and} \qquad D_e = \frac{2ab}{a+b} \qquad (2\text{-}30b)$$

For annular spaces, area $A_c = \frac{1}{4}\pi(D_2^2 - D_1^2)$, wetted perimeter = $D_2 + D_1$, and hence

$$D_e = \frac{4\pi(D_2^2 - D_1^2)}{4\pi(D_2 + D_1)} = D_2 - D_1 \qquad (2\text{-}30c)$$

where D_1 is the outside diameter of the inner pipe and D_2 is the inside diameter of the outside pipe.

The friction-factor relationships for the circular conduits apply also to the noncircular conduits, but the equivalent diameter must be used in calculating the Reynolds number.

Example 2-5. Find the hydraulic radius and equivalent diameter for the section shown in Fig. 2-8 if the water is flowing 4 ft 6 in deep in the channel.

Solution. The hydraulic radius is

$$r_h = \frac{\text{area of cross section of flow}}{\text{wetted perimeter}}$$

$$\text{Area of cross section} = \tfrac{1}{2}(a+b)h$$

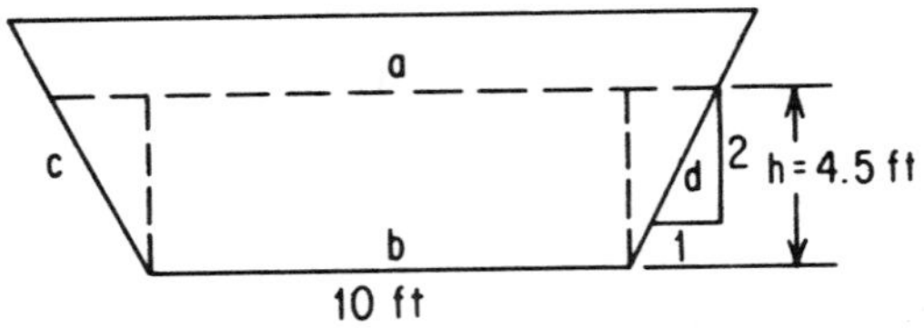

Figure 2-8. Data sketch for Example 2-5.

$$\text{Width of top at liquid level} = 10 + 2(4.5)(\tfrac{1}{2}) = 14.5 \text{ ft}$$

$$\text{Length of each side } (c \text{ and } d) = \sqrt{4.5^2 + \left(\frac{4.5}{2}\right)^2} = 5.031 \text{ ft}$$

$$\text{Wetted perimeter} = 2(5.031) + 10 = 20.062 \text{ ft}$$

$$\text{Area of cross section} = \tfrac{1}{2}(14.5 + 10)(4.5) = 55.125 \text{ ft}^2$$

Then

$$r_h = \frac{55.125}{20.062} = 2.75 \text{ ft}$$

and the equivalent diameter is

$$D_e = 4r_h = 4(2.75) = 11 \text{ ft}$$

Note: The friction factor used in the previous equations is the Fanning friction factor which is an arbitrary constant and should be used only in conjunction with the Fanning friction-factor chart of Fig. 2-7. Other charts are also in use. The most common of these is the Moody-Darcy chart. The Moody friction factor f' is related to the Fanning friction factor f by the following relation:

$$f' = 4f \tag{2-31}$$

Frictional Losses through Fittings and Valves. Frictional losses occur because of local disturbances in the flow-through conduits. The disturbances are caused by fittings such as bends and elbows and by sudden changes in the direction of the flow caused by obstructions. These losses are generally expressed in terms of a resistance coefficient K, an equivalent length L_e, and a flow coefficent C_v.

Resistance Coefficient *K*. The resistance coefficient K is defined as the number of the velocity heads that are lost because of a fitting or an obstruction. It is assumed to be independent of the friction factor irrespective of either the laminar or turbulent flow. Data on K values for the various fittings, obstructions, and valves are available,[2,1e] and some are given in Table 2-1.

TABLE 2-1 Equivalent Lengths and *K* Values (Turbulent Flow)

Type	Comment		L_e/D*	L_e/d*
	Valves			
Gate (disk or plug)	Fully open		13	1.1
Globe	Seat flat,bevel or plug		340	28.3
(conventional)	wing or pin guided disc		450	37.5
Angle	With no obstruction			
(conventional)	(seat flat, bevel, or plug)		145	12.1
	Wing or pin guided disk		200	16.7
Y-pattern globe	Stem at 60° from run of pipe		175	14.6
	Stem at 45° from run of pipe		145	12.1
Ball valve	Fully open (full port)		3	0.25
Butterfly	Sizes 2 to 8 in		45	3.8
	Sizes 10 to 14 in		35	2.9
	Sizes 14 to 24 in		25	2.1
Plug	Straight-way		18	1.5
	3-way (straight run)		30	2.5
	3-way through branch		90	7.5
Foot valve with	Poppet disk		420	35
strainer	Hinged disk		75	6.25
Swing check	Clearway		50	4.2
	Tilting seat		100	8.34
Tilting disk check	Disk angle, 5° (15°)	2 to 8 in	40 (120)	3.3 (10)
		10 to 14 in	20 (90)	2.5 (7.5)
		16 to 48 in	20 (60)	1.7 (5)
Lift or stop check	Globe lift or stop		450	37.5
	Angle lift or stop		200	16.7

*L_e and D are in feet and d is in inches.

TABLE 2.1 (*Continued*)

Type	Comment	L_e/D*	L_e/d*
	Flanged Fittings		
Standard elbow	90°	30	2.5
	45°	16	1.33
	90°long radius	20	1.67
Return bend	100° close pattern	50	4.17
Standard tee	Flow-through run	20	1.67
	Flow-through branch	60	5.0
Elbow 90°	90° $R/D = 1.0$	18	1.5
	$R/D = 1.5$	12	1.0
	$R/D = 2.0$	10	0.83
	Welded Fittings		
Elbow 45°	Multiply L_e/D for 90° elbow by 0.64		
Return bend, 180°	Multiply L_e/D for 90° elbow by 1.34		
Tees	100% flow-through run	18	0.7
	100% flow-out through branch	58	4.83
	100% flow-in through branch	43	3.6
Reducer		30	2.5

Miscellaneous Obstructions

Type	Comment	K
Pipe entrance	With inward projection	0.78
Pipe entrance	Sharp-edged	0.50
Pipe entrance	Slightly rounded	0.23
Pipe entrance	Well rounded	0.04
Exit from pipe	Projecting type, sharp edged, or rounded	1.0
Sudden contraction	β = small diam./large diam.	$0.5(1 - \beta^2)^2$
Sudden expansion	β = small diam./large diam.	$(1 - \beta^2)^2$

*L_e and D are in feet and d is in inches.

Equivalent Length L_e. The equivalent length of a fitting or an obstruction is the length of a pipe that effects the same pressure drop due to friction as the fitting or obstruction under consideration. It is usually expressed in terms of the equivalent pipe diameter L_e/D. Equivalent lengths are available from various sources.[3a] Some values are reproduced in Table 2-1.

Flow Coefficient. The flow coefficient C_v of a control valve is defined as the flow of water at 60°F in gallons per minute at a pressure drop of 1 lbf/in^2 across the valve when the valve is completely open.

Relation between K and L_e. The resistance coefficient K and the equivalent length are related by

$$K = 4f\frac{L_e}{D} \qquad \text{or} \qquad K = f'\frac{L_e}{D} \tag{2-32}$$

where f is the Fanning friction factor and f' the Moody-Darcy friction factor.

Example 2-6. Water is being pumped through the system shown in Fig. 2-9. The temperature of the water is 86°F and the flow rate 100 gpm. Calculate the total pressure drop through the system if the viscosity of water is 0.85 cP.

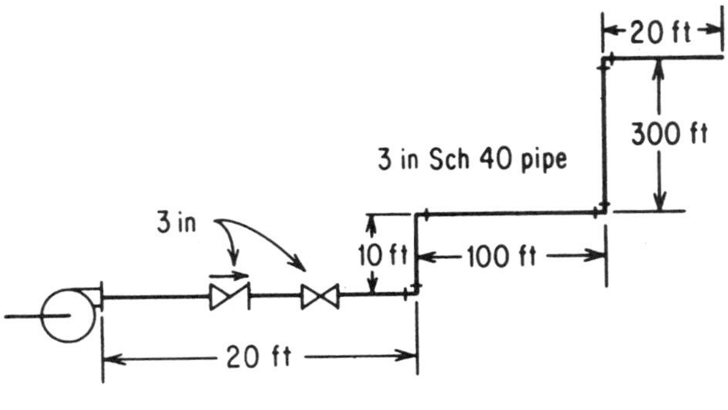

Figure 2-9. Pumping system (Example 2-6). Distances covered by fittings are not included.

Solution.

$$\text{ID} = 3.068 \text{ in} = 0.2557 \text{ ft}$$

$$\text{Total length of pipe} = 20 + 10 + 100 + 300 + 20 = 450 \text{ ft}$$

***K* Values**

90° bends	$K = 4(0.34) = 1.36$
3-in check valve	$K = 2.00$
3-in gate valve	$K = 0.23$
Exit enlargement	$K = 1.00$

$$\Delta P = \frac{\Sigma K u^2}{2g_c} + \frac{2fLu^2}{g_c D}$$

$$= \left(\Sigma K + \frac{4Lf}{D}\right)\frac{u^2}{2g_c}$$

$$\Sigma K = 1.36 + 2 + 0.23 + 1 = 4.59$$

$$u = \frac{110}{7.48(60)(0.7854)(0.2557)^2} = 4.773 \text{ ft/s}$$

$$\frac{Du\rho}{\mu} = \frac{0.2557(4.773)(62.4)}{0.85(0.000672)} = 133{,}327$$

For steel pipe

$$\epsilon = 0.00015 \text{ ft} \qquad \frac{\epsilon}{D} = \frac{0.00015}{0.2557} = 0.00059$$

$$f = 0.0048 \qquad \text{from Fig. 2-7}$$

$$\Delta P = \left[4.59 + \frac{4(450)(0.0048)}{0.2557}\right]\frac{4.773^2}{64.4} = 13.58 \text{ ft}$$

$$= (13.58 \text{ ft})(0.433 \text{ psi/ft}) = 5.88 \text{ psi}$$

Example 2-7. Solve Example 2-6 with the use of the equivalent lengths. Obtain the equivalent lengths from data sources[2,3a] or from Table 2-1.

Solution. Use the following equivalent lengths:

Item	Equivalent Length, L_e, ft
Four 3-in elbows	(4)(7.5) = 30
One 3-in check valve	35
One 3-in gate valve	3.5
Sudden enlargement	15
Straight pipe length	450
Total equivalent length	30 + 35 + 3.5 + 15 + 450 = 533.5

$$f = 0.0048 \qquad \text{as in Example 2-6}$$

$$\Delta P = \frac{2fL_e u^2}{g_c D}\left(\frac{\rho}{144}\right) = \frac{2(0.0048)(533.5)(4.773)^2(62.4)}{(32.17)(0.2557)(144)} = 6.10 \text{ psi}$$

Example 2-8. Soda ash liquor having 1250 kg/m^3 density and 1.2 cP viscosity flows through a 215-m length of a 150-mm-ID steel pipe. The equivalent length of the fittings may be taken as 70 pipe diameters. A venturimeter with a throat diameter of 75 mm installed in the line shows a differential column height of 26 mm on a mercury manometer. What is the flow rate in kilograms per second? What is the total pressure drop through the line if the roughness of the pipe inside surface is 0.0000457 m? Assume the flow coefficient of the venturi is 0.985.

Solution.

$$\rho = 1250 \text{ kg/m}^3 \qquad \text{Specific gravity} = 1.25$$

$$\mu = 1.2 \text{ cP} = 0.0012 \text{ N}\cdot\text{s/m}^2 \text{or kg/m}\cdot\text{s}$$

From Eq. (2-6),

$$\Delta H = h_m\left(\frac{s_m}{s} - 1\right)$$

Therefore,

$$\Delta H = 26\left(\frac{13.6}{1.25} - 1\right) = 257 \text{ mm of liquid} = 0.257 \text{ m}$$

$$\beta = \frac{D_T}{D_P} = \frac{75}{150} = 0.5 \qquad \beta^4 = 0.0625 \qquad 1 - \beta^4 = 0.9375$$

$$u_T = \text{velocity through the throat of the venturi}$$

$$= C\sqrt{\frac{2g_c\,\Delta H}{1-\beta^4}}$$

$$= 0.985\sqrt{\frac{2(9.81)(0.257)}{0.9375}} = 2.284 \text{ m/s}$$

$$A_T = \text{cross section of venturi} = \tfrac{1}{4}\pi(0.075)^2 = 0.004418 \text{ m}^2$$

$$\text{Flow rate} = u_T A_T$$

$$= (2.284 \text{ m/s})(0.004418 \text{ m}^2)(1250 \text{ kg/m}^3)$$

$$= 12.62 \text{ kg/s}$$

$$\text{Volumetric flow rate} = 2.284(0.004418) = 0.0101 \text{ m}^3\text{/s}$$

$$\text{Flow cross section of pipe} = \tfrac{1}{4}\pi(0.15)^2 = 0.01767 \text{ m}^2$$

$$\text{Velocity through pipe} = \frac{0.0101 \text{ m}^3\text{/s}}{0.01767 \text{ m}^2} = 0.572 \text{ m/s}$$

$$N_{\text{Re}} = \frac{Du\rho}{\mu} = \frac{0.15(0.572)(1250)}{0.0012} = 89{,}375$$

$$\text{Roughness factor} = \frac{\epsilon}{D} = \frac{0.0000457}{0.15} = 0.0003$$

From Fig. 2-7,

$$\text{Fanning friction factor} = 0.0048$$

$$\text{Equivalent length of fittings} = 70(150) = 10{,}500 \text{ mm} = 10.5 \text{ m}$$

$$\text{Total equivalent length} = 215 + 10.5 = 225.5 \text{ m}$$

$$\Delta P \text{ through pipe and fittings} = \frac{2fL_e u^2\rho}{g_c D}$$

$$= \frac{2(0.0048)(225.5)(0.572)^2(1250)}{1(0.15)}$$

$$= 5902 \text{ N/m}^2 = 5.9 \text{ kN/m}^2$$

Assume that the permanent ΔP through the venturi is 15 percent of the pressure differential across the venturi. On this basis, the differential head across the venturi is

$$\Delta P = \Delta H\, g_c \rho$$
$$= (0.257\text{ m})(9.81\text{ m/s}^2)(1250\text{ kg/m}^3)$$
$$= 3151.5\text{ N/m}^2 = 3.15\text{ kN/m}^2$$
$$\Delta P\text{ through venturi} = 0.15(3.15) = 0.47\text{ kN/m}^2$$
$$\text{Total }\Delta P\text{ in the line} = 5.9 + 0.47 = 6.37\text{ kN/m}^2$$
$$= \frac{6.37}{6.896} = 0.924\text{ psi}$$

Example 2-9. Water at 86°F is to flow through a horizontal pipe at the rate of 175 gpm. $\mu = 0.85$ cP. A 25-ft head is available. What should be the pipe diameter? Assume ϵ, the roughness of the pipe, is 0.00015 ft and the length of pipe is 1000 ft.

Solution. From Eq. (2-26),

$$h_L = \frac{2fLu^2}{g_c D}\text{ ft} \qquad \text{or} \qquad D = \frac{2fLu^2}{g_c h_L}$$

u can be calculated in terms of D as

$$u = \frac{175\text{ gpm}}{(7.48\text{ gal/ft}^3)(\frac{1}{4}\pi D^2)(60\text{ s/min})} = \frac{0.497}{D^2}\text{ ft/s}$$

Substitution in the equation for D gives

$$D = \frac{2fL}{g_c h_L}\left(\frac{0.497}{D^2}\right)^2 \qquad \text{or} \qquad D^5 = \frac{2(0.497)^2(1000)f}{32.2(25)} = 0.6137f$$

Since f is not known, the solution for D is to be obtained by trial and error. A first estimate could be obtained by using the guidelines in Table 2-2. Thus the estimated line size $= 0.25\sqrt{175} = 3.3$ in. Since 3.3 in is not a standard pipe size, assume 4-in schedule 40 pipe.

$$\text{ID} = 4.026\text{ in} = 0.3355\text{ ft}$$

$$u = \frac{0.497}{(0.3355)^2} = 4.415\text{ ft/s}$$

$$\frac{Du\rho}{\mu} = \frac{0.3355(4.415)(62.4)}{0.85(0.000672)} = 1.62 \times 10^5$$

TABLE 2-2 **Line Sizing Guideline**

	Typical ΔP_{100}, psi/100 ft	Typical Velocity, ft/s	Remarks
1. Liquid service			
a. Pump suction	0.05 to 0.5	1 to 6	Use lower ΔP_{100} and velocity for hydrocarbons and boiling liquids. For water and similar service $\Delta P_{100} = 0.5$ to 1 may be used. (Line size is often dictated by the NPSH required by the pump.) Rule of thumb: line size (in) = $\sqrt{\text{gpm}/10}$
b. Pump discharge	2 to 6	3 to 14	Final line size should be selected from an economic analysis, which may sometimes lead to $\Delta P_{100} \doteq 8$. Rule of thumb: line size (in) = $0.25\sqrt{\text{gpm}}$
c. Water header	0.5 to 1	2 to 10	
d. Water lateral	0.5 to 2	2 to 12	
e. Reboiler inlet			
i. Once-through	0.1 to 0.2		Usually dictated by the static head available and flow rate.
ii. Thermosyphon recirculation	0.8 to 1		Usually dictated by the static head available and the required circulation ratio. Rule of thumb: cross section of inlet pipe = (0.5) (cross section of all tubes).
f. Gravity flow			For self-venting, line size, in = $0.92Q^{0.4}$ where Q = flow in gpm.

2. Vapors and gases			
a. Steam			
15–28 inHg vacuum	0.05 to 0.2		Use lower ΔP_{100} at lower operating pressure.
15 inHg vacuum to 0 psig	0.2 to 0.5		
0–50 psig	0.5 to 1.0		
50–150 psig	1 to 1.5		
150–300 psig	1.5 to 2		
Over 300 psig	3		
b. Vapor and gas			
15–28 inHg vacuum	0.025 to 0.05		Use lower ΔP_{100} at lower operating pressure. For operating pressure below 15 inHg vacuum, total system pressure drop should not exceed 10 percent operating pressure (absolute). Line size is often dictated by the total allowable system pressure drop and economics.
15 inHg vacuum to 0 psig	0.05 to 0.1		
0–50 psig	0.1 to 0.25		
50–150 psig	0.25 to 0.75		
150–300 psig	0.75 to 1.5		
c. Kettle type reboiler outlet	0.1 to 0.2		
d. Compressor suction		25 to 100	Final line size should be selected from a pressure profile study and economics.
e. Compressor discharge	1 to 2	100 to 200	

TABLE 2-2 Line Sizing Guideline (*Continued*)

	Typical ΔP_{100}, psi/100 ft	Typical Velocity, ft/s	Remarks
3. Two-phase flow			
a. Slurry	....	4 to 10	Never size line on the basis of ΔP_{100}. Velocity should be greater than deposit velocity. For slurries containing solids that tend to stick to the surface when settled, velocities like 16 ft/s are not unusual.
b. Gas-liquid	....	35 to 75 maximum $u = 100/\sqrt{\rho_{mix}}$	ρ_{mix} = density of mixture in lb/ft^3. Prefer dispersed flow. Avoid slug flow. Rule of thumb: condensate line size is two sizes smaller than steam line size.
c. Air-solid air/solid (ft^3/lb)			
10 to 40	....	90 to 70	Use higher velocity for lower air/solid ratio.
40 to 100	....	70 to 60	
d. Thermosyphon reboiler outlet	0.1 to 1.0	....	Rule of thumb: cross section of outlet pipe = cross section of all tubes.

$$\frac{\epsilon}{D} = \frac{0.00015}{0.3355} \doteq 0.00045$$

$$f = 0.0047 \qquad \text{from Fig. 2.7}$$

$$D^5 = 0.6137(0.0047) = 0.002884$$

$$D = 0.3105 \text{ ft} = 3.73 \text{ in}$$

which is close to assumed 4.026-in ID. A 4-in-diam. schedule 40 pipe should be specified.

Example 2-10. Water is flowing through an annular channel at a rate of 25 gpm. The channel is made of $\frac{1}{2}$- and $1\frac{1}{4}$-in schedule 40 pipes. Calculate ΔP through an annular channel of length 20 ft, assuming $\epsilon/D = 0.0014$:

$$\mu = 0.9 \text{ cP} \qquad \rho = 62.4 \text{ lb/ft}^3$$

Solution. First calculate equivalent diameter. The outside diameter of the inner pipe is $D_1 = 0.84$ in $= 0.07$ ft, and the inside diameter of the outer pipe is $D_2 = 1.38$ in $= 0.115$ ft

$$\text{Equiv. diam.} = \frac{4\pi(D_2^2 - D_1^2)}{4\pi(D_2 + D_1)} = \frac{4(\text{cross-sectional area of flow})}{\text{wetted perimeter}}$$

$$= D_2 - D_1 = 0.115 - 0.07 = 0.045 \text{ ft}$$

$$\text{Cross section} = \tfrac{1}{4}\pi(D_2^2 - D_1^2) = 0.785(0.115^2 - 0.07^2) = 0.006535 \text{ ft}^2$$

$$\text{Velocity } u = \frac{25}{7.48(60)(0.006535)} = 8.52 \text{ ft/s}$$

$$N_{\text{Re}} = \frac{D_e u \rho}{\mu} = \frac{0.045(8.52)(62.4)}{0.9(0.000672)} = 39{,}557$$

$$\frac{\epsilon}{D} = 0.0014 \qquad \text{(given)}$$

$$f = 0.0068$$

from the Fanning friction chart (Fig. 2-7).

$$\Delta P = \frac{2fLu^2\rho}{g_c D_e(144)} = \frac{2(0.0068)(20)(8.52)^2(62.4)}{32.17(0.045)(144)} = 5.91 \text{ psi}$$

Pump Calculations

Various terms in connection with the pump calculations (Fig. 2-10) are defined in the following paragraphs.

Capacity

Capacity is the quantity of fluid discharged per unit time. In the fps system, this is expressed in gallons per minute (gpm) for liquids.

Static Head. Static head for a liquid being pumped is the difference in elevation, in feet, between the datum line and the liquid surface or the point of free delivery. For the horizontal centrifugal pumps, the datum line is the pump centerline. For the vertical pumps the datum line is taken at the eye of the first-stage impeller.

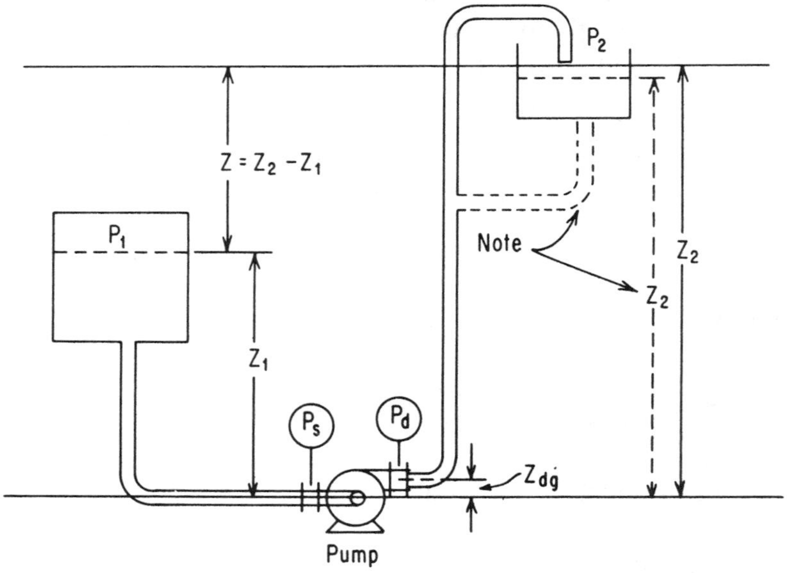

Figure 2-10. Pumping system terminology. (*Note:* If the discharge line of the pump is connected to the bottom of the receiver tank as shown by the dashed line, then Z_2 is the distance from the centerline of the pump to the liquid surface.)

In Fig. 2-10, Z_1 and Z_2 are the static heads. Note that Z_2 is measured up to the end of the discharge pipe because the pipe end is the point of free delivery. If the pipe were connected at the bottom, Z_2 would be measured up to the liquid surface.

Pressure Head. This is given in feet and defined as follows:

$$\text{Pressure head} = \frac{144P_1}{\rho} = \frac{2.31P_1}{s} \tag{2-33}$$

where P_1 = absolute pressure, psia
ρ = density of the fluid, lb/ft^3
s = specify gravity of the fluid with respect to water.

Velocity Head. This is given in feet and defined as follows:

$$\text{Velocity head} = \frac{u^2}{2g_c}$$

where u is the velocity, ft/s, and g_c is Newton's law proportionality factor = 32.17 ft · lb/lbf · s^2

Static Suction Head. This is the difference in elevation, in feet, between the centerline (or impeller eye) of the pump and the liquid surface in the suction vessel. The liquid surface is above the pump centerline. In Fig. 2-10, Z_1 is the static suction head.

Static Suction Lift. When the liquid level in the suction vessel is below the centerline (or impeller eye) of the pump, the difference in elevation, in feet, between the liquid surface of the suction vessel and the centerline of the pump is called the *suction lift*. In Fig. 2-11*c*, Z_1 is the static suction lift.

Total Suction Head or Lift. This is defined as total suction head (Fig. 2-11*b* and *c*) and is the absolute pressure head in the supply vessel plus the static suction head minus the friction head, or

$$\text{Total suction head} = Z_{pt} \pm Z_1 - Z_f \qquad (2\text{-}34)$$

Use the minus sign for Z_1 in the case of the suction lift.

If the total suction head (or lift) is measured from the reading of a pressure gauge at the suction flange of the pump, then

$$\text{Total suction head} = \frac{P_S(144)}{\rho} + \frac{u_s^2}{2g_c} \qquad (2\text{-}35)$$

where P_S = ±* gauge reading + barometric pressure, psia
ρ = density of liquid, lb/ft^3
u_s = suction velocity, ft/s
g_c = Newton's law proportionality factor
= 32.17 ft · lb/lbf · s^2

Static Discharge Head. This is the difference in elevation, in feet, between the point of the free delivery or the liquid surface in the discharge vessel and the centerline of the pump. In Fig. 2-10, Z_2 is the static discharge head.

Total Discharge Head. The total discharge head is defined as the absolute pressure head in the discharge vessel plus the static discharge head plus the friction head. If the total discharge head is determined from the reading of the pressure gauge at the discharge flange of the pump as shown in Fig. 2-10, then

$$\text{Total discharge head} = \frac{144P_d}{\rho} + Z_{dg} + \frac{u_d^2}{2g_c} \qquad (2\text{-}36)$$

where P_d = barometric pressure + gauge reading, psia
Z_{dg} = elevation of the discharge flange from the datum line, ft
u_d = discharge velocity, ft/s

*Use the minus sign when using the vacuum-gauge reading

Total Dynamic Head or Total Head (TDH). This is the energy, in ft · lbf/lb, of liquid that the pump has to impart to the liquid in order to transport it to the desired location. It can be calculated from

$$\text{TDH} = \text{total discharge head} - \text{total suction head}$$

or from

$$\text{TDH} = \frac{144P_2}{\rho} - \frac{144P_1}{\rho} + \frac{\Delta P_{f1}(144)}{\rho} + \frac{(144)\,\Delta P_{f2}}{\rho} + Z_2 - Z_1$$

$$= (P_2 - P_1 + \Delta P_{f1} + \Delta P_{f2})\,\frac{2.31}{s} + Z_2 - Z_1 \quad (2\text{-}37)$$

where P_2 = pressure in discharge vessel, psia
P_1 = pressure in suction vessel, psia
ΔP_{f1} = pressure drop in suction line, psi
ΔP_{f2} = pressure drop in discharge line, psi
s = gravity of liquid at pumping temperature with respect to water
Z_2 = elevation of liquid discharge point above pump centerline, ft
Z_1 = elevation of liquid level in suction vessel above pump centerline, ft (Z_1 is negative for suction lift.)

Shutoff Head. This is the head developed by a pump with discharge valve closed. It is added to the suction head to determine the maximum discharge pressure of the pump.

Net Positive Suction Head ($NPSH_A$). The net positive suction head available $NPSH_A$ is the total suction head, in feet, of liquid (absolute) that is available in excess of the liquid vapor pressure, also expressed in feet, of liquid at the pump suction flange. It is required to move the liquid into the eye of the impeller, for which the pump itself is not responsible. The liquid should be brought into the pump in the liquid state without vaporization.

Each pump requires a particular NPSH depending upon its design. Typical suction systems with formulas for $NPSH_A$ in each case are given below.

System 1A. In this case, the suction supply is open to atmosphere, the liquid level of the supply is above the pump centerline (Fig. 2-11*a*), and the $NPSH_A$ is

$$NPSH_A = Z_1 + Z_a - Z_v - Z_f + \frac{u^2}{2g_c} \tag{2-38}$$

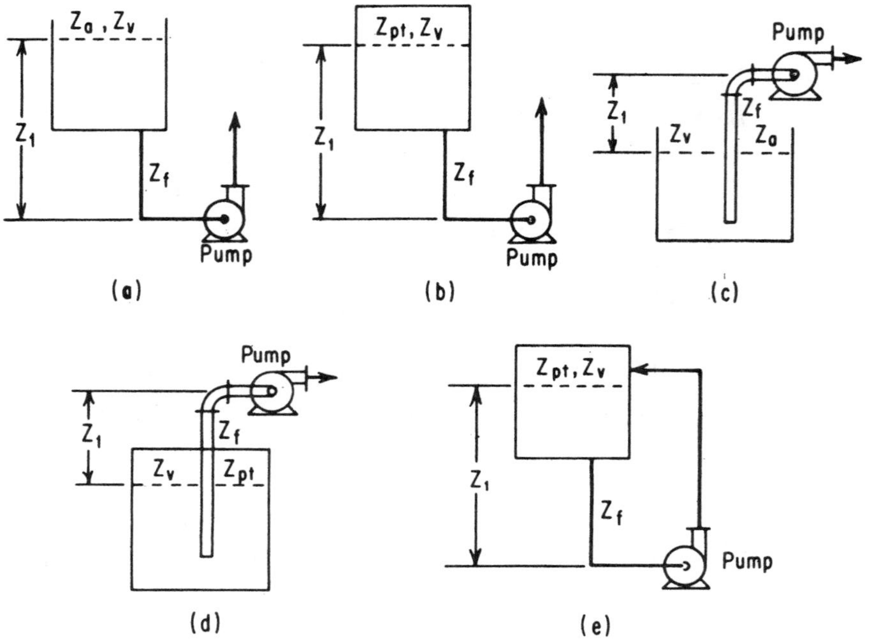

Figure 2-11. $NPSH_A$ for various suction conditions: (*a*) suction vessel open to atmosphere and liquid level above pump centerline; (*b*) suction vessel closed to atmosphere and above the centerline of pump; (*c*) suction vessel open to atmosphere but below the centerline of pump; (*d*) suction vessel closed to atmosphere and below the centerline of pump; (*e*) boiling liquid in a closed tank located above the pump.

where Z_1 = static suction head, ft
Z_a = absolute atmospheric pressure over the liquid level, ft
Z_v = absolute vapor pressure of liquid at the pumping temperature, ft
Z_f = frictional losses, ft
$u^2/2g_c$ = velocity head, ft (usually negligible)

System 1B. In this case, the suction vessel is closed to the atmosphere and is located above the pump centerline (Fig. 2-11*b*). For this system

$$\mathrm{NPSH}_A = Z_1 + Z_{pt} - (Z_v + Z_f) \tag{2-39}$$

where Z_{pt} is the total absolute pressure on the surface of the liquid, ft. If $Z_{pt} = Z_v$, the above reduces to

$$\mathrm{NPSH}_A = Z_1 - Z_f \tag{2-40}$$

System 2A. In this case, the suction vessel is open to atmosphere but is located below the centerline of the pump (Fig. 2-11*c*). For this system

$$\mathrm{NPSH}_A = Z_a - (Z_1 + Z_v + Z_f) \tag{2-41}$$

in which Z_1 is the suction lift, ft.

System 2B. In this case, the suction vessel is closed to atmosphere and below the centerline of the pump (Fig. 2-11*d*). For this case:

$$\mathrm{NPSH}_A = Z_{pt} - (Z_1 + Z_v + Z_f) \tag{2-42}$$

System 3. For a boiling liquid in a closed tank located above the pump suction line (e.g., liquid refrigerant),

$$\begin{aligned}\mathrm{NPSH}_A &= Z_1 + Z_{pt} - (Z_v + Z_f)\\ &= Z_1 + Z_{pt} - Z_v - Z_f\\ &= Z_1 - Z_f \qquad (2\text{-}43)\end{aligned}$$

since $Z_{pt} = Z_v$ in this case. Requirements of NPSH are usually determined on the basis of handling water.

In an existing system, the NPSH_A is determined by gauge reading at the pump suction and with the use of the following formula:

$$\mathrm{NPSH}_A = Z_{pt} - Z_v \pm R_G + Z_k \pm Z_{1p} \tag{2-44}$$

where R_G = gauge reading, ft (Use minus sign for vacuum gauge reading.)

Z_k = velocity head in suction pipe at the gauge connection, ft

Z_{1p} = static suction head or lift of pressure gauge with respect to centerline of the pump (Use plus sign for the suction head and minus sign for the suction lift.)

Cavitation. When there is no sufficient NPSH at the pump suction, the pressure of the liquid reduces to a value equal to or below its vapor pressure, which causes the liquid to vaporize resulting in the formation of small vapor bubbles. These bubbles collapse, when they reach a high-pressure area as they move along the impeller vanes. This is called cavitation of the pump. To prevent the adverse effects of cavitation (such as pump noise, loss of head, and impeller damage), it must be ensured that the available NPSH in the system is greater than the NPSH required by the pump.

Specific Speed. The specific speed of an impeller is defined as the revolutions per minute needed to produce 1 gpm at 1 ft head. The specific speed is related to the capacity, head, and impeller speed by

$$N_s = \frac{nQ^{1/2}}{H^{3/4}} \tag{2-45}$$

$$n_s = \frac{nQ_v^{1/2}}{H^{3/4}} \tag{2-46}$$

where N_s = specific speed of the pump impeller, rpm
n_s = specific speed of the blower or fan impeller, rpm
n = impeller speed, rpm
Q = flow, gpm
H = total dynamic head, ft of fluid flowing
Q_v = flow, cfm, in case of fans

Cavitation Parameter σ. The cavitation in the pump must be avoided for the sake of efficiency and for the prevention of impeller damage. For pumps, a cavitation parameter σ is given by

$$\sigma = \frac{P_1/\rho - P_v/\rho + Z_1 - Z_f}{H} \tag{2-47}$$

where H = total dynamic head of the pump, ft
P_v = vapor pressure, ft
P_1 = pressure upon the liquid surface in the suction vessel, ft

The critical value of the cavitation parameter σ_c is the value at which there is an observed change in the efficiency. σ_c and N_s are related as follows:

$$\sigma_c = \begin{cases} 6.3 \times 10^{-6} N_s^{1.33} & \text{for single suction pumps} \quad (2\text{-}48) \\ 4 \times 10^{-6} N_s^{1.33} & \text{for double suction pumps} \quad (2\text{-}49) \end{cases}$$

Suction Specific Speed. The suction specific speed S is given by

$$S = \frac{nQ^{1/2}}{\text{NPSH}^{3/4}} \tag{2-50}$$

σ_c depends upon both N_s and S. A relation among σ_c, N_s, and S is

$$\sigma_c = \left(\frac{N_s}{S_c}\right)^{4/3} \tag{2-51}$$

where S_c is the critical suction speed of the pump, rpm. The critical net positive suction head, NPSH_c is given by

$$\text{NPSH}_c = \sigma_c H \tag{2-52}$$

Performance Curves. A plot of the head developed vs. the pump capacity is called the *performance curve*. It is also called the *head-capacity* or the *HQ* curve (Fig. 2-12). The head developed is the net head obtained after subtraction of the vane and shock losses. For a given pump, the head-capacity curve is unique.

System Head. The total head a pump has to produce is the sum of all the work from the liquid source to the discharge. This includes the frictional resistance, the static head difference, and the pressure head difference that have to be overcome. For a given pump and piping arrangement, a system curve can be prepared and superimposed on the head-capacity curve (Fig. 2-12). The system head curve is a function of the system static head, pressure head, and frictional head. Only the frictional head varies with the flow.

The point of intersection of the system curve with the *HQ* curve is the operating point (*A* in Fig. 2-12). This is the only flow rate the pump will deliver. If a change in the flow rate delivered is desired, characteristics for the system must be changed (point *B* in Fig. 2-12). This is usually accomplished by throttling the discharge valve. Head efficiency and horsepower curves vary with specific speed.

Brake Horsepower. The brake horsepower (bhp) is the actual horsepower consumed by the pump in generating the required head and volumetric flow rate.

Centrifugal Pump Calculations. The expressions for calculation of various quantities in the case of the centrifugal pumps are summarized next:

$$Z = \Delta P \frac{2.31}{s} \quad \text{and} \quad \Delta P = 0.433 s Z \tag{2-53}$$

$$Q = \frac{W}{500s} \tag{2-54}$$

$$\text{TDH} = (P_2 - P_1 + \Delta P_{f1} + \Delta P_{f2}) \frac{2.31}{s} + Z_2 - Z_1 \tag{2-55}$$

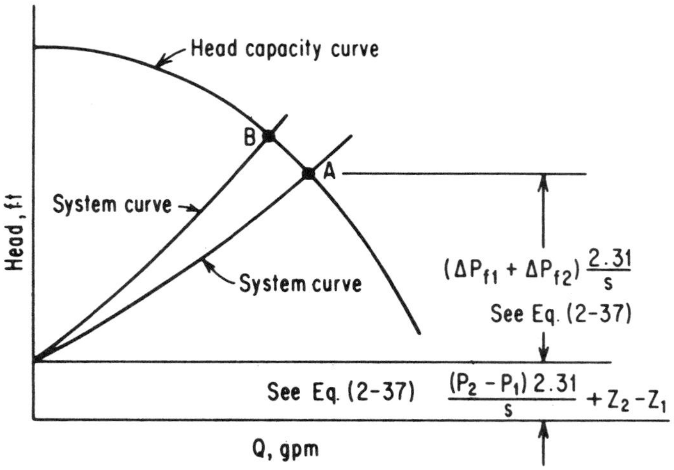

Figure 2-12. Operating characteristics and operating points for a centrifugal pump.

Net Positive Suction Head.

$$\text{NPSH} = (P_1 - P_v - \Delta P_{f1})\frac{2.31}{s} \pm Z_1 \qquad \text{ft} \tag{2-56}$$

$$P_s = P_1 \pm 0.433 Z_1 s - \Delta P_{f1} \tag{2-57}$$

$$P_d = P_s + 0.433(\text{TDH})s \tag{2-58}$$

$$\text{hhp} = \frac{W(\text{TDH})}{1.98 \times 10^6} = \frac{Q(\text{TDH})s}{3960} = \frac{Q(P_d - P_s)}{1714} \tag{2-59}$$

$$\text{bhp} = \frac{W(\text{TDH})}{1.98 \times 10^6 \epsilon} = \frac{Q(\text{TDH})s}{3960\epsilon} = \frac{Q(P_d - P_s)}{1714\epsilon} \tag{2-60}$$

$$\text{hp} = \frac{\text{bhp}}{\epsilon_m} \tag{2-61}$$

$$kW = \frac{0.745(\text{bhp})}{\epsilon_m}$$

where W = weight rate of flow, lb/h
TDH = required total dynamic head of the pump, ft

P_1 = pressure in suction vessel, psia
P_2 = pressure in discharge vessel, psia
ΔP_{f1} = total pressure drop in suction line, psi
ΔP_{f2} = total pressure drop in discharge line, psi
P_v = vapor pressure at pumping temperature, psia
Q = pump capacity, gpm
Z_1 = static suction head, ft
Z_2 = static discharge head, ft
hp = motor horsepower
hhp = hydraulic horsepower
bhp = brake horsepower
ϵ = efficiency of pump
ϵ_m = motor efficiency
P_s = suction pressure, psia
P_d = discharge pressure, psia
s = specific gravity

Actual motor size selection depends upon available standard motor sizes. This results in having motors of ratings 110 to 125 percent of the rated brake horsepower of the pump.

Pump Affinity Laws. These are the relationships among the capacity Q, head H, power bhp, impeller diameter D, and speed of revolution (in revolutions per minute) of centrifugal pumps and fans.

1. Effect of Speed Change When D Is Constant.

Capacity:
$$\frac{Q_2}{Q_1} = \frac{N_2}{N_1} \tag{2-62}$$

Head:
$$\frac{H_2}{H_1} = \left(\frac{N_2}{N_1}\right)^2 \tag{2-63}$$

Power:
$$\frac{\text{bhp}_2}{\text{bhp}_1} = \left(\frac{N_2}{N_1}\right)^3 \tag{2-64}$$

NPSH required:
$$\frac{\text{NPSH}_2}{\text{NPSH}_1} = \left(\frac{N_2}{N_1}\right)^2 \tag{2-65}$$

These laws can be used to determine the performance curve at another rpm level if the performance curve is available at a known rpm.

2. Effect of Impeller Diameter Change. Within the same pump at a constant speed, the following relations apply when the impeller diameter is changed.

Capacity:
$$\frac{Q_2}{Q_1} = \frac{D_2}{D_1} \tag{2-66}$$

Head:
$$\frac{H_2}{H_1} = \left(\frac{D_2}{D_1}\right)^2 \tag{2-67}$$

bhp:
$$\frac{\text{bhp}_2}{\text{bhp}_1} = \left(\frac{D_2}{D_1}\right)^3 \tag{2-68}$$

The laws relating to the impeller diameter are accurate within a certain range of the change of impeller diameter. In general, these laws are not as accurate as the laws relating to the rpm.

3. For Geometrically Similar Pumps with Different Impeller Diameter But Same Speed. Pumps are geometrically similar if they are of different sizes but of the same style, and the relationship between the casing and impeller dimensions is the same. For such pumps,

$$\begin{aligned} Q_2 &= Q_1\left(\frac{D_2}{D_1}\right)^3 & N &= \text{const.} \\ H_2 &= H_1\left(\frac{D_2}{D_1}\right)^2 & N &= \text{const.} \\ \text{bhp}_2 &= \text{bhp}_1\left(\frac{D_2}{D_1}\right)^5 & N &= \text{const.} \end{aligned} \tag{2-69}$$

These laws can be applied to develop a performance curve at a different diameter from a given performance curve at a known diameter.

When the diameter and rpm are both changed, the theorem of joint variation can be applied to estimate the effect of change. The above affinity laws apply to centrifugal fans and blowers, too. In the case of this equipment, NPSH does not apply.

Example 2-11. Determine the available NPSH from Fig. 2-13*a*.

Solution.

$$\begin{aligned}\text{NPSH} &= Z_1 + Z_a - Z_v - Z_f \\ &= 5 + \frac{(14.7 - 10 - 1)(2.31)}{0.5} = 22.1 \text{ ft}\end{aligned}$$

Example 2-12. The adjoining Fig. 2-13*b* shows a typical setup for NPSH requirement test. One set of data in one of the tests was

Pressure gauge reading = 9.12 psi (vacuum)

Vapor pressure of liquid = 0.507 psia

Specific gravity of liquid = 1

Velocity in suction line = 8.03 ft/s

Calculate the NPSH required for the specified flow.

Solution. From Eq. (2-44),

$$\begin{aligned}\text{NPSH} &= Z_{pt} \pm R_G - Z_v + Z_k \pm Z_{1p} \\ &= \frac{14.7 - 9.12 - 0.507}{0.433(1)} + \frac{8.03^2}{2(32.2)} - 0.5 = 12.22 \text{ ft}\end{aligned}$$

Example 2-13. A solution (specific gravity = 1.25, viscosity = 1.2 cP) is pumped through a 100-mm-ID stainless steel pipe of total length 200 m in the horizontal and vertical directions. The net elevation is 15 m. In the line, there are fifteen 90° standard elbows, five ball valves, a control valve, and a filter. The roughness factor of the inside surface is 0.0004 m. Flow rate is 0.0285 m^3/s. Maximum pressure drop through the filter is 210 cmHg, and the equivalent length of the control valve can be taken as 250 pipe diameters. The pipe discharges to an open tank.

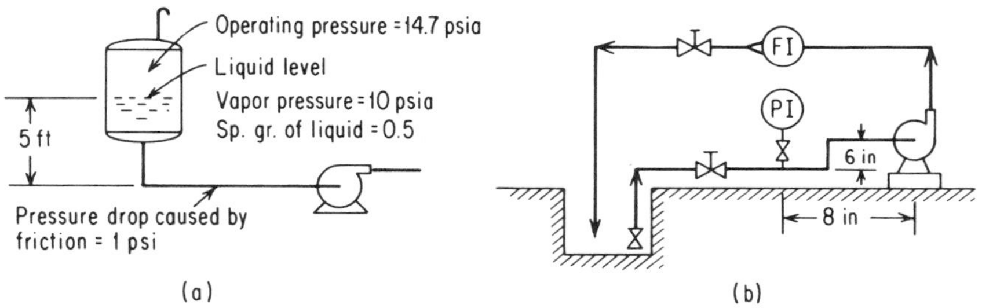

Figure 2-13. (*a*) Data sketch for Example 2-11; (*b*) data sketch for Example 2-12.

Calculate (**a**) pressure loss through the pipes and fittings, (**b**) total head loss in meters, (**c**) total head to be developed by the pump, and (**d**) power requirements of the pump if it is 65 percent efficient.

Solution.

a.

$$\text{Flow through pipe} = 0.0285 \text{ m}^3/\text{s}$$

$$\text{Area of pipe-flow cross section} = \tfrac{1}{4}\pi(0.1)^2 = 0.007854 \text{ m}^2$$

$$\text{Velocity through pipe} = \frac{0.0285}{0.007854} = 3.63 \text{ m/s}$$

$$\text{Viscosity of liquid } \mu = 1.2 \text{ cP} = 0.0012 \text{ N}\cdot\text{s/m}^2 \text{ or kg/m}\cdot\text{s}$$

$$N_{\text{Re}} = \frac{Du\rho}{\mu} = \frac{0.1(3.63)(1250)}{0.0012} = 3.78 \times 10^5$$

$$\text{Roughness factor} = \frac{\epsilon}{D} = \frac{0.00004}{0.1} = 0.0004$$

From Fig. 2-7, Fanning friction factor $f \doteq 0.0043$.

Item	Equivalent Length, m	
Straight pipe	200	
Fifteen 90° std. elbows	45	(3 m each)
Five ball valves	3.5	(0.7 m each)
Control valve	25	(250 pipe diam.)
Pipe exit	4.5	
Total	278	

ΔP through pipe, fittings, and control valve is

$$\Delta P = \frac{2fL_e u^2 \rho}{D}$$

$$= \frac{2(0.0043)(278)(3.63)^2(1250)}{0.1}$$

$$= 393{,}792 \text{ N/m}^2 = 393.8 \text{ kN/m}^2$$

b.

$$\Delta P \text{ through line including control valve} = 393.8 \text{ kN/m}^2$$

$$\Delta P \text{ through filter} = \frac{210(10 \text{ mmHg})(0.133 \text{ kN/m}^2)}{1 \text{ cmHg}}$$

$$= 279.3 \text{ kN/m}^2$$

Then

$$\text{Total } \Delta P = 393.8 + 279.3 = 673.1 \text{ kN/m}^2 = 673{,}100 \text{ N/m}^2$$

$$\text{Head loss} = \frac{673{,}100}{1250(9.81)} = 54.9 \text{ m}$$

c. Total head to be developed is the head loss plus the net elevation:

$$54.9 + 15 = 69.9 \text{ m}$$

d.

$$\text{Mass flow} = (0.0285 \text{ m}^3/\text{s})(1250 \text{ kg/m}^3) = 35.625 \text{ kg/s}$$

$$\text{Power required} = (35.625 \text{ kg/s})(69.9 \text{ m})(9.81 \text{ m/s}^2)$$

$$= 24{,}429 \text{ (kg}\cdot\text{m}^2/\text{s}^2)/\text{s}$$

$$= 24.43 \text{ kW}$$

$$= \frac{24.43}{0.746} = 32.8 \text{ hp}$$

Example 2-14. A pump takes brine from a tank and transports it to another tank through a 6-in schedule 40 line. A sketch of the system is shown in Fig. 2-14. The flow rate is 825 gpm. In the suction line there are: one gate valve, one strainer, two standard tees, and two 90° elbows. In the discharge line, there are: six standard tees, one gate valve, one check valve, and five 90° short elbows (not all shown in the figure). Two

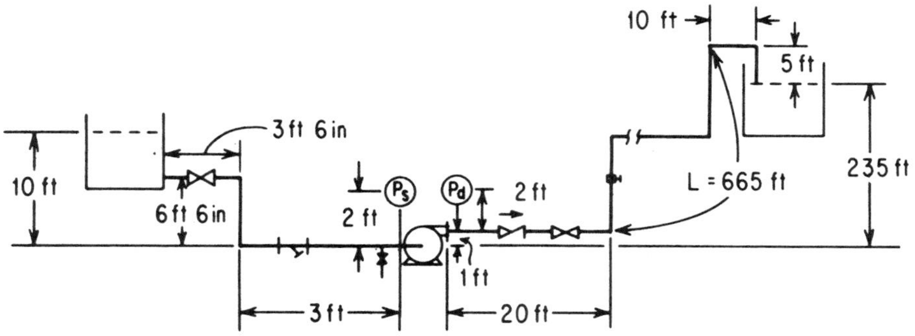

Figure 2-14. Data sketch for Example 2-14.

pressure gauges are installed at the suction and discharge of the pump as shown. Specific gravity of brine is 1.2, and its viscosity is 1.2 cP. Calculate the frictional pressure drops in the suction and discharge lines. What pressures are registered by the gauges?

Solution. Calculate the pressure drop in the suction line. The line contains the items shown in Table 2-3.

TABLE 2-3

	K
One sudden contraction	1(0.5) = 0.50
One gate valve	1(0.09) = 0.09
One strainer	1(0.88) = 0.88
Two standard tees	2(0.6) = 1.2
Two 90° short elbows	2(0.3) = 0.6
	ΣK = 3.27
Straight pipe length = 13 ft	

$$\text{ID} = 6.065 \text{ in} = 0.5054 \text{ ft}$$

$$\text{Velocity of brine} = \frac{825 \text{ gpm}}{7.48(60)(0.785)(0.5054)^2} = 9.163 \text{ ft/s}$$

$$N_{Re} = \frac{Du\rho}{\mu} = \frac{0.5054(9.163)(1.2)(62.4)}{1.2(0.000672)} = 4.3 \times 10^5$$

$$\frac{\epsilon}{D} = \frac{0.00015}{0.5054} = 0.0003$$

f from Fanning friction factor chart (Fig. 2-7) is 0.0041. The pressure drop on the suction side is

$$\Delta P_S = \left(\Sigma K + \frac{4fL}{D}\right)\frac{u^2}{2g_c}$$

$$= \left[3.27 + \frac{4(0.0041)(13)}{0.5054}\right]\frac{9.163^2}{64.4}$$

$$= 4.8 \text{ ft} = 4.8(0.433)(1.2) = 2.5 \text{ psi}$$

The suction static head is

$$10(0.433)(1.2) = 5.2 \text{ psi}$$

The suction velocity head is

$$\frac{9.163^2}{64.4}(0.433)(1.2) = 0.678 \doteq 0.7 \text{ psi}$$

Therefore the pressure at inlet of nozzle is

$$14.7 + 5.2 - 2.5 - 0.7 = 16.7 \text{ psia}$$

The pressure indicated by gauge is 16.7 minus the elevation of the gauge from the centerline at the suction nozzle and equals

$$16.7 - 2(0.433)(1.2) \doteq 15.7 \text{ psia} = 1 \text{ psig}$$

Calculate pressure drop in the discharge line. The line contains the data listed in Table 2-4.

TABLE 2-4

		K
Six standard tees	6(0.6) =	0.36
One gate valve		0.09
Five check valve		2.0
Five 90° short elbows	5(0.3) =	1.5
One sudden expansion		1.0
	ΣK =	4.95
Total pipe length = 700 ft		

The friction loss in the discharge line is therefore

$$\Delta P_{fd} = \left[4.95 + \frac{4(0.0041)(700)}{0.5054}\right]\frac{9.163^2}{64.4} = 36.1 \text{ ft} = 18.8 \text{ psi}$$

The discharge static head is

$$(235 \text{ ft})(0.433)(1.2) = 122.1 \text{ psi}$$

The pressure indicated by the gauge (neglecting the velocity head) is calculated to be

$$\begin{aligned} 14.7 + 122.1 + 18.8 - \text{elevation} &= 155.6 - 3(0.433)(1.2) \\ &= 154 \text{ psia} = 139.3 \text{ psig} \end{aligned}$$

where the elevation of the gauge from the centerline is in psi.

Example 2-15. In Example 2-14, calculate **(a)** NPSH available, **(b)** total head, **(c)** hydraulic horsepower, and **(d)** brake horsepower if efficiency is 74.7 percent.

The vapor pressure of water over the brine solution at 86°F is 0.6 psia.

Solution.

a. From Eq. (2-38) and considering the velocity head negligible,

$$\text{NPSH}_A = Z_1 + Z_a - Z_v - Z_f$$

From previous problem solution and the data of this example,

$$Z_1 = 10 \text{ ft}$$

$$Z_a = 14.7 \text{ psia} = 14.7\frac{2.31}{1.2} = 28.3 \text{ ft}$$

$$Z_f = 4.8 \text{ ft} \qquad \text{(calculated in Example 2-14)}$$

$$Z_v = (0.6 \text{ psi})\,\frac{2.31}{1.2} \doteq 1.2 \text{ ft}$$

$$\text{NPSH}_A = 10 + 28.3 - 1.2 - 4.8 = 32.3 \text{ ft}$$

b. *Total Head.*

$$\text{TDH} = (P_2 - P_1 + \Delta P_{f1} + \Delta P_{f2})\,\frac{2.31}{s} + Z_2 - Z_1$$

$$= (14.7 - 14.7 + 2.5 + 18.8)\frac{2.31}{1.2} + 235 - 10$$

$$= 266 \text{ ft} = 138.3 \text{ psi}$$

c. *Hydraulic Horsepower.*

$$\text{hhp} = \frac{(\text{gpm})(\text{psi})}{1714} = \frac{825(138.3)}{1714} = 66.6$$

d. *Brake Horsepower.*

$$\text{bhp} = \frac{\text{hhp}}{\text{efficiency}} = \frac{66.6}{0.747} = 89.2$$

Example 2-16. Water is pumped at 86°F through a piping system from one tank to another as shown in Fig. 2-15. The suction line consists of 30 ft of 3-in schedule 40 straight pipe and 260 ft of 2-in schedule 40 pipe in the discharge line. If the globe valves are completely open, determine the flow through piping in gallons per minute. Assume the viscosity of water is 0.85 cP. Data for the centrifugal pump to be used are given below.

Capacity, gpm	0	10	20	30	40	50	60	70	80
Developed head, ft	120	119.5	117	113	107.5	100	90	82	75
Efficiency, %	0	13	23.5	31.6	37.5	42.2	42.5	41.7	39.5

Calculate the brake horsepower.

Solution. Plot the pump-head curve as in Fig. 2-16. The flow rate through the piping will be given by the intersection of the system curve with the pump-head curve. To construct the system curve, the total head for a given flow through the piping system is given by

$$H = P_2 - P_1 + Z_2 - Z_1 + \Delta P_{f1} + \Delta P_{f2}$$

$$P_2 - P_1 = 0$$

$$Z_2 - Z_1 = 60 - 8 = 52 \text{ ft}$$

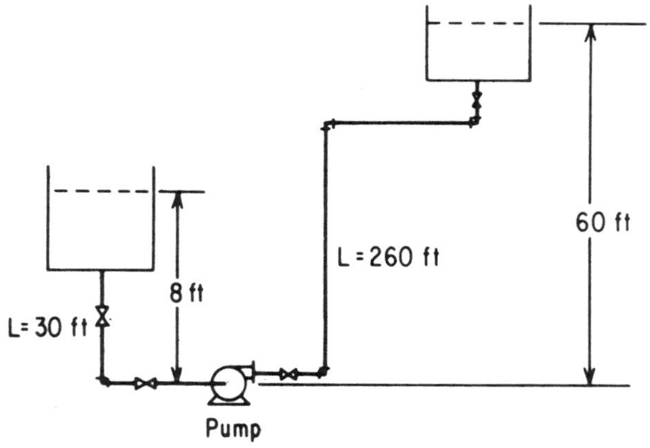

Figure 2-15. Data sketch for Example 2-16.

$$\Delta P_{f1} = h_1 = \frac{2f_1 L_1 u_1^2}{g_c D_1}$$

$$\Delta P_{f2} = h_2 = \frac{2f_2 L_2 u_2^2}{g_c D_2} \qquad \text{(both in feet)}$$

where L_1 and L_2 are total equivalent lengths. See Table 2-5.

TABLE 2-5

	Length, ft
Suction Line	
3-in pipe length	30
Contraction	7.5
Two open glove valves	180.0
One 90° elbow	7.5
Total	225
Discharge Line	
2-in pipe length	260
Three elbows	16
Two globe valves	120
Sudden expansion	1.3
Total	397.3

$$D_1 = \frac{3.068}{12} = 0.2557 \text{ ft} \qquad D_2 = \frac{2.067}{12} = 0.1723 \text{ ft}$$

Prepare a table of total head vs. capacity for the pump.

$$\Delta P_{f1} = \frac{2(225)f_1 u_1^2}{32.17(0.2557)} = 54.7 f_1 u_1^2$$

$$\Delta P_{f2} = \frac{2(397.3)f_2 u_2^2}{32.17(0.1723)} = 143.4 f_2 u_2^2$$

$$u_1 = \frac{\text{gpm}}{7.48(60)(0.785)(0.2557)^2} = 0.044 \text{ (gpm)}$$

$$u_2 = \frac{\text{gpm}}{7.48(60)(0.785)(0.1723)^2} = 0.096 \text{ (gpm)}$$

$$H = 52 + 54.7 f_1 u_1^2 + 143.4 f_2 u_2^2$$

$$\frac{\epsilon}{D_1} = \frac{0.00015}{0.2557} = 0.00059 \qquad \frac{\epsilon}{D_2} = \frac{0.00015}{0.1723} = 0.0009$$

The head H is plotted vs. capacity in gallons per minute in Fig. 2-16 to get the system curve. The system curve cuts the pump head curve at 70 gpm and 85 ft head. At 70 gpm, efficiency is 41.5 percent. Thus,

$$\Delta P = 85(0.433) = 36.84 \text{ psi}$$

$$\text{bhp} = \frac{70(36.84)}{1714(0.415)} = 3.62$$

Assuming 125 percent of the rated brake horsepower, the motor hp is

$$\text{hp} = 3.62(1.25) = 4.53$$

A 5-hp standard size motor will have to be specified.

Example 2-17. A liquid pumping system has the following data:

Capacity	42 gpm
TDH	50 ft
NPSH available	10 ft
Specific gravity	0.5

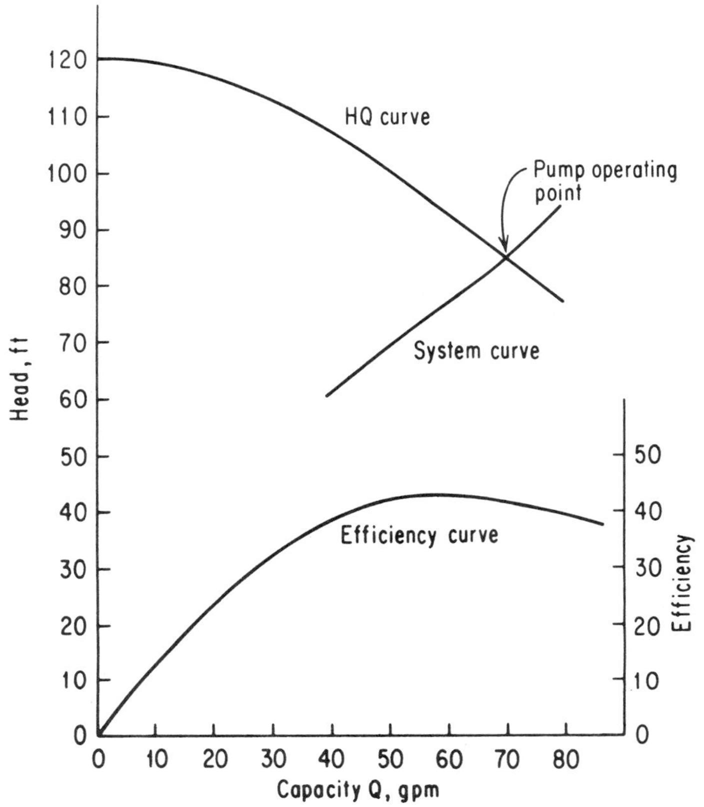

Figure 2-16. *HQ* and system curves (Example 2-16).

A pump with a $5\frac{1}{2}$-in impeller is available in storage. However, the performance curve is lost. The only information from the vendor is the performance curves of a 6-in impeller pump of the identical model number at various rpm as shown in Fig. 2-17. Also available in stock are electric motors with an 80 percent efficiency as given in Table 2-6. Justify whether the pump in storage can be used for the specified duty with or without any modification of the pump dimensions. Also determine which one of the above motors will meet the pumping requirement.

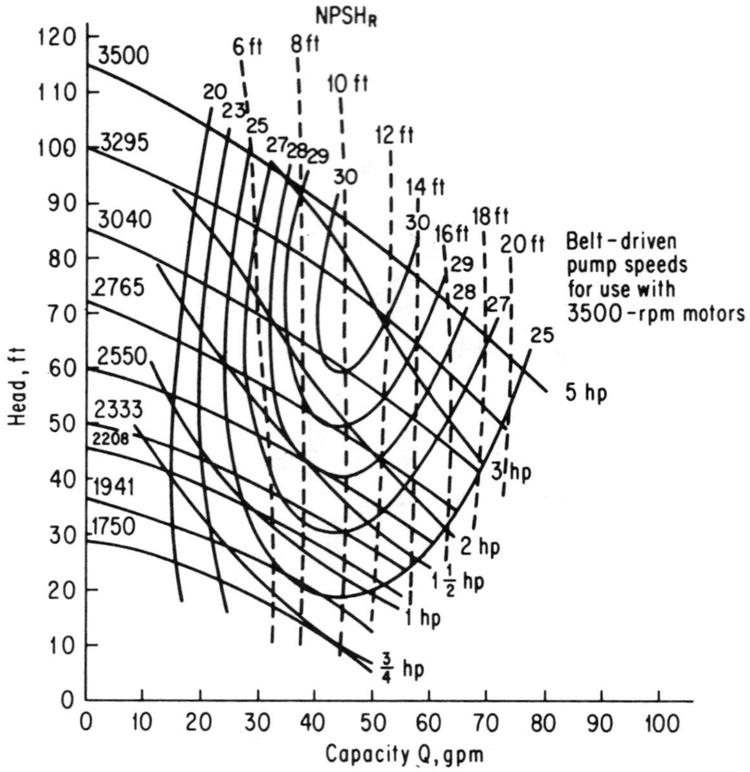

Figure 2-17. Pump characteristics (Example 2-17). Pump size = 1 × 1 – 6; eye area = 1.23in^2.

TABLE 2-6 Revolutions per Minute of Stock Electric Motors

hp		
1	**$1\frac{1}{2}$**	**2**
3500	3500	3500
3295	3295	3295
3040	3040	3040
2500	2500	2500
1750	1750	1750

Solution. The brake horsepower is given by

$$\text{bhp} = \frac{(\text{gpm})(\text{psi})}{1714\ (\text{pump efficiency})} = \frac{(\text{gpm})(\text{TDH})(0.433)(s)}{1714\ (\text{pump efficiency})}$$

$$= \frac{42(50)(0.433)(0.5)}{1714\ (\text{pump efficiency})} = \frac{0.265}{\text{pump efficiency}}$$

We do not know the pump efficiency. We ask the question: If a $5\frac{1}{2}$-in impeller pump has a capacity of 42 gpm and TDH of 50 ft, what are the corresponding capacity and TDH of a 6-in impeller pump at the same efficiency and same rpm?

From the affinity laws, *at an equal efficiency*, and *equal rpm*,

$$\frac{Q_6}{Q_{5(1/2)}} = \frac{D_6}{D_{5(1/2)}} = \frac{6}{5.5} = 1.091$$

$$Q_6 = 1.091(42) = 46 \text{ gpm}$$

$$\frac{H_6}{H_{5(1/2)}} = \left(\frac{D_6}{D_{5(1/2)}}\right)^2 = 1.19$$

$$H_6 = 1.19(50) = 59.5 \text{ ft}$$

Locating $Q_6 = 46$ gpm and TDH = 59.5 ft on the performance curve gives the impeller 3040 rpm and a pump efficiency of 30 percent. This is also the efficiency of a $5\frac{1}{2}$-in impeller at 42 gpm and 50 ft TDH and operating at 3040 rpm. Therefore the brake horsepower of the pump in storage is

$$\frac{0.265}{0.30} = 0.883 \qquad \text{Motor hp} = \frac{0.883}{0.8} = 1.10$$

Hence, choose a $1\frac{1}{2}$-hp, 3040-rpm motor, and the existing pump will work. Next, check the NPSH requirement. From the performance curve, the NPSH required is 10 ft, when the impeller diameter is 6 in at 3040 rpm. At the same rpm but with an impeller diameter of $5\frac{1}{2}$ in, estimate the NPSH required:

$$\frac{H_2}{H_1} = \left(\frac{D_2}{D_1}\right)^2 \qquad \text{when } N = \text{const.}$$

$$H_2 = H_1\left(\frac{D_2}{D_1}\right)^2 = 10\left(\frac{5.5}{6}\right)^2 = 8.4 \text{ ft}$$

The motor hp can also be estimated alternatively. From Fig. 2-17, the bhp for a 6-in impeller at the point of equivalence (i.e., at 46 rpm and 59.5 ft TDH) is 2.4 hp approximately. Hence bhp for a $5\frac{1}{2}$-in impeller at the same rpm (3040) by the affinity law is

$$\text{bhp}\left(\frac{D_{5(1/2)}}{D_6}\right)^3(s) = 2.4\left(\frac{5.5}{6}\right)^3(0.5) = 0.92$$

$$\text{Motor hp} = \frac{0.92}{0.8} = 1.15$$

So, choose a $1\frac{1}{2}$-hp motor of 3040 rpm.

Note: The bhp's shown on the pump performance curves are generally based on water. Hence, when estimating the bhp for any other liquid, correction of specific gravity needs to be made, as shown above.

Control Valves

A control valve is a variable opening used to regulate the flow of a process fluid as required by the process. The three important aspects of a control valve are capacity, characteristics, and rangeability.

Capacity. This is expressed in terms of C_v, the flow coefficient, which is the flow of water at 60°F in gallons per minute at a pressure drop of 1 psi across the valve when it is completely open.

Characteristics. This is the relationship between the change in the valve opening and the change in the flow through the valve, viz., equal percentage, linear and nonlinear.

Rangeability. This is the ratio of the maximum controllable flow to the minimum controllable flow. In terms of C_v, the rangeability R is given by

$$R = \frac{\text{rated } C_v}{\text{minimum controllable } C_v}$$

The formulas used for control valve sizing are as follows:

$$\text{Volumetric flow:} \quad C_v = \begin{cases} Q\sqrt{\dfrac{s}{\Delta P}} & \text{liquid service} \\[2ex] \dfrac{Q}{963}\sqrt{\dfrac{s_g T}{\Delta P(P_1 + P_2)}} & \text{gas service} \end{cases} \tag{2-70}$$

$$\text{Mass flow:} \quad C_v = \begin{cases} \dfrac{W}{500\sqrt{s\ \Delta P}} & \text{liquid service} \\[2ex] \dfrac{W}{3.22\sqrt{\Delta P(P_1 + P_2)s_{gf}}} & \text{gas service} \end{cases} \tag{2-71}$$

where C_v = valve coefficient
Q = flow rate, gpm for liquid, ft^3/min for gases and vapors
s = specific gravity at flowing temperature (water s at 60°F = 1)
ΔP = pressure drop across valve, psi
W = flow rate, lb/h
s_g = specific gravity of gas relative to air
= $\frac{1}{29}$ the molecular weight of gas
$s_{gf} = s_g(520/T)$
T = flow temperature, °R
P_1 = upstream pressure, psia
P_2 = downstream pressure, psia

In terms of ΔP and Q (gpm) the rangeability is given by

$$R = \frac{Q_{max}}{Q_{min}}\sqrt{\frac{\Delta P_{min}}{\Delta P_{max}}} \tag{2-72}$$

where Q_{max} = maximum flow
Q_{min} = minimum flow
ΔP_{min} = pressure drop at the minimum flow (not the minimum pressure drop)
ΔP_{max} = pressure drop at the maximum flow (not the maximum pressure drop)

Equations (2-70) and (2-71) apply to subcritical flow. For liquids, when the ΔP across the control valve is less than the ΔP that would cause flashing, the flow is subcritical. For gases, the flow is subcritical when the ΔP across the control valve is roughly less than 0.5 times upstream pressure in absolute units. The sizing calculation of the control valve for critical and supercritical flows requires additional information which is beyond the scope of this book.

Control Valve Sizing. For sizing a control valve for liquid service, a flow coefficient C_{vc} is calculated with the design flow rate in gallons per minute from

$$C_{vc} = Q\sqrt{\frac{s}{\Delta P}} \tag{2-73}$$

For a good range of control, a value of $C_v > C_{vc}$ is required. The normal practice is to select a valve such that[3b]

$$\frac{C_v}{C_{vc}} = 1.25 \text{ to } 2 \tag{2-74}$$

where C_v is the coefficient of the selected valve from the manufacturer's catalog. The pressure drop across the control valve is generally taken as 50 percent of the frictional drop in the system excluding the control valve or 5 to 10 psi minimum. Thus if the frictional drop in the line, heat exchanger, fittings, etc., at the design flow rate is 30 psi, the drop across the control valve should be 15 psi. For a very high-pressure drop line, the percentage drop across the valve may be 15 to 20 percent of the frictional drop through the system.

The minimum pressure drop at the fully open condition is given by

$$\Delta P_{\min} = \left(\frac{Q}{C_v}\right)^2 s \qquad \text{psi} \tag{2-75}$$

For intermediate positions, the pressure drop can be

calculated by

$$\Delta P = \left[\frac{Q}{(C_{vc}/C_v)C_v}\right]^2 s \qquad \text{psi} \tag{2-76}$$

where C_v is the valve coefficient from the manufacturer's catalog and C_{vc}/C_v is the intermediate flow condition.

When sizing control valves for a gas, steam, or vapor service, the valve coefficients may be calculated with the use of either Eq. (2-70) or (2-71). For good control, the ratio of the selected valve coefficient to the calculated coefficient should be according to Eq. (2-74) in the case of gases also.

Example 2-18. Select the control valve size from the following data assuming subcritical flow of 100 gpm, an available pressure drop of 10 psi, and a specific gravity of 2.5.

Valve size, in	1	$1\frac{1}{2}$	2	3	4
C_v	7.6	19	35	76	130

After the final selection, estimate the maximum flow through the valve for the specified pressure drop.

Solution.

$$C_{vc} = Q\sqrt{\frac{s}{\Delta P}} = 100\sqrt{\frac{2.5}{10}} = 50$$

With a 3-in valve,

$$C_v = 76$$

$$\frac{C_v}{C_{vc}} = \frac{76}{50} = 1.52$$

This ratio is between 1.25 and 2 and acceptable. Therefore,

$$76 = Q\sqrt{\frac{2.5}{10}} = Q(0.5)$$

$$Q = 152 \text{ gpm} \qquad \text{(maximum)}$$

Example 2-19. An existing pumping system has a flow control valve working at 50 percent of the rated flow coefficient. The system study indicates that the valve is working with a pressure drop of 20 psi and specific gravity of 0.9. Because of the expansion of the plant capacity, the flow from the pump is to be increased from 100 to 130 gpm and, consequently, the pressure drop available for the control valve drops to 15 psi. If the pump is retained, will the control valve need replacement?

Solution. For existing operation,

$$C_{vc} = Q\sqrt{\frac{s}{\Delta P}} = 100\sqrt{\frac{0.9}{20}} = 21.2$$

$$\text{Rated } C_v = \frac{21.2}{0.5} = 42.4$$

For the new case,

$$C_{vc} = 130\sqrt{\frac{0.9}{15}} = 31.8 \qquad \frac{C_v}{C_{vc}} = \frac{42.4}{31.8} = 1.33$$

This ratio is within the range of 1.25 to 2, and hence the control valve would work and would need no replacement.

Example 2-20. In a system, the frictional drop excluding the pressure drop across the control valve for the maximum controllable flow of 100 gpm is 30 psi. The pressure drop available for the control valve for this maximum flow is 5 psi. If the rangeability of the control valve is 50 and the terminal pressures do not change, estimate the minimum controllable flow of the valve.

Solution.

$$R = \frac{Q_{\max}}{Q_{\min}}\sqrt{\frac{\Delta P_{\min}}{\Delta P_{\max}}}$$

$$\Delta P_f = \text{frictional drop} = \Delta P_1 + \Delta P_2$$

$$P_1 - P_2 = \text{constant} = \Delta P_f + \Delta P$$

$$\Delta P = \text{control valve pressure drop}$$

ΔP_1 = frictional drop upstream of the control valve

ΔP_2 = frictional drop downstream of the control valve

P_1 = upstream terminal pressure

P_2 = downstream terminal pressure

Any lowering in the frictional drop would increase the available ΔP for the control valve.

$$(\Delta P_f)_{max} = 30 \text{ psi}$$

when the flow is

$$Q_{max} = 100 \text{ gpm} \qquad \text{and} \qquad \Delta P_{max} = 5 \text{ psi}$$

$$(\Delta P_f)_{min} = \left(\frac{Q_{min}}{Q_{max}}\right)^2 (30)$$

The decrease in the frictional drop at the minimum flow is

$$30 - \left(\frac{Q_{min}}{Q_{max}}\right)^2 (30) = 30\left[1 - \left(\frac{Q_{min}}{Q_{max}}\right)^2\right]$$

$$\Delta P_{min} = 5 + 30\left[1 - \left(\frac{Q_{min}}{Q_{max}}\right)^2\right]$$

$$R = \frac{Q_{min}}{Q_{min}}\sqrt{\frac{\Delta P_{min}}{\Delta P_{max}}}$$

$$50 = \frac{100}{Q_{min}}\sqrt{\frac{5 + 30[1 - (Q_{min}/100)^2]}{5}}$$

$$1.25\, Q_{min}^2 = 5 + 30 - 0.003\, Q_{min}^2$$

Solving for Q_{min} gives $Q_{min} = 5.3$ gpm.

Flow of Compressible Fluids

A fluid is compressible when its density changes with pressure. The following terms are often used in connection with the flow of compressible fluids.

Mach Number N_{Ma}. This is the ratio of the fluid velocity to the velocity of sound in the same fluid under identical temperature and pressure.

$$N_{Ma} = \frac{u}{a} \tag{2-77}$$

where u = fluid velocity
a = velocity of sound in the same fluid

$$N_{Ma} \begin{cases} < 1, \text{ subsonic flow} \\ = 1, \text{ sonic flow} \\ > 1, \text{ supersonic flow} \end{cases}$$

Critical Pressure Ratio. For a constant pressure, the flow rate of a compressible fluid increases as the downstream pressure is lowered until a downstream pressure is reached at which the flow rate is the maximum and the fluid velocity equals the sonic velocity. At this condition, the ratio of downstream pressure to the upstream pressure in absolute scale is called the *critical pressure ratio*. For gases obeying ideal-gas law, the critical pressure ratio is given by

$$\frac{P_{\text{downstream}}}{P_{\text{upstream}}} = \left(\frac{2}{k+1}\right)^{k/(k-1)} \tag{2-78}$$

$$k = \frac{C_p}{C_v} = \frac{C_p}{C_p - 1.987} \tag{2-78a}$$

where k = specific heat ratio
C_p = molar specific heat at constant pressure, Btu/lb · mol · °F
C_v = molar specific heat at constant volume, Btu/lb · mol · °F

If the downstream pressure is less than what is calculated from the above equation, the flow rate becomes independent of the downstream pressure.

Critical Velocity. The critical velocity of a gas through a pipe or an orifice is the sonic velocity. It is given by

$$u_s = \sqrt{kg_cRT} \qquad \text{ft/s} \tag{2-79}$$

where k = ratio of specific heats at constant pressure and volume, respectively

g_c = 32.2 lb · ft/lbf · s^2

$R = \dfrac{1545}{M}$ ft · lbf/lb · °R

T = temperature, °R

M = molecular weight

For the estimation of the flow of compressible fluids, the following rules may be used:

1. When the pressure drop is less than 10 percent of inlet pressure, use the density based on either the inlet or outlet conditions.
2. When the pressure drop is more than 10 percent but less than 40 percent of the inlet pressure, use the average density based on the inlet and outlet conditions.
3. For a pressure drop more than 40 percent of the inlet pressure, both the inlet and outlet densities should be considered.
4. At a very high pressure, the flow of gases may be treated the same way as the flow of incompressible fluid since the change in the density is small.

Isothermal Flow and Adiabatic Flow of Compressible Fluid Obeying Ideal-Gas Law. For these topics, see other texts.

Compression Equipment. The objective of compression is to deliver a gas in a required quantity at a pressure higher than the initial. The compression equipment is basically of two types:

1. Positive displacement, e.g., reciprocating
2. Velocity or dynamic compressor, e.g., centrifugal

Compression is of three types: (1) adiabatic (2) isothermal, and (3) polytropic. In the adiabatic compression there is no heat exchange with the surroundings and the equation $PV^k = C_1$ applies, where k is the ratio of specific heats C_p/C_v for the gas.

During an isothermal compression, the heat of compression is removed from the gas by cooling to maintain the gas temperature constant. The equation $PV = C_2$ applies in this case.

The polytropic compression is neither adiabatic nor isothermal. In this case the equation $PV^n = C_3$ applies. If ϵ_p is the polytropic compression efficiency, k and n are related by the equation

$$\frac{n}{n-1} = \epsilon_p \frac{k}{k-1} \tag{2-80}$$

where n is the polytropic coefficient. An average value of 0.725 for ϵ_p is usually assumed for estimation purposes. The actual efficiency will be different depending upon the speed, wheel design, compression ratio, and other factors.

A compressor operates on a predetermined performance curve which shows the relationship between the total head and horsepower requirement as a function of the volumetric flow rate. As in the case of pumps, the exact operating point of a compressor is determined by the intersection of the system and performance curves. Compressor calculations are done either with the use of simplified equations (Table 2-7) for the adiabatic or polytropic paths or by the use of the Mollier diagrams where such diagrams are available or by using the thermodynamic property data. The Mollier diagrams are easier to use but they are available mostly for the pure components, especially for the refrigerants.

Ratio of Specific Heats *k*. Values of the specific heat ratios k are available for some gases at 1 atm. When the k value is not available, it can be calculated from molar specific heats at constant pressure by the formula

$$k = \frac{C_p}{C_p - 1.987} \tag{2-81}$$

For gas mixtures, an average k_{av} can be calculated on the basis of their mole fractions.

Multistage Compression. The multistage operation allows interstage cooling of the gas which reduces work of compression. The minimum work is obtained when the compression ratio in each stage is equal to

$$\frac{P_n}{P_{n-1}} = \left(\frac{P_n}{P_1}\right)^{1/N_s} \tag{2-82}$$

where N_s = number of stages
P_n = pressure after n stages
P_1 = pressure to first stage

Fans. These are used for low pressure and usually for pressure heads less than 1.5 psi. They are either centrifugal or axial flow type. Air horsepower of a fan is given by

$$\text{Air hp} = \frac{144(P_2 - P_1)Q_V}{33{,}000} = 0.00436\,(P_2 - P_1)Q_V \tag{2-83}$$

where P_2 and P_1 are in psia. If $P_2 - P_1$ is expressed in inches of water, the air horsepower is given by

$$\text{Air hp} = 0.0001575 Q_V\,\Delta P \tag{2-84}$$

where ΔP is the developed head across the fan in inches of water and Q_V is flow in cubic feet per minute at the inlet conditions.

Fan Static Pressure P_s. This is the total pressure rise ΔP minus the velocity pressure in the fan outlet. Thus

TABLE 2-7 Calculation of Power Requirements in Compression

Type of Compression	Adiabatic
Equation of state	$PV^k = C_1$
Exponent	$k = \dfrac{C_p}{C_v}$
Theoretical discharge temperature	$T_2 = T_1\left(\dfrac{P_2}{P_1}\right)^{(k-1)/k}$
Actual discharge temperature	$T_2 = T_1 + \dfrac{T_1[(P_2/P_1)^{(k-1)/k} - 1]}{\epsilon_{ad}}$
Head, ft · lbf/lb or ft	$H_{ad} = \dfrac{k}{k-1}\dfrac{RT_1}{M}\left[\left(\dfrac{P_2}{P_1}\right)^{(k-1)/k} - 1\right]Z_{av}$
Gas horsepower, ghp (using cfm)	$\text{ghp} = \dfrac{0.00437Q_1P_1[k/(k-1)][(P_2/P_1)^{(k-1)/k} - 1]Z_{av}}{\epsilon_{ad}}$
Gas horsepower, ghp (using weight)	$\text{ghp} = \dfrac{WH_{ad}}{33{,}000\ \epsilon_{ad}}$
Brake horsepower, bhp	$\text{bhp} = \dfrac{\text{ghp}}{\epsilon_m}$

C_p, C_v = molar specific heats at constant pressure and constant volume, respectively, Btu/lb mol · °R
T_1, T_2 = inlet and outlet temperature, °R
P_1, P_2 = inlet and discharge pressure, psia
ϵ = efficiency (ad = adiabatic, P = polytropic, iso = isothermal, m = mechanical)
W = lb/min, Q_1 = cfm at suction conditions
H_{ad} = adiabatic head, ft; H_p = polytropic head, ft; H_{iso} = isothermal head, ft
R = gas constant = 1545 ft · lbf/lb · mol · °R
M = molecular weight
Z_{av} = $\frac{1}{2}(Z_1 + Z_2)$, Z_1, Z_2 compressibilities at suction and discharge

Polytropic	Isothermal
$PV^n = C_2$	$PV = C_3$
$\dfrac{n-1}{n} = \dfrac{k-1}{k}\dfrac{1}{\epsilon_P}$	1
$T_2 = T_1\left(\dfrac{P_2}{P_1}\right)^{(n-1)/n}$	$T_2 = T_1$
$T_2 = T_1\left(\dfrac{P_2}{P_1}\right)^{(n-1)/n}$	$T_2 = T_1$
$H_p = \dfrac{n}{n-1}\dfrac{RT_1}{M}\left[\left(\dfrac{P_2}{P_1}\right)^{(n-1)/n} - 1\right]Z_{av}$	$H_{iso} = Z_{av}\dfrac{RT_1}{M}\ln\dfrac{P_2}{P_1}$
$\text{ghp} = \dfrac{0.00437Q_1P_1[(P_2/P_1)^{(n-1)/n} - 1]Z_{av}}{\epsilon_P}$	$\text{ghp} = \dfrac{0.00437Q_1P_1\ln(P_2/P_1)}{\epsilon_{iso}}$
$\text{ghp} = \dfrac{WH_P}{33{,}000}$	$\text{ghp} = \dfrac{WH_{iso}}{33{,}000\ \epsilon_{iso}}$
$\text{bhp} = \dfrac{\text{ghp}}{\epsilon_m}$	$\text{bhp} = \dfrac{\text{ghp}}{\epsilon_m}$

$P_s = \Delta P - P_v$. P_v = velocity head expressed in inches of water. It is given by

$$P_v = \frac{u_2^2}{2g_c}\left[\frac{\rho_{air}(12)}{62.4}\right] \qquad \text{in}H_2O$$

where u_2 is the fan outlet velocity, ft/s.

$$u_2 = 18.3\sqrt{\frac{P_v}{\rho_{air}}} \qquad \rho_{air} = \text{air density} \qquad \text{lb/ft}^3 \qquad (2\text{-}85)$$

Example 2-21. A centrifugal fan operating at 1740 rpm has charactertistics as shown in Fig. 2-18. It is connected to a duct system which offers a static resistance of 1.5 inH_2O when handling 2500 cfm of air. (**a**) At what flow, static pressure, and bhp will the fan and the duct system operate when connected together? (**b**) The flow through the duct is to be 4500 cfm of air by changing pulley ratios. What speed, static pressure, and bhp would be required to do this?

Solution.

a. A fan or a pump will always operate at the intersection of its head-capacity curve and the system curve. Therefore, a system curve has to be prepared first.

ΔP at 2500 cfm is given to be 1.5 inH_2O. Assume turbulent flow and f, the friction factor, constant. Then $\Delta P \propto u^2$, i.e., as Q^2, if the duct cross section is constant. Calculate values as follows:

cfm	2500	3000	3250	3500	4000
ΔP, in H_2O	1.5	2.16	2.54	2.94	3.84

These points are plotted on the same graph to give the system curve which cuts the HQ curve. This is the operating point for the system (Fig. 2-18). At this point,

$$\text{Flow} = 3520 \text{ cfm}$$

$$\text{bhp} = 3.28$$

$$\text{Static pressure} = 2.98 \text{ in}H_2O$$

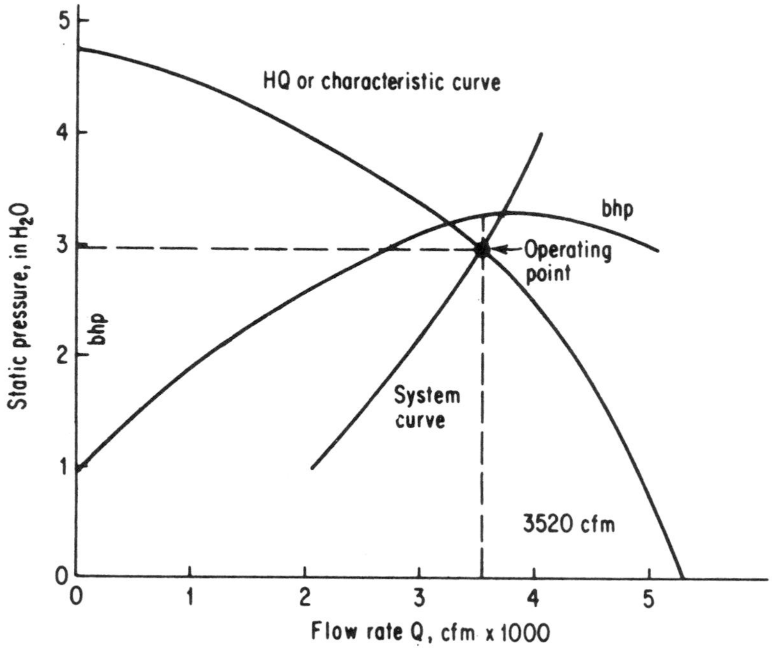

Figure 2-18. Data and solution (Example 2-21).

b. To obtain the solution to part **b**, make use of the affinity laws:

$$\Delta P_s \propto Q^2 \qquad \text{New } \Delta P_s = 2.98\left(\frac{4500}{3520}\right)^2 = 4.87 \text{ inH}_2\text{O}$$

$$\text{bhp} \propto Q^3 \qquad \text{New bhp} = 3.28\left(\frac{4500}{3520}\right)^3 = 6.85 \text{ bhp}$$

$$\text{rpm} \propto Q \qquad \text{New rpm} = 1740\,\frac{4500}{3520} = 2225 \text{ rpm}$$

References

1. *Chemical Engineers' Handbook*, 5th ed., R. H. Perry (ed.), McGraw-Hill, New York, 1973; (*a*) p. 5-7; (*b*) p. 5-11; (*c*) p. 5-13; (*d*) p. 5-10; (*e*) p. 5-36.
2. L. L. Simpson and M. L. Weirick, *Chem. Eng.*, vol. 85, desk book issue/April 3, 1978, pp. 35–42.
3. R. Kern, *Chem. Eng.*, vol. 82, 1975: (*a*) Jan. 6, pp. 115–120; (*b*) April 14, p. 88.

Chapter

3

Heat Transfer

There are three modes of heat transfer: conduction, convection, and radiation. The basic laws governing these are covered in this chapter.

Thermal Conduction

Fourier's law states that

$$q_k = -kA \frac{dT}{dx} \tag{3-1}$$

where q_k = heat transferred by conduction, Btu/h
k = thermal conductivity of material, Btu/(h · ft^2 · °F/ft)
A = area of cross section perpendicular to direction of heat flow, ft^2
dT/dx = temperature gradient in the direction of heat flow, °F/ft

If k is independent of temperature,

$$q_k = \frac{T}{L/Ak} \tag{3-2}$$

$$R_k = \frac{L}{Ak} \qquad \text{thermal resistance} \tag{3-2a}$$

$$K_k = \frac{Ak}{L} \qquad \text{thermal conductance} \tag{3-2b}$$

where L is the distance between the hot and cold surfaces in the direction of heat flow, in feet.

Thermal Radiation

By the Stefan-Boltzmann law, the heat transfer by radiation for a blackbody is

$$q_r = \sigma A_1 T_1^4 \tag{3-3}$$

where q_r = heat transfer by radiation, Btu/h
σ = Stefan-Boltzmann constant
= 0.173×10^{-8} Btu/h · ft^2 · °R^4
= 5.67×10^{-8} W/m^2 · K^4
A_1 = area of radiating surface, ft^2
T_1 = radiating surface temperature, °R

Real bodies emit radiation at lower rates than blackbodies. Gray bodies emit a constant fraction of the blackbody emission at each wavelength at a given temperature. The net rate of the radiation heat transfer from a gray body at T_1 to a surrounding blackbody at T_2 is given by

$$q_r = \sigma A_1 \epsilon_1 (T_1^4 - T_2^4) \tag{3-4}$$

where ϵ_1 is the emissivity of the gray surface.

Heat transfer between two gray bodies is given by

$$q_r = \sigma A_1 F_{1\to 2} (T_1^4 - T_2^4) \tag{3-5}$$

where $F_{1\to 2}$ is a view factor which accounts for the emissivity and relative geometry of the gray bodies. Radiation coefficient is given by

$$h_r = \frac{\sigma F_{1\to 2}(T_1^4 - T_2^4)}{T_1 - T_2'} \tag{3-6}$$

where T_2' is a convenient reference temperature. Usually the surface temperature is taken as the reference temperature.

Convection

The convection heat transfer rate can be expressed as

$$q_c = h_c A \; \Delta T \tag{3-7}$$

where q_c = rate of heat transfer by convection, Btu/h
A = area of heat transfer, ft^2
ΔT = temperature difference between the surface and the fluid, $t_s - t_f$, °F
h_c = convection coefficient of heat transfer, $Btu/h \cdot ft^2 \cdot °F$

Example 3-1. A 6-in-thick wall is 12 ft high and 16 ft long. One face is at 1500°F and the other at 300°F. Find the heat loss in Btu/h.

Solution.

$$A = 12(16) = 192 \text{ ft}^2$$

$$\Delta T = 1500 - 300 = 1200°\text{F}$$

$$\Delta X = \frac{6}{12} = 0.5 \text{ ft}$$

$$Q = -kA \frac{\delta T}{\delta X} = kA \frac{\Delta T}{\Delta X} = 0.15(192)\frac{1200}{0.5} = 69{,}120 \text{ Btu/h}$$

Composite Walls

For resistances in series (Fig. 3-1a), the rate of heat transfer is given by

$$q = \frac{\Delta T}{R_T} = \frac{t_0 - t_n}{(X_A/k_a A) + (X_B/k_b A) + (X_C/k_c A) + \cdots + (X_n/k_n A)} \tag{3-8}$$

where ΔT is the overall temperature difference, °F.

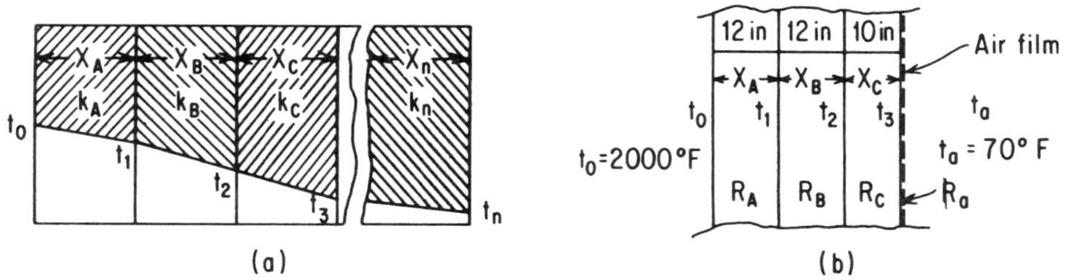

Figure 3-1. (*a*) Composite wall resistances in series; (*b*) composite wall resistances in series for Example 3-21.

Example 3-2a. The inside temperature of a composite wall is maintained at 2000 °F, and the outside temperature is 70 °F. The thicknesses from the hotter to colder surfaces are 12, 12, and 10 in, respectively, and the corresponding thermal conductivities are 0.4, 0.2, 0.1 Btu/(h · ft^2 · °F/ft), respectively. The outside surface coefficient of heat transfer is 2 Btu/h · ft^2 · °F. Calculate the heat loss and outside surface temperature. (Refer to Fig. 3-1*b*.)

Solution. Heat loss q_L is given by

$$q_L = \frac{t_0 - t_1}{R_A} = \frac{t_1 - t_2}{R_B} = \frac{t_2 - t_3}{R_C} = \frac{t_3 - t_a}{R_a} = \frac{t_0 - t_a}{R}$$

$$R_A = \frac{X_A}{k_a A} = \frac{12/12}{0.4(1)} = 2.5$$

$$R_B = \frac{X_B}{k_b A} = \frac{12/12}{0.2(1)} = 5.0$$

$$R_C = \frac{X_C}{k_c A} = \frac{10/12}{0.1(1)} = 8.33$$

$$R_A = \frac{1}{h_a A} = \frac{1}{2(1)} = 0.5$$

$$R_T = \Sigma R = 16.33 \qquad \Delta T = 2000 - 70 = 1930°\text{F}$$

$$\text{Heat loss} = q_L = \frac{1930}{16.33} = 118 \text{ Btu/h} \cdot \text{ft}^2$$

$$q_L = \frac{t_3 - t_a}{R_a} = 118\ \text{Btu/h} \cdot \text{ft}^2$$

$$t_3 - t_a = 118(0.5) = 59°\text{F}$$

Therefore $t_3 = t_a + 59 = 70 + 59 = 129°\text{F}$

Example 3-2b. A furnace is constructed with 0.3 m of fire brick, 0.15 m of insulating brick, and 0.25 m of ordinary building brick. The inside surface temperature is 1530 K, and the outside surface temperature is 325 K. Calculate the heat loss per unit area and the temperatures at the junctions of the bricks. The thermal conductivities of the fire, insulating, and building bricks are 1.4, 0.21, and 0.7 W/m · K, respectively.

Solution. Calculate resistance based on 1 m^2 area.

$$\begin{aligned}\text{Total resistance} &= \frac{0.3}{1.4(1)} + \frac{0.15}{0.21(1)} + \frac{0.25}{0.7(1)} \\ &= 0.2143 + 0.7143 + 0.3571 \\ &= 1.2857\ \text{K/W}\end{aligned}$$

Therefore
$$\text{Heat loss} = \frac{1530 - 325}{1.2857} = 937.2\ \text{W}$$

$$\text{Total temperature drop} = 1530 - 325 = 1205\ \text{K}$$

$$\text{Temperature drop over fire brick} = \frac{0.2143}{1.2857}(1205) = 201\ \text{K}$$

$$\text{Temperature drop over insulating brick} = \frac{0.7143}{1.2857}(1205) = 669\ \text{K}$$

$$\text{Temperature drop over building brick} = \frac{0.3571}{1.2857}(1205) = 335\ \text{K}$$

Temperature at junction of fire brick and insulating brick

$$= 1530 - 201 = 1329\ \text{K}$$

Temperature at junction of insulating and building bricks

$$= 1329 - 669 = 660\ \text{K}$$

Heat Flow Through a Pipe Wall. The differential equation for heat flow through a pipe wall (Fig. 3-2*a*) is given by

$$q = 2\pi rk\left(-\frac{dt}{dr}\right) \qquad \text{Btu/h}\cdot\text{ft} \tag{3-9}$$

The boundary conditions are when $r = r_i$, $t = t_i$, and when $r = r_o$, $t = t_o$, where i and o refer to the inside and outside surfaces. The solution of the differential equation is

$$q = \frac{2\pi k(t_i - t_o)}{2.3 \log (r_o/r_i)} = \frac{2\pi k(t_i - t_o)}{\ln (r_o/r_i)} = \frac{2\pi k(t_i - t_o)}{\ln (D_o/D_i)} \qquad \text{Btu/h}\cdot\text{ft} \tag{3-10}$$

If $D_i > 0.75D_o$, the arithmetic mean of the two areas can be used without much error. Then

$$q = \frac{\Delta T}{R} = \frac{\Delta T}{\dfrac{l_w}{k_a A_m}} = \frac{t_i - t_o}{\dfrac{\frac{1}{2}(D_o - D_i)}{\frac{1}{2}\pi k_a(D_o + D_i)}} = \frac{t_i - t_o}{\dfrac{1}{\pi k_a}\dfrac{D_o - D_i}{D_o + D_i}} \qquad \text{Btu/h}\cdot\text{ft} \tag{3-11}$$

where $l_w = \frac{1}{2}(D_o - D_i)$ is the wall thickness of the pipe.

Composite Cylindrical Resistance. Referring to Fig. 3-2*b*, the total temperature drop is given by

$$t_1 - t_3 = \frac{q}{2\pi k_a} \ln \frac{D_2}{D_1} + \frac{q}{2\pi k_b} \ln \frac{D_3}{D_2} \tag{3-12}$$

Heat Loss from an Insulated Pipe to Air. For an insulated pipe in which steam is flowing, there are four resistances to be considered:

1. The steam film resistance due to the condensation of steam is

$$q = h_s \pi D_i(t_s - t_s') \tag{3-13a}$$

where h_s is the steam film heat transfer coefficient, $\text{Btu/h}\cdot\text{ft}^2\cdot{}^\circ\text{F}$.

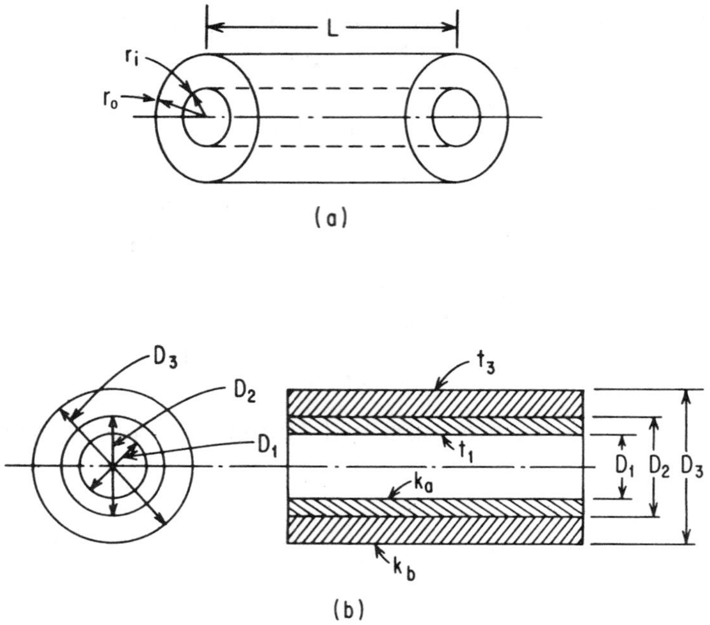

Figure 3-2. (*a*) Heat flow through pipe wall; (*b*) composite cylindrical resistances.

2. The pipe wall resistance is

$$q = \frac{2\pi k_p}{\ln (D_o/D_i)}(t_s' - t_s'') \tag{3-13b}$$

3. The insulation resistance is

$$q = \frac{2\pi k_b}{\ln (D_s/D_o)}(t_s'' - t_o) \tag{3-13c}$$

4. The air film resistance is

$$q = h_a \pi D_s (t_o - t_a) \tag{3-13d}$$

Combination of the four equations above gives

$$t_s - t_a = q\left[\frac{1}{h_s \pi D_i} + \frac{\ln (D_o/D_i)}{2\pi k_p} + \frac{\ln (D_s/D_o)}{2\pi k_b} + \frac{1}{h_a \pi D_s}\right] \tag{3-14}$$

where q = heat loss, Btu/h · ft
t_s = temperature of steam, °F
t_a = temperature of air, °F
k_p = thermal conductivity of pipe wall, Btu/(h · ft^2 · °F/ft)
k_b = thermal conductivity of insulation, Btu/(h · ft^2 · °F/ft)
h_a = surface coefficient of heat transfer, Btu/h · ft^2 · °F

Generally the steam-film and pipe-wall resistances are negligible and can be ignored in comparison with the other resistances.

Critical Radius and Maximum Heat Loss. The addition of the insulation to the outside of small-diameter tubes or pipes does not always reduce the heat loss. The expression for the value of the critical radius at maximum heat loss can be derived from Eq. (3-14) and is given by

$$(r_s)_c = \frac{k_b}{h_a} \tag{3-15}$$

where $(r_s)_c$ is the critical radius.

Long, Hollow Cylinder. Heat-transfer rate through a long, hollow cylinder is given by

$$q = \frac{k_m 2\pi L \ \Delta t}{\ln (r_2/r_1)} \tag{3-16}$$

where $\Delta t = t_i - t_o$ and t_i and t_o are the inside and outside surface temperatures, respectively. To simplify Eq. (3-16), log mean area is defined as

$$A_m = \frac{2\pi r_2 L - 2\pi r_1 L}{\ln (2\pi r_2 L / 2\pi r_1 L)} = \frac{A_2 - A_1}{\ln (A_2/A_1)} \tag{3-17}$$

The arithmetic mean area is given by

$$A_a = \frac{A_1 + A_2}{2} \tag{3-18}$$

When $A_2/A_1 \ngtr 2$, A_a is within 4 percent of A_m . If the log mean areas are used, the expression for q [Eq. (3-14)] can also be written as

$$q = \frac{t_s - t_a}{1/(h_i A_i) + l_w/(k_w A_{mp}) + x_i/(k_b A_{mb}) + 1/(h_o + h_r)A_o} \qquad (3\text{-}19)$$

where A_{mp} = log mean area for pipe
A_{mb} = log mean area for insulation
x_i = thickness of insulation

Hollow Sphere. For heat transfer through a hollow sphere, the mean area is

$$A_m = 4\pi r_1 r_2 = \sqrt{A_1 A_2} \qquad (3\text{-}20)$$

Radiation. The relation among the absorbed, reflected, and transmitted energies is

$$\alpha + R + \tau = 1 \qquad (3\text{-}21)$$

where α, R, and τ are the absorptivity, reflectivity, and transmittivity, respectively.

For opaque bodies $\tau = 0$, and therefore

$$\alpha + R = 1 \qquad (3\text{-}21a)$$

Kirchhoff's Law

At thermal equilibrium, the ratio of the total emissive power to the absorptivity for all bodies is the same, or

$$\frac{\epsilon_1}{\alpha_1} = \frac{\epsilon_2}{\alpha_2} = \epsilon_b \qquad (3\text{-}22)$$

where subscript b refers to blackbody. Emissivity of an actual body is given by

$$\epsilon = \frac{\text{emissive power of body } \epsilon_a}{\text{blackbody emissive power } \epsilon_b} \qquad (3\text{-}23)$$

Radiation to a Completely Absorbing Receiver

When a radiating source is small by comparison with the receiving enclosure, it is assumed for simplification that the heat radiated by the source is not reflected back to it. In this case, the heat loss by the source is given by

$$\frac{q_r}{A_1} = \epsilon_1 \sigma (T_1^4 - T_2^4) \qquad (3\text{-}24)$$

where q_r = heat loss due to radiation, Btu/h
ϵ_1 = emissivity of radiating source
σ = Stefan-Boltzmann constant
T_1 = temperature of the radiating source, °R
T_2 = temperature of the receiving surface, °R

Example 3-3. A 2-in IPS steel pipe carries steam at 325°F through a room at 70°F. What decrease in radiation occurs if the bare pipe is coated with 26% aluminum paint?

Solution. Assume negligible resistance of the steam film and metal wall.

Surface temperature t_s = 325°F = 785°R

Air temperature t_a = 70°F = 530°R

OD of pipe = 2.38 in

Area A_0 = 0.622 ft²/ft of pipe

Emissivity of oxidized steel = 0.79

Emissivity of painted steel (26% Al) = 0.3

$$\text{Heat loss from bare pipe} = 0.173 \times 10^{-8}(0.79)[(785)^4 - (530)^4]$$
$$= 0.173(0.79)[(\tfrac{785}{100})^4 - (\tfrac{530}{100})^4]$$
$$= 411.1 \text{ Btu/h} \cdot \text{ft}^2$$

$$\text{Heat loss from painted pipe} = 0.173(0.3)[(\tfrac{785}{100})^4 - (\tfrac{530}{100})^4]$$
$$= 156.1 \text{ Btu/h} \cdot \text{ft}^2$$

$$\text{Decrease in radiation} = 411.1 - 156.1 = 255 \text{ Btu/h} \cdot \text{ft}^2$$

Heat Exchange Between Source and Receiver

In general, the radiation received by a receiver from a source is given by

$$Q = \begin{cases} F_A A_1 \sigma(T_1^4 - T_2^4) & \text{for blackbodies} \quad (3\text{-}25) \\ F_A F_\epsilon A_1 \sigma(T_1^4 - T_2^4) & \text{for gray surfaces} \quad (3\text{-}26) \end{cases}$$

where F_ϵ = emissivity correction[5j]
F_A = correction for relative geometry of the bodies or view factor[5j]
σ = Stefan-Boltzmann constant
T_1 = temperature of source, °R
T_2 = temperature of receiver, °R

Nongray Enclosures

For a small nongray body of area A_1 and at T_1 in black surroundings at T_2, the radiation interchange is given by

$$q_{1\to2} = A_1\sigma(\epsilon_1 T_1^4 - \alpha_{1\to2} T_2^4) \quad (3\text{-}26a)$$

where ϵ_1 is the emmissivity of small nongray body of area A_1 and $\alpha_{1\to2}$ is the absorptivity of surface A_1 at T_1 for blackbody radiation at T_2.

Example 3-4. Calculate the heat transfer by radiation between an oxidized nickel tube of 3 in OD at a temperature of 1000°F and an enclosing chamber maintained at 2000°F if **(a)** the chamber is very large relative to the tube diameter and **(b)** 12-in-square inside. The chamber is lined with glazed silica brick having an emissivity of 0.85.

Solution.

a. If the chamber is very large, it is not necessary to correct for emissivity of silica brick, because the surroundings from the enclosed small body appear black.

ϵ of nickel at 1000°F = 0.463 (interpolated value)

$$\epsilon \text{ of nickel at } 2000°\text{F} = 0.62 \text{ (extrapolated value)}$$

$$= \text{absorptivity at } 2000°\text{F}$$

$$\frac{\text{Surface area}}{\text{Tube length (ft)}} = \pi D(1) = \tfrac{3}{12}\pi(1) = 0.7854 \text{ ft}^2$$

$$q_r = 0.173(0.7854)\left[0.463\left(\frac{1000+460}{100}\right)^4 - 0.62\left(\frac{2000+460}{100}\right)^4\right]$$

$$= -27{,}993 \text{ Btu/h}\cdot\text{ft}$$

b. In this case, it is necessary to allow for ϵ of the silica brick, since enclosure is small. For this, use the following equation given for two nonblack source-sink surfaces of areas A_1 and A_2:

$$\frac{1}{A_1 F_{1\to2}} = \frac{1}{A_1}\left(\frac{1}{\epsilon_1} - 1\right) + \frac{1}{A_2}\left(\frac{1}{\epsilon_2} - 1\right) + \frac{1}{A_1 \overline{F}_{1\to2}} \qquad (3\text{-}26b)$$

where $F_{1\to2}$ is the overall interchange factor for gray surfaces and $\overline{F}_{1\to2}$ is the view factor with allowance for the refractory surface. Since the temperature of the brick is higher than that of the tube, $F_{1\to2}$ is more important than $\overline{F}_{1\to2}$. Therefore, use $\alpha_{1\to2}$ instead of ϵ_1 of the tube. Thus $\epsilon_1 = \alpha_{1\to2} = 0.62$. $\overline{F}_{1\to2} = 1$, since all radiation emitted by A_1 is received by A_2. $\alpha_{1\to2} = 0.62$ and $\epsilon_2 = 0.85$.

$$A_1 = 0.7854 \text{ ft}^2\text{/ft of tube}$$

$$A_2 = (4)(1)(1) = 4 \text{ ft}^2\text{/ft of enclosing chamber}$$

From Eq. (3-26*b*),

$$\frac{1}{F_{1\to2}} = \left(\frac{1}{\epsilon_1} - 1\right) + \frac{A_1}{A_2}\left(\frac{1}{\epsilon_2} - 1\right) + 1$$

$$= \left(\frac{1}{0.62} - 1\right) + \frac{0.7854}{4}\left(\frac{1}{0.85} - 1\right) + 1 = 1.6476$$

from which

$$F_{1\to2} = 0.607$$

Therefore

$$q_r = 0.173(0.7854)(0.607)\left[\left(\frac{1460}{100}\right)^4 - \left(\frac{2460}{100}\right)^4\right]$$

$$= -26{,}457 \text{ Btu/h}\cdot\text{ft}$$

Example 3-5. A 4-in NPS steel pipe carrying steam at 450°F is insulated with 1 in of kapok surrounded by 1 in of magnesite. The surrounding air is at 70°F. What is the heat loss per linear foot given the following data:

$$k \text{ of kapok} = 0.02 \quad \text{Btu/(h}\cdot\text{ft}^2\cdot{}^\circ\text{F/ft)}$$

$$k \text{ of magnesite} = 0.35 \quad \text{Btu/(h}\cdot\text{ft}^2\cdot{}^\circ\text{F/ft)}$$

$$\epsilon \text{ of plaster} = 0.9$$

Solution. Neglect the steam-film and pipe-wall resistances. Here t_s, the surface temperature, is not known. Therefore, a trial-and-error solution is required. q, the heat loss in Btu/h · ft, is given by

$$q = \frac{t_i - t_a}{(1/2\pi k_k)\ln(D_3/D_2) + (1/2k_m\pi)\ln(D_s/D_3) + (1/h_a\pi D_s)}$$

Since the outside diameter of 4-in NPS pipe is 4.5 in,

$$D_2 = \frac{4.5}{12} = 0.375 \text{ ft}$$

$$D_3 = \frac{6.5}{12} = 0.5417 \text{ ft}$$

$$D_s = \frac{8.5}{12} = 0.7084 \text{ ft}$$

$$t_a = 70^\circ\text{F} = 530^\circ\text{R}$$

Assuming $t_s = 130^\circ\text{F} = 590^\circ\text{R}$,

$$h_r = 0.173(0.9)\frac{\left(\frac{590}{100}\right)^4 - \left(\frac{530}{100}\right)^4}{130 - 70} = 1.1 \text{ Btu/h}\cdot\text{ft}^2\cdot{}^\circ\text{F}$$

$$h_c = 0.5\left(\frac{\Delta t}{d_0}\right)^{0.25} = 0.5\left(\frac{60}{8.5}\right)^{0.25} = 0.82 \text{ Btu/h}\cdot\text{ft}^2\cdot{}^\circ\text{F}$$

Note: d_0 is in inches in the formula for h_c.

$$h_a = h_c + h_r = 1.92 \text{ Btu/h}\cdot\text{ft}^2\cdot{}^\circ\text{F}$$

$$q = \frac{450 - 70}{\frac{1}{2\pi(0.02)}\ln\left(\frac{0.5417}{0.375}\right) + \frac{1}{2\pi(0.35)}\ln\left(\frac{0.7084}{0.5417}\right) + \frac{1}{1.92\pi(0.7084)}}$$

$$= \frac{380}{3.049 + 0.234} = 115.8 \text{ Btu/h} \cdot \text{ft}$$

Checking t_s for correctness,

$$q = \frac{t_s - 70}{1/1.92\pi(0.7084)} = 115.8 \text{ Btu/h} \cdot \text{ft}$$

$$t_s = 97.1°\text{F} \qquad \text{no check}$$

Another trial with $t_s = 100°\text{F}$, $h_r = 1.01$, $h_c = 0.69$, and $h_a = 1.7$ gives $t_s = 100.3°\text{F}$ which is close to the assumed value of 100°F. Hence $t_s = 100.3°\text{F}$, and the heat loss = 114.7 Btu/h · ft.

Heat Transfer by Convection to Fluids Flowing Inside and Outside of Pipes

Overall Coefficient of Heat Transfer. The overall coefficient of heat transfer is expressed in terms of the individual coefficients of heat transfer. U_o, the overall coefficient based on the outside area, is given by

$$\frac{1}{U_o} = \frac{1}{h_i}\frac{D_o}{D_i} + f_{di}\frac{D_o}{D_i} + \frac{l_w}{k_w}\frac{D_o}{D_{av}} + f_{do} + \frac{1}{h_o} \tag{3-27}$$

where U_o = overall heat transfer coefficient, Btu/h · ft² · °F, based on the outside area
h_i = inside film coefficient of heat transfer, Btu/h · ft² · °F
f_{di} = fouling resistance on the inside, h · ft² · °F/Btu
l_w = wall thickness of pipe, ft
k_w = thermal conductivity of pipe wall, Btu/(h · ft² · °F/ft)
f_{do} = fouling resistance on outside of pipe, h · ft² · °F/Btu
h_o = outside film coefficient of heat transfer, Btu/h · ft² · °F

D_i, D_o, D_{av} = inside, outside and average diameters respectively, ft

Based on inside area, U_i is given by

$$\frac{1}{U_i} = \frac{1}{h_i} + f_{di} + \frac{l_w}{k_w}\frac{D_i}{D_{av}} + f_{do}\frac{D_i}{D_o} + \frac{1}{h_o}\frac{D_i}{D_o} \tag{3-27a}$$

Logarithmic Mean Temperature Difference. The logarithmic mean temperature difference is given by the relation

$$\Delta T_{LM} = \frac{\Delta T_2 - \Delta T_1}{\ln(\Delta T_2/\Delta T_1)} \tag{3-28}$$

ΔT_1 and ΔT_2 are the terminal temperature differences.

The expression for log mean difference is the same for both the parallel and countercurrent flows. When one of the fluids is isothermal, the numerical value of ΔT_{LM} is the same for both the parallel and countercurrent flows.

Caloric Fluid Temperature.[5h] When U is not constant along the length of the exchanger because of a change in the viscosity of the fluid, the fluid properties should be evaluated at the caloric temperature for the calculation of the film coefficients.

Pipe-Wall Temperature (when *U* is not constant). If the caloric temperature and the film coefficient h_i and h_o are known, and if the pipe-wall resistance is neglected, the wall temperature t_w can be calculated by the following relations. When hot fluid is on the outside of the pipe,

$$t_w = \begin{cases} t_c + \dfrac{h_o}{h_{io} + h_o}(T_c - t_c) & \text{(3-29a)} \\ T_c - \dfrac{h_{io}}{h_{io} + h_o}(T_c - t_c) & \text{(3-29b)} \end{cases}$$

where T_c and t_c are the caloric temperatures of the hot and cold fluids respectively, and $h_{io} = h_i(D_i/D_o)$. When the hot fluid is inside the pipe, t_w is given by

$$t_w = \begin{cases} t_c + \dfrac{h_{io}}{h_{io} + h_o}(T_c - t_c) & \text{(3-30a)} \\ T_c - \dfrac{h_o}{h_{io} + h_o}(T_c - t_c) & \text{(3-30b)} \end{cases}$$

If h_i and h_o are not known, a trial-and-error calculation to fix the caloric temperature is required.

Film Coefficients for Fluids in Pipes and Tubes (No Change of Phase).

Streamline Flow. Sieder and Tate[1] correlation for laminar flow ($N_{Re} < 2100$) is

$$\frac{h_i D_i}{k} = 1.86\left[\left(\frac{D_i G}{\mu}\right)\left(\frac{C_p \mu}{k}\right)\left(\frac{D_i}{L}\right)\right]^{1/3}\left(\frac{\mu}{\mu_w}\right)^{0.14} \tag{3-31}$$

Equation (3-31) simplifies to

$$\frac{h_i D_i}{k} = 1.86\left(\frac{4Wc_p}{\pi k L}\right)^{1/3}\left(\frac{\mu}{\mu_w}\right)^{0.14} \tag{3-31a}$$

For turbulent flow ($N_{Re} > 10{,}000$), the Sieder and Tate[1] equation is

$$\frac{h_i D_i}{k} = 0.027\left(\frac{D_i G}{\mu}\right)^{0.8}\left(\frac{c_p \mu}{k}\right)^{1/3}\left(\frac{\mu}{\mu_w}\right)^{0.14} \tag{3-32}$$

The bulk temperature for streamline flow and the caloric temperature for turbulent flow are to be used to evaluate the physical properties of the fluid, except μ_w which is taken at the wall temperature. If the caloric temperature factor chart for the fluid is not available, the average bulk fluid temperature is used to evaluate the physical properties when: (1) viscosity at the cold temperature is low (~5 cP), (2) temperature range is moderate (~100°F), and (3) temperature difference is low (~75°F). Also $(\mu/\mu_w)^{0.14}$ is assumed equal to ~1.

At moderate Δt values, the Dittus-Boelter[2] equation for turbulent flow is

$$\frac{h_i D_i}{k} = \begin{cases} 0.023 \left(\dfrac{D_i G}{\mu_b}\right)^{0.8} \left(\dfrac{c_p \mu}{k}\right)_b^{0.4} & \text{for heating} \quad (3\text{-}33a) \\ 0.023 \left(\dfrac{D_i G}{\mu_b}\right)^{0.8} \left(\dfrac{c_p \mu}{k}\right)^{0.3} & \text{for cooling} \quad (3\text{-}33b) \end{cases}$$

where h_i = inside heat-transfer coefficient, $\text{Btu/h} \cdot \text{ft}^2 \cdot {}^\circ\text{F}$
D_i = inside diameter, ft
k = thermal conductivity of fluid, $\text{Btu} \cdot \text{ft/h} \cdot \text{ft}^2 \cdot {}^\circ\text{F}$
G = mass velocity, $\text{lb/h} \cdot \text{ft}^2$
μ = viscosity, $\text{lb/h} \cdot \text{ft}$
c_p = specific heat of fluid = $\text{Btu/lb} \cdot {}^\circ\text{F}$
L = length of pipe, ft
W = flow rate, lb/h

The subscript b indicates that the fluid physical properties are to be evaluated at the bulk temperature of the fluid.

Evaluation of Wall Temperature t_w (when U is fairly constant). When the overall heat-transfer coefficient U is fairly constant over the length of the exchanger, a trial-and-error calculation is required to evaluate t_w.

$$\Delta T_i = \frac{(1/h_i A_i)\, \Delta T}{(1/h_i A_i) + (1/h_o A_o) + (l_w/k_w A_{\text{av}})} \qquad (3\text{-}34)$$

where ΔT is the difference between the average temperature of the hot fluid and the average temperature of the cold fluid.

Preliminary estimates of h_i and h_o are made, and then the wall temperature is obtained by the equations

$$t_w = \begin{cases} T + \Delta T_i & \text{for heating} \qquad (3\text{-}35a) \\ T - \Delta T_i & \text{for cooling} \qquad (3\text{-}35b) \end{cases}$$

where T is the average temperature of the fluid.

Equivalent Diameter for Heat Transfer. When a fluid flows through a circular annulus, the wetted area for the heat transfer is not the same as that for the fluid flow. Therefore, values of the equivalent diameter are different and are given by

$$D_e = \begin{cases} \dfrac{D_2^2 - D_1^2}{D_1} & \text{for heat transfer} \quad (3\text{-}36a) \\ D_2 - D_1 & \text{for pressure drop} \quad (3\text{-}36b) \end{cases}$$

Flow Across Banks of Tubes. The Reynolds number for flow over tube banks is based on the minimum free flow area. For in-line tube arrangement (Fig. 3-3*a*), the minimum free flow area between two adjacent tubes is given by

$$A_{\min} = S_T - D_o \qquad \text{ft}^2/\text{unit length of tube} \qquad (3\text{-}37)$$

where S_T is center-to-center distance of the tubes in adjacent longitudinal rows or the transverse pitch, ft.

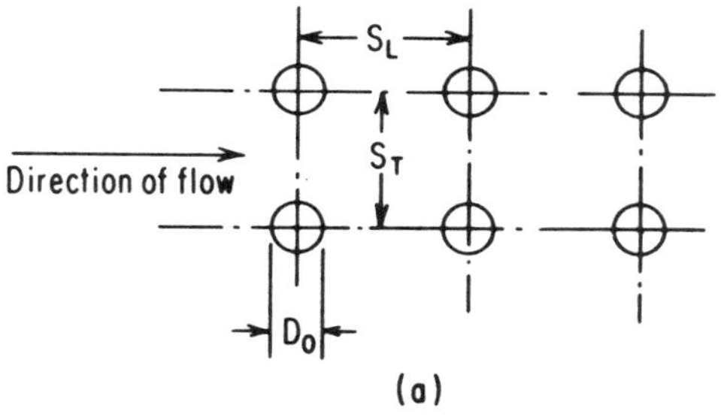

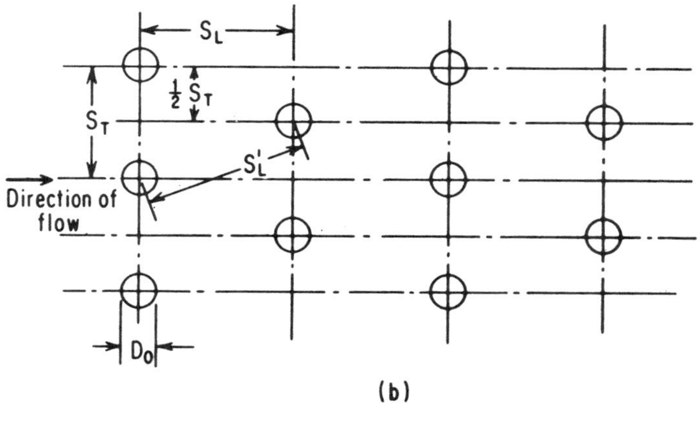

Figure 3-3. Tube arrangements in cross-flow exchangers: (*a*) in-line, (*b*) staggered.

For a shell containing N transverse rows of in-line tubes perpendicular to the direction of flow, the minimum flow area (total) is given by

$$A_{\min} = (N - 1)(S_T - D_o) + 2C \quad \text{ft}^2/\text{unit length of bundle} \tag{3-38}$$

where C is the clearance between the shell and the outermost row of tubes, ft.

For the staggered tube arrangement (Fig. 3-3*b*), the minimum flow area may occur between the adjacent tubes in a row or between the diagonally opposed tubes. In the former case, the minimum flow area is given by

$$A_{\min} = S_T - D_o \tag{3-39a}$$

S_T is as shown in Fig. 3-3*b*. In the latter case, if S_L/S_T is very small,

$$\sqrt{S_T^2 + S_L^2} < (S_T + \tfrac{1}{2}D_o) \tag{3-39b}$$

For the diagonally opposed tubes, the maximum velocity is then given by

$$V_{\max} = V_f \frac{S_T}{\sqrt{S_T^2 + S_L^2} - D_o} \tag{3-40}$$

where $V_{\max}$ = maximum velocity, ft/s
V_f = free flow velocity based on shell area without tubes, ft/s
S_L = center-to-center spacing between transverse tube rows or longitudinal pitch, ft

For gases flowing normal to the staggered tubes, Colburn[3] recommended the equation

$$\frac{h_m D_o}{k_f} = 0.33\left(\frac{c_p\mu}{k}\right)_f^{1/3}\left(\frac{D_o G_{\max}}{f}\right)^{0.6} \tag{3-41}$$

The subscript f indicates that the fluid properties are to be evaluated at the film temperature t_f.

For air, the above reduces to

$$\frac{h_m D_o}{k_f} = 0.3\left(\frac{D_o G_{\max}}{\mu_f}\right)^{0.6} \tag{3-42}$$

For the in-line tube arrangement, the constant in the above equation should be taken as 0.26 instead of 0.33. The fluid properties are to be evaluated at

$$t_f = \tfrac{1}{2}(t_s + t) \tag{3-43}$$

where t_s is the surface temperature and t is the bulk temperature.

The ratios of h_m for N rows deep to h_m for 10 rows deep are given in Table 3-1.

TABLE 3-1 h_m for N Rows/h_m for 10 Rows[4e]

N	1	2	3	4	5	6	7	8	9	10
Ratio for staggered tubes	0.68	0.75	0.83	0.89	0.92	0.95	0.97	0.98	0.99	1
Ratio for in-line tubes	0.64	0.80	0.87	0.90	0.92	0.94	0.96	0.98	0.99	1

A simplified equation for gases flowing normal to a 10-row bank of staggered tubes is

$$h_m = \frac{0.133 c_p G_{max}^{0.6}}{D_o^{0.4}} \tag{3-44}$$

Free Convection of Air Outside Tubes. Correlations for the heat-transfer coefficient for free convection of air outside tubes are

$$h_c = \begin{cases} 0.5\left(\dfrac{\Delta t}{d_o}\right)^{0.25} & \text{horizontal pipes} \\ 0.4\left(\dfrac{\Delta t}{d_o}\right)^{0.25} & \text{long vertical pipes} \\ 0.28\left(\dfrac{\Delta t}{Z}\right)^{0.25} & \text{vertical plates} < 2 \text{ ft high} \\ 0.3(\Delta t)^{0.25} & \text{vertical plates} > 2 \text{ ft high} \\ 0.38(\Delta t)^{0.25} & \text{horizontal plates facing downward} \\ 0.20(\Delta t)^{0.25} & \text{horizontal plates facing upward} \end{cases} \tag{3-45}$$

In the above equations, d_o is the outer diameter of the tube in inches, Δt is the temperature difference between the hot surface and the cold air in °F, Z is the height of the plate in feet, and h_c is in Btu/h · ft^2 · °F.

Free Convection of Fluids Outside Horizontal Pipes.

Single Horizontal Pipes. The heat-transfer coefficient for both liquids and gases outside a single horizontal pipe is given by

$$\frac{h_c D_o}{k_f} = \alpha \left[\left(\frac{D_o^3 \rho_f^2 g \beta \, \Delta t}{\mu_f^2} \right) \left(\frac{c_p \mu_f}{k_f} \right) \right]^{0.25} \tag{3-46}$$

where the film temperature t_f is

$$t_f = \frac{t_w + t_b}{2} \qquad t_b = \text{bulk temperature} \tag{3-46a}$$

The constant α is 0.47 for small pipes and 0.53 for large pipes.

In Eq. (3-46), the three groups are dimensionless. Therefore consistent units of the variables are to be used. For example, g, acceleration due to gravity, equals 4.18×10^{8} ft/h^2 and β is the coefficient of thermal expansion, 1/°F.

Banks of Tubes. For free convection outside banks of tubes (liquids and gases),[5a] the convection heat-transfer coefficient is given by

$$h_c = 116 \left[\left(\frac{k_f^3 \rho_f^2 c_f \beta}{\mu'_f} \right) \left(\frac{\Delta t}{d_o} \right) \right]^{0.25} \tag{3-47}$$

For ideal gases, $\beta = 1/T$ where T = average temperature in absolute scale. For liquids,

$$\beta = \frac{1/s_2 - 1/s_1}{s_{av}(t_2 - t_1)} \tag{3-47a}$$

where s_2 and s_1 are specific gravities at t_2 and t_1 and s_{av} equals

$\frac{1}{2}(s_1 + s_2)$. In Eq. (3-47), μ_f' is in centipoise and d_o is outside diameter in inches.

Condensation Outside Tubes.

Vertical Tubes. Some correlations for the condensation coefficients are given below. For vertical tubes[5b] (length L of tube is known), the condensation coefficient is given by

$$\bar{h} = 0.943 \left(\frac{k_f^3 \rho_f^2 \lambda g}{\mu_f L \, \Delta t_f} \right)^{0.25} \tag{3-48}$$

where $t_f = \frac{1}{2}(t_{sv} + t_w)$
$\Delta t_f = t_f - t_w$
t_{sv} = temperature of saturated vapor

λ is the latent heat of condensation. Consistent units of the variable are to be used.

Inclined Tubes.[5c] For the inclined tubes $\bar{h}$ is given by

$$\bar{h} = 0.943 \left(\frac{k_f^3 \rho_f^2 \lambda g \sin \alpha}{\mu_f L \, \Delta t_f} \right)^{0.25} \tag{3-49}$$

where α is the angle that the gravity component of the condensate weight makes with the line perpendicular to the tube, and λ is the latent heat, Btu/lb.

When the length of the vertical tube is not known, the following equation can be used for the vertical single or multiple tubes.[5d]

$$\bar{h} = 1.47 \left(\frac{4G'}{\mu_f} \right)^{-1/3} \left(\frac{k_f^3 \rho_f^2 g}{\mu_f^2} \right)^{1/3} \tag{3-50}$$

$$G' = \frac{W'}{\pi D_o} \qquad \text{lb/h} \cdot \text{ft} \tag{3-51}$$

where W' is the condensate loading/tube = W/N_t and N_t is the number of tubes.

Horizontal Tubes.[5d] For horizontal tubes $\bar{h}$ is given by

$$\bar{h} = \begin{cases} 0.725\left(\dfrac{k_f^3\rho_f^2 g\lambda}{\mu_f D_o\,\Delta t_f}\right)^{0.25} & \text{for single tube} \quad (3\text{-}52) \\ 1.51\left(\dfrac{4G''}{\mu_f}\right)^{-1/3}\left(\dfrac{k_f^3\rho_f^2 g}{\mu_f^2}\right)^{1/3} & \text{for multiple tubes} \quad (3\text{-}53) \end{cases}$$

where

$$G'' = \frac{W}{LN_t^{2/3}} \qquad (3\text{-}54)$$

The equations for estimating the condensation coefficients outside the vertical and horizontal tubes hold good for the streamline flow of the condensate, i.e., for $N_{\mathrm{Re}} < 2100$ where $N_{\mathrm{Re}} = DG'/\mu$ for vertical tubes and $N_{\mathrm{Re}} = DG''/\mu$ for horizontal tubes. The effects of the high velocity, turbulence, etc., can be accounted for.[4b]

Condensation Inside Tubes.

Horizontal Tubes. Equation (3-53) is to be used with

$$G'' = \frac{W}{0.5LN_t} \qquad (3\text{-}55)$$

Vertical Tubes. $\bar{h}$ is given in terms of j factor by

$$j = \bar{h}\left(\frac{\mu_f^2}{k_f^3\rho_f^2 g}\right)^{1/3} = \phi\left(\frac{4G'}{\mu_f}\right) \qquad (3\text{-}56)$$

where j is to be obtained from a plot[5e] of j versus N_{Re}.

Film Temperature for Condensation. Kern[5f] recommends the relation

$$t_f = \tfrac{1}{2}(t_{sv} + t_w) \qquad {}^\circ\mathrm{F} \qquad (3\text{-}57)$$

where t_{sv} = temperature of saturated vapor, °F
t_w = wall temperature, °F
$\Delta t_f = t_f - t_w$

McAdams[4c] recommends the relation

$$t_f = t_{sv} - 0.75\ \Delta t \tag{3-57a}$$

where $\Delta t = t_{sv} - t_w$

Tube Wall Temperature. This may be estimated by the equation

$$t_w = \frac{h_i t_i + h_o t_o}{h_i + h_o} \tag{3-58}$$

where t_i is the bulk temperature of the fluid inside the tube and t_o is the bulk temperature of the fluid outside the tube.

Note that if t_f is calculated as suggested by McAdams,[4c] Δt should be used in the place of Δt_f in Eqs. (3-48), (3-49), and (3-52).

Example 3-6. A bare horizontal 12-in IPS steam pipe carries saturated steam at 240°F. The temperature of ambient air is 70°F. Calculate the rate of heat loss per foot of pipe length.

Solution. Assume the steam-film resistance and pipe-wall resistance negligible. Therefore, the surface temperature is 240°F. For horizontal pipe,

$$h_c = 0.5\left(\frac{\Delta t}{d_o}\right)^{0.25}$$

$$= 0.5\left(\frac{170}{12.75}\right)^{0.25} = 0.96\ \text{Btu/h}\cdot\text{ft}^2\cdot{}^\circ\text{F}$$

$$h_r = \frac{0.173(0.8)\ [(\frac{700}{100})^4 - (\frac{530}{100})^4]}{240 - 70}$$

$$= 1.31\ \text{Btu/h}\cdot\text{ft}^2\cdot{}^\circ\text{F}$$

$$h_c + h_r = 0.96 + 1.31 = 2.27\ \text{Btu/h}\cdot\text{ft}^2\cdot{}^\circ\text{F} = h_a$$

$$\frac{\text{Outer area of pipe}}{\text{Foot length}} = \pi D_o L = \pi\frac{12.75}{12}(1) = 3.338\ \text{ft}^2\text{/ft length}$$

Therefore, heat loss is

$$Q_L = h_a(A_o)(\Delta t)$$
$$= 2.27(3.338)(240 - 70)$$
$$= 1288 \text{ Btu/h} \cdot \text{ft}$$

Example 3-7. An oil is flowing at velocity of 5 ft/s through a 10-ft-long, 1-in-diam. 18 BWG tube. On the outer surface of the tube, steam is condensing at 220°F. The tube is clean. Oil enters the tube at 86°F and leaves at 104°F. The physical-property data (all assumed constant) for oil are:

$$\text{Oil density } \rho = 55 \text{ lb/ft}^3$$
$$\text{Specific heat } c_p = 0.48 \text{ Btu/lb} \cdot °\text{F}$$
$$\text{Thermal conductivity } k = 0.08 \text{ Btu/(h} \cdot \text{ft}^2 \cdot °\text{F/ft)}$$

The viscosity μ varies with temperature as follows:

t, °F	80	90	100	110	130	140	220
μ, cP	20	18	16.2	15	12	11	3.6

Calculate h_i, the inside oil film coefficient of heat transfer.

Solution. The inside diameter of the tube is

$$0.902 \text{ in} = \frac{0.902}{12} = 0.0752 \text{ ft}$$

Neglect the steam-film resistance. The oil leaves the tube at 104°F.

$$t_w = 220°\text{F}$$

$$t_b = \text{bulk temperature of oil} = \frac{86 + 104}{2} = 95°\text{F}$$

At $t_b = 95°$F, by interpolation $\mu = 17.1$ cP $= 41.38$ lb/ft · h.

$$N_{\text{Re}} = \frac{Du\rho}{\mu} = \frac{0.0752(5)(55)(3600)}{41.38} = 1799 < 2100$$

Therefore, Eq. (3-31) for the streamline flow applies.

$$\frac{h_i D_i}{k} = 1.86\left[\left(\frac{DG}{\mu}\right)\left(\frac{c_p\mu}{k}\right)\left(\frac{D}{L}\right)\right]^{1/3}\left(\frac{\mu}{\mu_w}\right)^{0.14}$$

$$\frac{DG}{\mu} = 1799 \qquad \text{(as calculated before)}$$

$$\frac{c_p\mu}{k} = \frac{0.48(41.38)}{0.08} = 248.3$$

$$\frac{D}{L} = \frac{0.0752}{10} = 0.00752$$

$$\frac{h_i D_i}{k} = 1.86[(1799)(248.3)(0.00752)]^{1/3}\left[\frac{41.38}{(3.6)(2.42)}\right]^{0.14}$$

$$= 34.65 \text{ Btu/h} \cdot \text{ft}^2 \cdot {}^\circ\text{F}$$

$$h_i = 34.65 \text{ x } .08 / .752$$
$$= 36.86 \text{ Btu / h.ft}^{2\circ}\text{F}$$

Example 3-8. Aniline is flowing at a velocity of 8 ft/s in a $\frac{3}{4}$-in 18 BWG tube. On the outer surface, steam is condensing at 220°F. The tube is clean. Using the Sieder-Tate equation, find h_i, the inside film coefficient of heat transfer if aniline enters the tube at 80°F and leaves at 120°F. Also determine the film coefficient using the Dittus-Boelter equation. The physical-property data for aniline are given in Table 3-2.

TABLE 3-2 Physical-Property Data for Aniline

Temp.,°F	μ, cP	k, Btu/(h · ft² · °F/ft)	s	c_p, Btu/lb · °F
60	4.84	0.10	1.026	0.43
100	2.53	0.10	1.001	0.49
150	1.44	0.098	0.986	0.505
200	0.91	0.096	0.962	0.515
300	0.48	0.093	0.922	0.540

Solution. The bulk temperature is

$$t_b = \frac{80 + 120}{2} = 100^\circ\text{F}$$

Assuming that the caloric temperature $= t_b = 100°F$,

$$\rho_c = 1.001(62.4) = 62.5 \text{ lb/ft}^3$$

$$(c_p)_c = 0.49 \text{ Btu/lb}\cdot°F$$

$$k_c = 0.10 \text{ Btu/(h}\cdot\text{ft}^2\cdot°F/\text{ft)}$$

$$\mu_c = 2.53(2.42) = 6.123 \text{ lb/h}\cdot\text{ft}$$

$$\mu_{220} = 0.91 - \tfrac{20}{100}(0.91 - 0.48) = 0.824 \text{ cP}$$

Assume that the metal wall resistance is negligible. Check for Reynolds number.

$$D_i = \frac{0.652}{12} = 0.05434 \text{ ft}$$

$$\frac{D_i u \rho_c}{\mu_c} = \frac{0.05434(8)(62.5)}{2.53(0.000672)} = 15{,}981$$

The flow is in the turbulent region, and therefore the Sieder-Tate equation for the turbulent flow applies.

$$h_i = 0.027\frac{k}{D_i}\left(\frac{D_i G}{\mu_c}\right)^{0.8}\left(\frac{c_p\mu_c}{k_c}\right)^{1/3}\left(\frac{\mu_c}{\mu_w}\right)^{0.14}$$

$$\frac{DG}{\mu_c} = 15{,}981$$

$$\frac{c_p\mu_c}{k_c} = \frac{0.49(6.123)}{0.10} = 30$$

$$\frac{\mu_c}{\mu_w} = \frac{2.53}{0.824} = 3.07$$

$$h_i = 0.027\left(\frac{0.1}{0.05434}\right)(15{,}981)^{0.8}(30)^{1/3}(3.07)^{0.14}$$

$$= 416.6 \text{ Btu/h}\cdot\text{ft}^2\cdot°F$$

The Dittus-Boelter equation gives

$$h_i = 0.023\left(\frac{DG}{\mu_b}\right)^{0.8}\left(\frac{c_p\mu_b}{k_b}\right)^{0.4}\frac{k_b}{D_i}$$

$$= 0.023(15{,}981)^{0.8}(30)^{0.4}\frac{0.1}{0.05434} = 380.5 \text{ Btu/h}\cdot\text{ft}^{2}°F$$

Example 3-9. Water flowing inside a 1-in-diam. 16 BWG horizontal copper tube is being heated from 90 to 200°F by saturated steam condensing on the outside of the tube at 250°F. If the average tube wall temperature is 210°F, calculate the condensation coefficient based on the inside area.

Solution.

$$t_{sv} = 250°\text{F} \qquad t_w = 210°\text{F}$$

The condensation film coefficient on the outside of the tube is calculated by Eq. (3-52) with Δt calculated as suggested by McAdams.

$$h = 0.725\left(\frac{k_f^3 \rho_f^2 g \lambda_f}{\mu_f D_o \, \Delta t}\right)^{0.25}$$

From Eq. (3-57*a*),

$$\Delta t = t_{sv} - t_w = 250 - 210 = 40°\text{F}$$

$$t_f = t_{sv} - 0.75\,\Delta t$$

$$= 250 - 0.75(250 - 210)$$

$$= 220°\text{F}$$

$$\mu_f = 0.654 \text{ lb/h}\cdot\text{ft} \qquad \rho_f = 59.63 \text{ lb/ft}^3$$

$$k_f = 0.395 \text{ Btu/(h}\cdot\text{ft}^2\cdot°\text{F/ft)} \qquad \lambda_f = 965.2 \text{ Btu/lb}$$

$$D_o = \tfrac{1}{12} = 0.0834 \text{ ft}$$

$$h_o = 0.725\left[\frac{0.395^3(59.63)^2(965.2)(4.18 \times 10^8)}{(0.654)(0.0834)(40)}\right]^{0.25}$$

$$= 1829 \text{ Btu/h}\cdot\text{ft}^2\cdot°\text{F}$$

h_{oi} based on inside area is

$$\frac{h_o D_o}{D_i} = 1829\left(\frac{1}{0.87}\right) = 2102 \text{ Btu/h}\cdot\text{ft}^2\cdot°\text{F}$$

For the calculation of h_o using the Kern's relationship for the film temperature,

$$h = 0.725\left(\frac{k_f^3 \rho_f^2 \lambda g}{\mu_f D_o \, \Delta t_f}\right)^{0.25}$$

From Eq. (3-57),

$$t_f = \tfrac{1}{2}(t_{sv} + t_w) = \tfrac{1}{2}(250 + 210) = 230°\text{F}$$

$$\Delta t_f = t_f - t_w = 230 - 210 = 20°\text{F}$$

$$\rho_f = 59.38 \text{ lb/ft}^3$$

$$\mu_f = 0.6195 \text{ lb/h} \cdot \text{ft}$$

$$\lambda = 958.8 \text{ Btu/lb}$$

$$k_f = 0.3955 \text{ Btu/(h} \cdot \text{ft}^2 \cdot °\text{F/ft)}$$

$$h_o = 0.725\left[\frac{(0.3955)^3(59.38)^2(958.8)(4.18 \times 10^8)}{0.6195(0.0834)(20)}\right]^{0.25}$$

$$= 2199 \text{ Btu/h} \cdot \text{ft}^{2}°\text{F}$$

$$h_{oi} = 2199\left(\frac{1}{0.87}\right) = 2528 \text{ Btu/h} \cdot \text{ft}^2 \cdot °\text{F}$$

Example 3-10. Water at an average temperature of 160°F is flowing inside a horizontal 1-in-diam. 14 BWG clean copper tube at a velocity of 6 ft/s. Saturated steam at 15 psig is condensing on the outside of the tube. Calculate the overall coefficient of heat transfer based on the outside area.

Solution. Use the Dittus-Boelter equation [Eq. (3-33*a*)] for the calculation of the water film coefficient.

$$\rho_b = 61.0 \text{ lb/ft}^3 \qquad \mu_b = 0.97 \text{ lb/h} \cdot \text{ft} \qquad c_p = 1 \text{ Btu/lb} \cdot °\text{F}$$

$$k_b = 0.385 \text{ Btu/(h} \cdot \text{ft}^2 \cdot °\text{F/ft)} \qquad D_i = \frac{0.834}{12} = 0.0695 \text{ ft}$$

$$h_i = 0.023\left(\frac{D_i u \rho}{\mu}\right)_b^{0.8}\left(\frac{c_p \mu}{k}\right)_b^{0.4}\left(\frac{k_b}{D_i}\right)$$

$$= 0.023\left[\frac{0.0695(6)(61)(3600)}{0.97}\right]^{0.8}\left[\frac{1(0.97)}{0.385}\right]^{0.4}\left(\frac{0.385}{0.0695}\right)$$

$$= 1761 \text{ Btu/h} \cdot \text{ft}^2 \cdot °\text{F}$$

Steam Film Coefficient. As a first approximation, assume

$$t_w = \frac{t_{sv} + t_b}{2} = \frac{250 + 160}{2} = 205°\text{F}$$

Use t_f as suggested by McAdams:

$$\begin{aligned} t_f &= t_{sv} - 0.75(t_{sv} - t_w) \\ &= 250 - 0.75(250 - 205) \\ &= 216.3°\text{F} \qquad \text{say } 220°\text{F} \end{aligned}$$

$\Delta t = 250 - 205 = 45°\text{F}$ $\rho_f = 59.63\ \text{lb/ft}^3$ $D_o = 0.0834\ \text{ft}$

$\lambda = 965.2\ \text{Btu/lb}$ $k_f = 0.395\ \text{Btu/(h} \cdot \text{ft}^2 \cdot °\text{F/ft)}$ $\mu_f = 0.654\ \text{lb/ft} \cdot \text{h}$

From Eq. (3-52),

$$h_o = 0.725\left[\frac{0.395^3(59.63)^2(965.2)(4.18 \times 10^8)}{0.654(0.0834)(45)}\right]^{0.25} = 1776\ \text{Btu/h} \cdot \text{ft}^2 \cdot °\text{F}$$

Check the assumed wall temperature. Take areas per linear foot. The water-side resistance is

$$\frac{1}{h_i A_i} = \frac{1}{1761\pi(0.0695)} = 0.002601$$

The metal-wall resistance is

$$\frac{l_w}{k_w}\frac{1}{A_{\text{av}}}$$

$$\begin{aligned} A_{\text{av}} &= \frac{A_o - A_i}{\ln (A_o/A_i)} = \frac{(D_o - D_i)\pi}{\ln (D_o/D_i)} \\ &= \frac{(0.0834 - 0.0695)\pi}{\ln (0.0834/0.0695)} = 0.2395\ \text{ft}^2 \end{aligned}$$

$$\text{Metal-wall resistance} = \frac{0.0834 - 0.0695}{226(2)}\ \frac{1}{0.2395} = 0.000128$$

$$\text{Steam-side resistance} = \frac{1}{1776\pi(0.0834)} = 0.00215$$

$$\begin{aligned} \text{Total resistance} &= \frac{1}{U_o A_o} \\ &= 0.002601 + 0.00215 + 0.000128 = 0.00488 \end{aligned}$$

$$\Delta T_{\text{total}} = 250 - 160 = 90°\text{F}$$

$$\Delta t_i = \frac{[1/(h_i A_i)]\,\Delta T}{1/(h_i A_i) + l_w/(k_w A_{\text{av}}) + 1/(h_o A_o)}$$

$$= \frac{0.002601(90)}{0.00488} = 48°\text{F}$$

$$t_w = t_b + \Delta t_i = 160 + 48 = 208°\text{F}$$

and $$t_f = 250 - 0.75(250 - 208) = 219°\text{F} \quad \text{(assumed 220°F)}$$

which is close. h_o was calculated with the physical properties at 220°F. Therefore, correct h_o for Δt only.

$$\text{New } \Delta t = t_{sv} - t_w = 250 - 208 = 42°\text{F}$$

$$h_o = 1776(\tfrac{45}{42})^{0.25} = 1807 \text{ Btu/h} \cdot \text{ft}^2 \cdot °\text{F}$$

The overall coefficient based on the outside area is

$$\frac{1}{U_o} = \frac{1}{h_o} + \frac{l_w}{k_w}\frac{A_o}{A_{\text{av}}} + \frac{1}{h_i}\frac{A_o}{A_i}$$

$$= \frac{1}{1807} + \frac{0.00695}{228}\,\frac{0.262}{0.2395} + \frac{1}{1761}\,\frac{0.262}{0.2183} = 0.001268$$

$$U_o = 788.5 \text{ Btu/h} \cdot \text{ft}^2 \cdot °\text{F}$$

The following example illustrates the calculation for a double pipe exchanger in which the equivalent diameter has to be considered.

Example 3-11. Three thousand pounds per hour of acetic acid is to be cooled from 250 to 150°F by heating 6000 lb/h of butyl alcohol which is available at 90°F in a double pipe exchanger with countercurrent flow. Acetic acid flows through a 1-in-diam 14 BWG tube which is surrounded by an outer pipe of 2.067 in ID. Calculate the length of the tube required. Thermal conductivity of the tube material is 10 Btu/(h · ft² · °F/ft). The specific heats of acetic acid and butyl alcohol are 0.55 and 0.65 Btu/lb · °F, respectively. A fouling factor of 0.001 is to be allowed for each stream.

Solution. *Overall Heat Balance.* The heat loss by acetic acid is

$$3000(250 - 150)(0.55) = 165{,}000 \text{ Btu/h}$$

The temperature rise of butyl alcohol is

$$\frac{165{,}000}{0.65(6000)} = 42.3°\text{F}$$

The outlet temperature of butyl alcohol is

$$90 + 42.3 = 132.3°\text{F}$$

Log mean temperature difference is expressed as follows:

Acetic acid 250 ⟶ 150
132.3 ⟵ 90 Butyl alcohol

$$\Delta t_1 = 250 - 132.3 = 117.7°\text{F} \qquad \Delta t_2 = 150 - 90 = 60°\text{F}$$

$$\Delta t_{LM} = \frac{117.7 - 60}{\ln(117.7/60)} = 85.63°\text{F}$$

Use the Dittus-Boelter equation which requires the evaluation of the physical properties at the bulk temperatures. Bulk temperature of acetic acid is

$$t_B = \frac{250 + 150}{2} = 200°\text{F}$$

Bulk temperature of butyl alcohol is

$$\frac{90 + 132.3}{2} = 111.2°\text{F}$$

The properties of the streams at the bulk temperatures are listed in Table 3-3.

TABLE 3-3

	Acetic Acid	**Butyl Alcohol**
μ, cP	0.5	2.7
ρ, lb/ft^3	65.5	50.5
c_p, Btu/lb · °F	0.55	0.65
k, Btu/(h · ft^2 · °F/ft)	0.098	0.096

The inside film coefficient of heat transfer when cooling is given by Eq. (3-33*b*):

$$h_i = 0.023\left(\frac{D_iG}{\mu}\right)_b^{0.8}\left(\frac{c_p\mu}{k}\right)_b^{0.3}\left(\frac{k}{D_i}\right)$$

$$D_i = \frac{0.834}{12} = 0.0695 \text{ ft}$$

$$A_c = 0.003794 \text{ ft}^2 \qquad G = \frac{3000}{0.003794} = 790{,}722 \text{ lb/h}\cdot\text{ft}^2$$

$$h_i = 0.023\left[\frac{0.0695(790{,}722)}{0.5(2.42)}\right]^{0.8}\left[\frac{0.55(0.5)(2.42)}{0.098}\right]^{0.3}\left(\frac{0.098}{0.0695}\right)$$

$$= 306 \text{ Btu/h}\cdot\text{ft}^2\cdot{}^\circ\text{F}$$

Outside Coefficient of Heat Transfer. For heat transfer, equivalent diameter D_e is

$$D_e = \frac{D_2^2 - D_o^2}{D_o}$$

$$= \frac{(2.067^2 - 1^2)}{12} = 0.2727 \text{ ft}$$

$$\text{Flow area} = \frac{\pi(2.067^2 - 1^2)}{4(144)} = 0.01785 \text{ ft}^2$$

$$G = \frac{6000}{0.01785} = 336{,}135 \text{ lb/h}\cdot\text{ft}^2$$

$$\frac{D_eG}{\mu} = \frac{0.2727(336{,}135)}{(2.7)(2.42)} = 14{,}029$$

$$\frac{k}{D_e} = \frac{0.096}{0.2727} = 0.352$$

$$\frac{c_p\mu}{k} = \frac{0.65(2.7)(2.42)}{0.096} = 44.23$$

From Eq. (3-33*a*) the outside coefficient of heat transfer is

$$h_o = 0.023\left(\frac{D_eG}{\mu_b}\right)^{0.8}\left(\frac{c_p\mu}{k}\right)^{0.4}\left(\frac{k_b}{D_e}\right)$$

$$= 0.023(14{,}029)^{0.8}(44.23)^{0.4}(0.352) = 76.6 \text{ Btu/h}\cdot\text{ft}^2\cdot{}^\circ\text{F}$$

$$\frac{1}{U_o} = \frac{1}{h_o} + f_{do} + \frac{l_w}{k_w}\frac{D_o}{D_{av}} + f_{di}\frac{D_o}{D_i} + \frac{1}{h_i}\frac{D_o}{D_i}$$

$$= \frac{1}{76.6} + 0.001 + \frac{0.006917}{10}\frac{1}{0.917} + 0.001\frac{1}{0.834} + \frac{1}{306}\frac{1}{0.834} = 0.01993$$

$$U_o = 50.2 \text{ Btu/h} \cdot \text{ft}^2 \cdot {}^\circ\text{F}$$

$$A_o = \frac{Q}{U\,\Delta T} = \frac{165{,}000}{50.2(85.63)} = 38.38 \text{ ft}^2$$

$$\text{Length of tube} = \frac{A}{\pi D_o} = \frac{38.38}{\pi(0.0834)} \doteq 147 \text{ ft}$$

Example 3-12. The data in Table 3-4 were obtained for a condenser in water flowing through the tubes and steam condensing on the outside. The steam film coefficient on the outside of the tube is constant and equal to 2000 Btu/h · ft^2 · °F. What is the scale resistance? Neglect the pipe-wall resistance.

TABLE 3-4 Overall Coefficients for a Condenser

u, ft/s	U_o, Btu/h · ft^2 · °F	$\frac{1}{U_o} \times 10^3$	$\frac{1}{u^{0.8}}$
6.91	485.4	2.060	0.213
6.35	473.3	2.113	0.228
5.68	452.1	2.212	0.249
4.90	421.2	2.374	0.280
2.93	333.2	3.001	0.423
7.01	480.5	2.081	0.211
2.95	325.1	3.076	0.421
4.12	364.6	2.743	0.322
6.76	400.3	2.498	0.217
2.86	298.0	3.356	0.431
6.27	452.7	2.209	0.230

Solution.

$$\frac{1}{U_o} = \frac{1}{h_o} + R_d + \frac{1}{au^{0.8}}$$

Since $h_i \propto u^{0.8}$, a plot of $1/U_o$ versus $1/u^{0.8}$ will be a straight line with intercept of $(1/h_o) + R_d$ which is constant (Wilson's plot).[4b] Values of $1/U_o$ and $1/u^{0.8}$ are given in the last two columns of Table 3-4. $1/U_o$ is plotted against $1/u^{0.8}$ in Fig. 3-4.

From the graph, the intercept is

$$0.00083 = \frac{1}{h_o} + R_d$$

$$R_d = 0.00063 - \frac{1}{h_o} = 0.00083 - \frac{1}{2000} = 0.00033$$

The scale resistance, or fouling factor, is 0.00033 h · ft² · °F/Btu.

Example 3-13. A tube bank consists of 1-in-OD tubes on 2-in centers in a square in-line arrangement of 16 rows in each direction. The tube wall is at 220°F and air at 80°F flows across the tube bank at a mean velocity of 8000 lb/h · ft² based on the minimum area of flow. Determine the heat-transfer coefficient for air. At what temperature does the air leave the tube bank? Assume ½-in clearance between the wall and the tubes in the extreme longitudinal rows.

Solution. The surface temperature is 220°F, and the air outlet temperature is not known. Assume it is 120°F. From Eq. (3-43),

$$t_f = \tfrac{1}{2}[220 + \tfrac{1}{2}(120 + 80)] = 160°\text{F}$$

The properties of air at 160°F are

$$\mu_f = 0.02 \text{ cP}(2.42) = 0.0484 \text{ lb/h} \cdot \text{ft}$$

$$D_o = 0.0834 \text{ ft}$$

$$k_f = 0.0174 \text{ Btu/(h} \cdot \text{ft}^2 \cdot °\text{F/ft)}$$

$$\rho_f = 0.0643 \text{ lb/ft}^3$$

$$h_o = 0.33\frac{k_f}{D_o}\left(\frac{D_o G_{\max}}{\mu_f}\right)^{0.6}\left(\frac{c_p \mu_f}{k_f}\right)^{1/3}$$

$$= 0.33\frac{0.0174}{0.0834}\left[\frac{0.0834(8000)}{0.0484}\right]^{0.6}\left[\frac{0.24(0.0484)}{0.0174}\right]^{1/3}$$

$$= 18.3 \text{ Btu/h} \cdot \text{ft}^2 \cdot °\text{F}$$

$$Q_o = 18.3(220 - 100) = 2196 \text{ Btu/h} \cdot \text{ft}^2$$

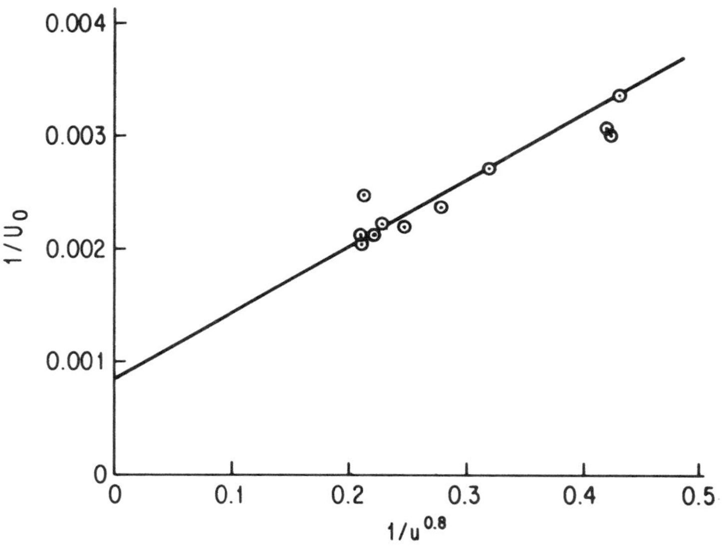

Figure 3-4. Wilson plot (Example 3-12).

The flow per square foot of minimum area is 8000 lb/h · ft².

$$\Delta T_{LM} = \frac{(220 - 80) - (220 - 120)}{\ln (140/100)} = 119°\text{F}$$

Consider a 1-ft length of bundle with 16 rows. The flow area is

$$(16 - 1)(\tfrac{2}{12} - \tfrac{1}{12}) + (2)(\tfrac{1}{2})(\tfrac{1}{12}) = \tfrac{16}{12}\ \text{ft}^2$$

The flow per foot of bundle is

$$(8000\ \text{lb/h} \cdot \text{ft}^2_{\text{m}})(\tfrac{16}{12}) = 10{,}667\ \text{lb/h} \cdot \text{ft}$$

The surface area per foot of bundle is

$$\frac{\pi(1)}{12}(16)(16) = 67.02\ \text{ft}^2$$

Heat balance gives

$$10{,}667(0.24)\ (t_o - 80) = 18.3(67.02)(119)$$

from which

$$t_o = 137°\text{F} \qquad (\text{assumed } 120°\text{F})$$

Reassume

$$t_o = 137°\text{F} \qquad \text{then } t_f = \tfrac{1}{2}\left(220 + \frac{80 + 137}{2}\right)$$

$$= 164°\text{F} \qquad \text{(previously calculated 160°F)}$$

Because of a very small change in temperature, there will not be much change in the physical properties of air.

$$h_o = 18.3\ \text{Btu/h}\cdot\text{ft}^2\cdot°\text{F} \qquad \text{as before}$$

$$\Delta T_{LM} = \frac{57}{\ln\frac{140}{83}} \doteq 109°\text{F}$$

Then,

$$10{,}667(0.24)\,(t_o - 80) = 67.02(18.3)\,\Delta T_{LM} = 67.02(18.3)(109.0)$$

$$t_o = 132.2 \text{ (137°F assumed)}$$

Another trial with $t_o = 133°\text{F}$ yields

$$\Delta T_{LM} = 114.4°\text{F}$$

and

$$t_o = 133.4°\text{F} \qquad \text{(assumed 133°F)}$$

$$h_i = 18.3\ \text{Btu/h}\cdot\text{ft}^2\cdot°\text{F} \qquad \text{and} \qquad t_o = 133.4°\text{F}$$

Shell and Tube Exchangers

For $2000 < N_{\text{Re}} < 10^6$, the heat-transfer coefficient[5g] on the shell side is given by

$$\frac{h_o D_e}{k} = 0.36\left(\frac{D_e G_s}{\mu}\right)^{0.55}\left(\frac{c_p \mu}{k}\right)^{1/3}\left(\frac{\mu}{\mu_w}\right)^{0.14} \tag{3-59}$$

$$a_s = \frac{D_{si} C' B}{144 P_T} \qquad \text{and} \qquad G_s = \frac{W}{a_s} \tag{3-60}$$

where d_o = tube outside diameter, in
D_e = shell side equivalent diameter, ft

a_s = cross-flow area, ft^2
D_{si} = inside diameter of shell, in
P_T = tube pitch, in
C' = clearance between tubes measured along tube pitch, in
$= P_T - d_o$
B = baffle spacing, in
W = weight flow of fluid, lb/h

(The baffle spacing should range from $D_{si}/5$ to D_{si}.) For square pitch tube arrangement,

$$D_e = \frac{P_T^2 - \frac{1}{4}\pi d_o^2}{3\pi d_o} \quad \text{ft} \tag{3-61}$$

For 60° triangular equilateral pitch arrangement,

$$D_e = \frac{2(0.43\, p_T^2 - \frac{1}{8}\pi d_o^2)}{3\pi d_o} \quad \text{ft} \tag{3-62}$$

Logarithmic Temperature Differences

In calculating the log mean temperature difference for the shell-tube exchangers, a correction factor F_T must be applied because the flows are not truly countercurrent or parallel, but a combination of the countercurrent and parallel. Thus

$$\Delta T_{\text{true mean}} = \text{LMTD}(F_T) \tag{3-63}$$

Values of F_T are available in the form of graphs in other texts.[5,6]

Example 3-14. Seventy thousand pounds per hour of an organic liquid is to be cooled by countercurrent flow of the water in a shell and tube exchanger. The data available are listed in Table 3-5. Estimate the required area of the exchanger.

TABLE 3-5

	Shell Side	Tube Side
Fluid circulated	Water	Organic
Total liquid, lb/h	198,330	70,000
Temperature in, °F	85	200
Temperature out, °F	100	115
Specific gravity	1 at 92°F	1.5 at 158°F
Viscosity, cP	1.0 at 92 °F	2 at 158°F
Specific heat, Btu/lb · °F	1 at 92°F	0.5 at 158°F
Thermal conductivity, Btu/(h · ft^2 · °F/ft)	0.36 at 92°F	0.1 at 158°F
Fouling factor, h · ft^2 · °F/Btu	0.002	0.002
Material of construction	Carbon steel	Carbon steel

Solution.

Step 1. Estimate a preliminary area and assume a geometry for the exchanger. Heat balance gives

$$Q = WC_p(t_2 - t_1) = 70{,}000(0.5)(200 - 115) = 2{,}975{,}000 \text{ Btu/h}$$

Alternatively,

$$Q = 198{,}330(1)(100 - 85)$$
$$= 2{,}974{,}950 \doteq 2{,}975{,}000 \text{ Btu/h}$$

Calculate LMTD:

$$\text{LMTD} = \frac{100 - 30}{\ln \frac{100}{30}} = 58.14\text{°F}$$

$$\begin{array}{lr} 85 & \xrightarrow{\hspace{6em}} 100 \\ 115 & \xleftarrow{\hspace{6em}} 200 \end{array}$$

For the cooling of organics, U varies between 50 to 150 Btu/h · ft^2 · °F. Assume for the first trial, U = 75 Btu/h · ft^2 · °F. The estimated area is

$$\frac{2{,}975{,}000}{75(58.14)} = 682 \text{ ft}^2$$

Assume $\frac{3}{4}$-in-OD 14 BWG tubes and a tube-side velocity of 5 ft/s. From

Table 11-2 of Ref. 6, the flow cross section is

$$0.2679\ \text{in}^2 = 0.00186\ \text{ft}^2/\text{tube}$$

The number of tubes N_t is

$$N_t = \frac{W}{Au\rho} = \frac{70{,}000}{0.00186(5)(3600)(62.4)(1.5)} = 22.34$$

say 22 tubes per pass. A tube of $\frac{3}{4}$ in OD has an outside surface of 0.1963 ft^2/ft of tube. Assume a tube length of 20 ft. Then N_p, the number of tube passes, is

$$N_p = \frac{682}{22(0.1963)(20)} = 7.9$$

say 8 passes. Therefore, the actual area in the exchanger = $8(22)(0.1963)(20) = 691\ \text{ft}^2$.

Step 2. Calculate the tube-side coefficient. The total tube-side flow area is $22(0.00186) = 0.04092\ \text{ft}^2$.

$$G_t = \frac{70{,}000}{0.04092} = 1.71 \times 10^6\ \text{lb/h}\cdot\text{ft}^2$$

The bulk temperature of fluid is

$$\tfrac{1}{2}(200 + 115) = 158°\text{F}$$

Use the Dittus-Boelter, Eq. (3-33*b*), for fluid cooling to calculate h_i:

$$\frac{h_i D_i}{k} = 0.023(N_{\text{Re}})^{0.8}(N_{\text{Pr}})^{0.3}$$

where N_{Pr} is the Prandtl number $c_p\mu/k$.

$$D_i = \frac{0.584}{12} = 0.04867\ \text{ft}$$

$$h_i = 0.023\left(\frac{0.1}{0.04867}\right)\left[\frac{0.04867(1.71 \times 10^6)}{2(2.42)}\right]^{0.8}\left[\frac{0.5(2)(2.42)}{0.1}\right]^{0.3}$$

$$= 301.6\ \text{Btu/h}\cdot\text{ft}^2\cdot°\text{F}$$

Step 3. Calculate shell-side coefficient. Assume $\frac{3}{4}$-in-OD tubes on $\frac{15}{16}$-in triangular pitch. With a $17\frac{1}{4}$-in-ID shell, an eight-tube pass exchanger can accommodate 178 tubes.[5k] Therefore, choose $17\frac{1}{4}$-in-ID shell. The baffle spacing chosen may vary from 17.25/5 = 3.45 to 17.25 in. The shell-side flow area is

$$a_s = \frac{D_{si}C'B}{144P_T} \qquad \text{ft}^2$$

where $D_{si} = 17.25$ in

$$C' = \frac{15}{16} - \frac{3}{4} = 0.1875 \text{ in}$$

B = baffle spacing (unknown)

$$P_T = \frac{15}{16} = 0.9375 \text{ in}$$

$$\text{Mass velocity} = G_S = \frac{W}{a_s} = u\rho(3600)$$

$$a_s = \frac{W}{u\rho(3600)}$$

Therefore

$$\frac{D_{si}C'B}{144P_T} = \frac{W}{u\rho(3600)}$$

From which

$$u = \frac{W(144)P_T}{B\rho(3600)(D_{si})(C')} = \frac{198{,}330(144)(0.9375)}{B(62.4)(3600)(17.25)(0.1875)} = \frac{36.85}{B}$$

If B is chosen as 6 in, $u = 36.85/6 = 6.14$ ft/s; this is reasonable since B lies between 3.45 and 17.25 in as required. The equivalent diameter is

$$D_e = \frac{0.86P_T^2 - \frac{1}{4}\pi d_o^2}{3\pi d_o} = \frac{0.86(0.9375)^2 - \pi(0.75)^2/4}{3\pi(0.75)} = 0.044 \text{ ft}$$

$$G_s = \frac{W}{a_s} = u\rho(3600) = 6.14(62.4)(3600) = 1{,}379{,}290 \text{ lb/h} \cdot \text{ft}^2$$

$$N_{\text{Re}} = \frac{D_eG_s}{\mu} = \frac{0.044(1.379 \times 10^6)}{1(2.42)} = 2.5 \times 10^4$$

$$N_{\text{Pr}} = \frac{c_p\mu}{k} = \frac{1(2.42)}{0.36} = 6.72$$

$$\frac{h_o D_e}{k} = 0.36(N_{\text{Re}})^{0.55}(N_{\text{Pr}})^{1/3}\left(\frac{\mu}{\mu_w}\right)^{0.14}$$

Assuming $(\mu/\mu_w)^{0.14} = 1.0$,

$$h_o = 0.36\frac{0.36}{0.044}(2.5 \times 10^4)^{0.55}(6.72)^{1/3} = 1458 \text{ Btu/h}\cdot\text{ft}^2\cdot{}^\circ\text{F}$$

Step 4. Calculate overall heat transfer coefficient U_o.

$$\frac{1}{U_o} = \frac{1}{h_o} + f_{do} + \frac{l_w}{k_w}\left(\frac{D_o}{D_{\text{av}}}\right) + f_{di}\frac{D_o}{D_i} + \frac{1}{h_i}\frac{D_o}{D_i}$$

$$= \frac{1}{1458} + 0.002 + \frac{0.083}{12(26)}\frac{0.75}{0.667} + 0.002\frac{0.75}{0.584}$$

$$+ \frac{1}{301.6}\frac{0.75}{0.584} = 0.00982$$

$U_o = 101.8 \text{ Btu/h}\cdot\text{ft}^2\cdot{}^\circ\text{F}$

Step 5. Calculate corrected LMTD and area.

$$R = \frac{T_1 - T_2}{t_2 - t_1} = \frac{85 - 100}{115 - 200} = 0.176$$

$$S = \frac{t_2 - t_1}{T_1 - t_1} = \frac{115 - 200}{85 - 200} = 0.74$$

From F_T chart for the 1-8 exchanger,[6] $F_T = 0.93$. The corrected $\Delta T_{LM} = F_T(\Delta T_{LM}) = 0.93(58.14) = 54.07^\circ\text{F}$

$$A = \frac{Q}{U_o\,\Delta T} = \frac{2{,}975{,}000}{101.8(54.07)} = 540.5 \text{ ft}^2$$

The area available in the exchanger is 691 ft^2. So the percent excess area is,

$$\frac{691 - 540.5}{540.5}(100) = 27.8 \text{ percent}$$

Therefore, the assumed area is too high an estimate. A second trial with a tube length of 18 ft (other dimensions being the same) gives a more practical estimate of the heat-exchanger surface required. The area of an exchanger with a tube length of 18 ft is $\frac{18}{20}(691) = 622\ \text{ft}^2$. It gives [(622 − 540.5)/540.5] (100) = 15.0 percent excess area.

In actual practice, the allowable pressure drop through the exchanger may dictate the selection of the geometry of the exchanger. In the example above, it is assumed that the allowable pressure drops are not exceeded by the selected geometry of the exchanger.

Example 3-15. A heat-exchanger specification sheet contains the information from Table 3-6.

TABLE 3-6

Tubes	$\text{OD} = \frac{3}{4}$ in ID = 0.62 in
Heat-transfer surface	91.1 ft^2
Heat exchanged	42,300 Btu/h
Corrected MTD	21.9°F
Transfer rates	
Service	21.2 $\text{Btu/h}\cdot\text{ft}^2\cdot{}^\circ\text{F}$
Clean	25.3 $\text{Btu/h}\cdot\text{ft}^2\cdot{}^\circ\text{F}$
Fouling resistances	
Shell side	0.002 $\text{h}\cdot\text{ft}^2\cdot{}^\circ\text{F/Btu}$
Tube side	0.002 $\text{h}\cdot\text{ft}^2\cdot{}^\circ\text{F/Btu}$

Check consistency of the data and calculate the percent extra surface included in the specific area.

Solution.

$$Q = U_o A_o (\Delta T)_{\text{LMTD}}(F_T)$$

$$= 21.2(91.1)(21.9) = 42{,}296\ \text{Btu/h}$$

This checks with the specified heat load of 42,300 Btu/h. The calculated dirty transfer rate may be obtained by

$$\frac{1}{U_D} = \frac{1}{U_C} + f_{do} + f_{di}\frac{D_o}{D_i}$$

$$\frac{1}{U_D} = \frac{1}{25.3} + 0.002 + 0.002\frac{0.75}{0.62} = 0.043945$$

$$U_D = 22.76 \text{ Btu/h} \cdot \text{ft}^2 \cdot {}^\circ\text{F}$$

$$\text{Calculated required area} = \frac{Q}{U_D \, \Delta T_{\text{MTD}}} = \frac{42{,}300}{22.76(21.9)} = 84.9 \text{ ft}^2$$

$$\text{Actual area} = 91.1 \text{ ft}^2$$

$$\text{Percent excess area} = \frac{91.1 - 84.9}{84.9}(100) = 7.34 \text{ percent}$$

Example 3-16. Benzene is to be condensed in a 1-2 vertical condenser at a rate of 6000 kg/h at essentially atmospheric pressure. Benzene condenses on the shell side while the water flows through the tubes. A unit having the following data is available:

Number of tubes	118
Tube OD	25 mm, ID = 22 mm
Length of tube	3 m
Thermal conductivity of tube material	0.045 kW/m · K

Cooling water is available at 303 K. The scale factors may be taken as

Shell side	$0.176 \text{ m}^2 \cdot \text{K/kW}$
Water side	$0.260 \text{ m}^2 \cdot \text{K/kW}$

Check whether this condenser is adequate for the required duty.

Solution. Latent heat of vaporization of benzene = 394 kJ/kg at 353 K and 1 atm. The density of water is 1000 kg/m^3. The condenser heat load is

$$(6000 \text{ kg/h})(\tfrac{1}{3600} \text{ h/s})(394 \text{ kJ/kg}) = 656.7 \text{ kJ/s} = 656.7 \text{ kW}$$

Assume a rise in the temperature of water of 5 K. By heat balance,

$$Mc_p(t_2 - t_1) = 656.7$$

Therefore, mass flow rate of water is

$$M = \frac{656.7}{5(4.1868)}$$

$$= 31.37 \text{ kg/s}$$

The water flow through tubes is

$$\frac{31.37}{1000} = 0.03137 \text{ m}^3\text{/s}$$

and the number of tubes per pass is

$$\frac{118}{2} = 59$$

The area of cross section for flow is

$$59(0.785)(0.022)^2 = 0.02243 \text{ m}^2$$

Water velocity on the tube side is

$$\frac{0.03137}{0.02243} = 1.3986 \text{ m/s}$$

This is a reasonable velocity consistent with the fouling factor suggested, and, therefore, a 5-K rise in the water temperature is adequate for the first trial. (Alternatively, a tube-side velocity of 1.5 m/s consistent with the scale factor could be assumed as a first trial and ΔT computed.)

Log Mean Temperature Difference.

$$\Delta T_1 = 353 - 303 = 50 \text{ K}$$

$$\Delta T_2 = 353 - 308 = 45 \text{ K}$$

$$\Delta T_{LM} = \frac{50 - 45}{\ln \frac{50}{45}} = 47.46 \text{ K}$$

Water Film Coefficient.
Use the Dittus-Boelter equation (3-33*a*), since water is heated.

$$h_i = 0.023\,\frac{k}{D_i}\left(\frac{D_iG}{\mu}\right)^{0.8}\left(\frac{c_p\mu}{k}\right)_b^{0.4}$$

$$\text{Bulk temperature } t_b = \tfrac{1}{2}(303 + 308) = 305.5\ \text{K}$$

Properties of water at this temperature are

$$\mu = 0.8\ \text{cP} = 0.0008\ \text{kg/m}\cdot\text{s}$$

$$c_p = 4186.8\ \text{J/ kg K} = 4.1868\ \text{kJ /kg K}$$

$$k = 0.623\ \text{W/m}\cdot\text{K}$$

$$\text{Mass velocity } G = \frac{31.37}{0.02243} = 1398.6\ \text{kg/m}^2\cdot\text{s}$$

$$h_i = 0.023\frac{0.623}{0.022}\left[\frac{0.022(1398.6)}{0.0008}\right]^{0.8}\left[\frac{4.1868(0.8)}{0.623}\right]^{0.4}$$

$$= 5943\ \text{W/m}^2\cdot\text{K} = 5.943\ \text{kW/m}^2\cdot\text{K}$$

Benzene Condensation Coefficient.
From Eq. (3-50),

$$h_o = 1.47\left(\frac{4G'}{\mu_f}\right)^{-0.33}\left(\frac{k^3\rho_f^2g}{\mu_f^2}\right)^{0.33}$$

The approximate wall temperature is

$$t_w = \frac{353 + 305.5}{2} = 329\ \text{K}$$

and the film temperature from Eq. (3-57) is

$$t_f = \frac{353 + 329}{2} = 341\ \text{K}$$

Properties of Benzene at the Film Temperature.

$$\mu_f = 0.35\ \text{cP} = 0.00035\ \text{N}\cdot\text{s/m}^2$$

$$k_f = 0.151\ \text{W/m}\cdot\text{K}$$

$$\rho_f = 880\ \text{kg/m}^3$$

$$\text{Benzene condensed} = \tfrac{6000}{3600} = 1.67\ \text{kg/s}$$

$$\text{Benzene condensed per tube} = \frac{1.67}{118} = 0.01415 \text{ kg/s}$$

$$G' = \frac{0.01415}{\pi(0.025)} = 0.1802 \text{ kg/m}\cdot\text{s}$$

$$h_o = 1.47\left[\frac{4(0.1802)}{0.00035}\right]^{-0.33}\left[0.151^3(880)^2\frac{9.81}{0.00035^2}\right]^{0.33}$$

$$= 853.04 \text{ W/m}^2\cdot\text{K} = 0.853 \text{ kW/m}^2\cdot\text{K}$$

The overall coefficient based on outside area from Eq. (3-27) is

$$\frac{1}{U_o} = \frac{1}{5.943}\left(\frac{25}{22}\right) + 0.26\left(\frac{25}{22}\right) + \frac{0.0015}{0.045}\left(\frac{25}{23.5}\right) + 0.176 + \frac{1}{0.853}$$
$$= 1.8705$$

Therefore $\qquad U_o = 0.5346 \text{ kW/m}^2\cdot\text{K}$

Check on Film Temperature.

$$t_w = \frac{5.943(305.5) + (0.853)(353)}{5.943 + 0.853} = 312 \text{ K}$$

$$t_f = 0.5(353 + 312) = 333 \text{ K (assumed 341 K)}$$

There will not be much change in the properties of benzene; hence a recalculation for benzene condensation is not necessary. The calculated value of U_o is therefore adequate. The area of heat transfer required is

$$\frac{656.7}{(0.5346)(47.46)} = 25.9 \text{ m}^2$$

The heat-transfer surface per tube is

$$\pi D_o L = \pi(0.025)(3) = 0.2356 \text{ m}^2$$

The total heat-transfer surface is

$$0.2356(118) = 27.8 \text{ m}^2$$

The percent excess area available is

$$\frac{(27.8 - 25.9)(100)}{25.9} = 7.3 \text{ percent}$$

Therefore, the available condenser is adequate for the specified duty.

Heating and Cooling of Liquid Batches

A detailed treatment of the various cases of batch heating and cooling is available elsewhere.[5h] An example is given below to illustrate the method of solution.

Example 3-17. A batch of 20,000 lb of a dilute water solution of a salt is to be concentrated from 5 to 25 percent solids by evaporation in a batch heater consisting of a coil in a tank which is equipped with an agitator. The heater coil remains submerged in the solution until the end of the evaporation process. The heating medium is a 40,000 lb/h stream of hot oil entering at 300°F. The specific heat of the oil is 0.5 Btu/lb · °F and can be assumed constant. U_o is 150 Btu/h · ft^2 · °F; assume U_o also is constant. The coil outside heat-transfer surface is 225 ft^2. The initial temperature of the solution is 60°F, and its boiling point is 212°F. The boiling-point elevation caused by the dissolved salts is negligible.

Solution. Since the solution is initially below its boiling point, it must be heated first to its boiling point (212°F). During this period both the batch liquid and the heating medium are nonisothermal.

Initial time θ_1 for heating the solution comes under the following category: coil in a tank heater, nonisothermal heating medium. Therefore, Eq. (18-9) of Ref. 5*i* applies. Now

$$M = 20{,}000 \text{ lb}$$

$$C_c = 1 \qquad C_H = 0.5$$

$$T_1 = 300°\text{F} \qquad t_1 = 60°\text{F} \qquad t_2 = 212°\text{F}$$

$$K_1 = \exp\left(\frac{UA}{WC_H}\right) = \exp\frac{150(225)}{40{,}000(0.5)} = 5.406$$

Then
$$\theta_1 = \frac{20{,}000(1)}{40{,}000(0.5)}\left[\frac{5.406}{5.406-1}\ln\left(\frac{300-60}{300-212}\right)\right] = 1.23 \text{ h}$$

During the evaporation period, the temperature of the solution is constant at 212°F. The heat-transfer rate is constant since U and A are constant. Therefore, the evaporation is a steady-state operation. First calculate water to be evaporated and the heat of evaporation. The solids in solution are

$$0.05(20{,}000) = 1000 \text{ lb}$$

The water in solution after evaporation is

$$1000\frac{0.75}{0.25} = 3000 \text{ lb}$$

Therefore, water evaporation is

$$19{,}000 - 3000 = 16{,}000 \text{ lb}$$

$$Q_T = \text{heat of evaporation} = 16{,}000(970.3) = 15.525 \times 10^6 \text{ Btu}$$

Now

$$q = UA\ \Delta T_{LM} = WC_H(T_1 - T_2)$$

$$UA\,\frac{(T_1 - t) - (T_2 - t)}{\ln[(T_1 - t)/(T_2 - t)]} = WC_H(T_1 - T_2)$$

from which

$$\frac{UA}{WC_H} = \ln\frac{T_1 - t}{T_2 - t}$$

$$\frac{T_1 - t}{T_2 - t} = \exp\left(\frac{UA}{WC_H}\right) = \exp\left[\frac{150(225)}{40{,}000(0.5)}\right] = 5.406$$

$$\frac{300 - 212}{T_2 - 212} = 5.406 \qquad \text{and} \qquad T_2 = 228.3°\text{F}$$

The rate of heat transfer is

$$40{,}000(0.5)(300 - 228.3) = 1.434 \times 10^6 \text{ Btu/h}$$

Evaporation time is

$$\frac{Q_T}{q} = \frac{15.525 \times 10^6}{1.434 \times 10^6} = 10.83 \text{ h}$$

Total time for heating and evaporation is 1.23 + 10.83 = 12.06 h.

References

1. E.N. Sieder and G.E. Tate, *Ind. Eng. Chem.* vol. 28, 1936, pp. 1429–1436.
2. F.W. Dittus and L.M.K. Boelter, *Univ. Calif. Pubs. Eng.* vol. 2, 1930, p. 443.
3. A.P. Colburn, *Trans. Am. Inst. Chem. Engrs.*, vol. 29, 1933, pp. 174–220.

4. W.H. McAdams *Heat Transmission*, 3d ed., McGraw-Hill, New York, 1954: (*a*) pp. 237–246; (*b*) pp. 335–336; (*c*) p. 330; (*d*) p. 345; (*e*) p. 274.
5. D.Q. Kern *Process Heat Transfer*, McGraw-Hill, New York, 1950: (*a*) p. 217; (*b*) p. 260; (*c*) p. 261; (*d*) p. 265; (*e*) p. 270; (*f*) p. 260; (*g*) pp. 136–139; (*h*) pp. 624–633; (*i*) p. 627; (*j*) p. 82; (*k*) pp. 841–842; (*l*) pp. 828–833.
6. *Chemical Engineers' Handbook*, 5th ed., R.H. Perry (ed.), McGraw-Hill, New York, 1973: pp. 10-24 and 10-25.

Chapter

4

Evaporation

Evaporation is the removal of a solvent, usually water, by its vaporization from relatively nonvolatile solute, usually a solid. The evaporators may be single-effect or multieffect, the latter for steam economy. The multieffect evaporators may be classified as forward feed, backward feed, parallel feed, and mixed feed. A further classification of the evaporators is natural circulation and forced circulation. Application of the various types of the evaporators is dependent upon the solution characteristics and the capacity required.

In general, the evaporation problems involve the calculation of the material and heat balances, steam economy, and required heat-transfer surfaces.

Evaporator Capacity

The heat-transfer capacity of an evaporator is given by

$$q = UA \, \Delta T \tag{4-1}$$

where q = total heat transferred, Btu/h

U = overall heat-transfer coefficient, Btu/h · ft^2 · °F
ΔT = overall temperature difference, °F

For a given heat input, the evaporation capacity is reduced if sensible heat is to be provided to bring the feed to its boiling point. The temperature difference that is available for heat transfer depends upon (1) the solution to be evaporated, (2) the liquid head in the tubes, and (3) the difference between pressure in the steam chest and that over the vapor space above the boiling liquid.

Boiling-Point Elevation

The boiling point of a solution is higher than the boiling point of the solvent because of the dissolved salts. The boiling-point elevation is small for dilute solutions but may be very high for the solutions of inorganic salts. Duhring's rule states that the boiling-point elevation of a given solution is a linear function of the boiling point of water. The other factor affecting the boiling point of a solution in an evaporator is the liquid head or height of the liquid level in the tube (hydrostatic head).

Heat-Transfer Coefficients

Heat-transfer coefficients are generally high for evaporators. Correlations[1a] are available for some cases.

Steam Economy in Evaporation

One advantage of the multieffect evaporation is the steam economy (pounds of water evaporated per pound of steam) that can be obtained. This is because the live steam is used only in one unit, whereas the vapor from one effect is used in the chest of the next effect. This requires using a vacuum on the system in order to have appreciable temperature differences. An optimum number of the effects for a given application is a function of the economy of steam savings weighted against the investment in the added number of effects.

Calculations

The usual assumptions made in making the calculations of the evaporations or the evaporator heat balances are: (1) there is no leakage or entrainment; (2) noncondensable content of the steam or vapor is negligible; (3) superheating the steam or vapor from each effect as well as subcooling of the condensate are considered negligible.

Where the enthalpies of the solutions as a function of concentration are available, a heat balance can be made in terms of the enthalpies. When the heat of dilution is negligible, the enthalpies can be calculated with the use of the specific heats of the solution. Usually the temperature of the thick liquor is taken as the reference temperature. The enthalpy of the feed is given by

$$h_F = c_{pf}(t_F - t) \tag{4-2}$$

where h_F = enthalpy of feed, Btu/lb
c_{pf} = specific heat, Btu/lb · °F
t_F, t = temperatures of the feed and thick liquor respectively, °F

Single-Effect Evaporator. Use of the enthalpy composition diagram to solve a problem of a single-effect evaporator is illustrated by the following example.

Example 4-1. A single-effect evaporator is used to concentrate 25,000 lb/h of a NaOH solution from 10 to 50% concentration. The evaporator is supplied with 15 psig steam. The feed is at 100°F. It operates under a vacuum of 26 inHg referred to a 30-in barometer. If $U = 450$ Btu/h · ft^2 · °F, calculate: (1) heat-transfer surface, (2) steam consumption, and (3) water requirements of a countercurrent baromet-

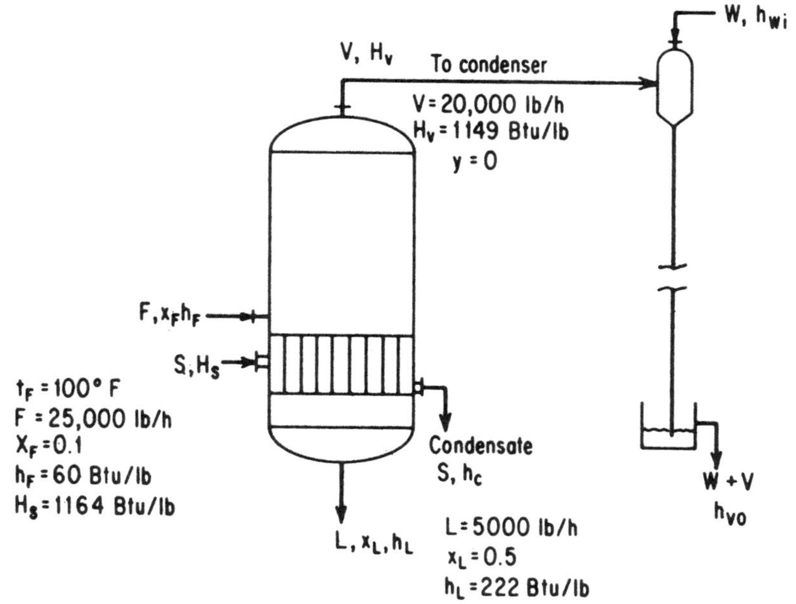

Figure 4-1. Material flows and enthalpies (Example 4-1).

ric condenser if the water is available at 86°F and if the outlet water temperature is not to exceed 120°F.

Solution. Refer to Fig. 4-1. Given quantities are

$$F = 25{,}000 \text{ lb/h} \qquad x_F = 0.1$$

$$x_L = 0.5 \qquad t_F = 100°\text{F}$$

$$y = 0$$

where y is the concentration of solute in overhead vapor. From steam tables,

$$\text{Steam pressure} = 14.7 + 15 = 29.7 \text{ psia}$$

$$t_s = 250°\text{F} \qquad H_s = 1164 \text{ Btu/lb} \qquad \lambda_s = 945.5 \text{ Btu/lb}$$

The boiling point of water under a 26-in Hg vacuum is to be obtained from steam tables.

$$26\text{-inHg vacuum} = 4\text{-inHg pressure} = \tfrac{4}{30}(14.7) = 1.96 \text{ psia}$$

From the steam tables, the boiling point of water at 1.96 psia (by interpolation) is

$$\text{bp} = \frac{1.960 - 1.942}{2.222 - 1.942}(130 - 125) + 125 = 125.3°\text{F} \doteq 125°\text{F}$$

The enthalpy of saturated steam at 125°F = 1116 Btu/lb (from steam tables). The boiling points of thick liquor (concentrated solution) are to be calculated.

For the enthalpies of NaOH solutions, refer to the enthalpy composition diagram[1b] for NaOH. To get the boiling points, refer to the Duhring's lines[2] for NaOH.

$$t_B = 198°\text{F}$$

$$h_F = 60 \text{ Btu/lb}$$

$$h_L = 222 \text{ Btu/lb}$$

Material Balance. Amounts of the thick liquor and evaporation are obtained by material balance as

$$Fx_F = Lx_L + Vy$$

$$25{,}000(0.1) = 0.5\text{L} + 0$$

Solving,

$$L = 5000 \text{ lb/h}$$

$$V = 20{,}000 \text{ lb/h}$$

Enthalpy Balance. The equation for the enthalpy balance is written as

$$Fh_F + S(H_s - h_c) = VH_v + Lh_L$$

where H_v is the enthalpy of the superheated vapor at 198°F.

$$\begin{aligned} H_v &= h_b + c_p(t_s - t_b) \\ &= 1116 + 0.46(198 - 125) \\ &= 1149.0 \text{ Btu/lb} \end{aligned}$$

where t_s is the temperature of superheated steam and t_b is the boiling point of the solution, °F.

$$25{,}000(60) + S(1164 - 218.5) = 20{,}000(1149) + 5000(222)$$

from which steam consumption S = 23,892 lb/h.

Heating Surface.

$$Q = UA\ \Delta T \qquad A = \frac{Q}{U\ \Delta T}$$

$$23{,}892(1164 - 218.5) = 450A(250 - 198)$$

$$A = \frac{23{,}892(1164 - 218.5)}{450(250 - 198)} = 965.4\ \text{ft}^2$$

$$\text{Steam economy} = \frac{\text{water evaporated}}{\text{live steam used}} = \frac{20{,}000}{23{,}892}$$
$$= 0.837\ \text{lb/lb of steam}$$

Condenser Water Requirements. An enthalpy balance on the condenser gives (refer to Fig. 4-1)

$$VH_v + Wh_{wi} = (W + V)h_{vo}$$

$$20{,}000(1149) + W(54.03) = (W + 20{,}000)(87.97)$$

$$W = 625{,}239\ \text{lb/h} = 1250.5\ \text{gpm}$$

In many cases, the enthalpy composition diagrams are not available. In such cases, approximate methods are used. These approximations are: (1) the latent of vaporization of water at the boiling point of the solution rather than the equilibrium temperature is used; (2) heats of dilution are neglected; and (3) the specific heats of the solution, if known, can be used to calculate the enthalpies of the feed and thick liquor. An example is given next.

Example 4-2. A solution is to be concentrated from 10 to 65% solids in a vertical long-tube evaporator. The solution has a negligible elevation of boiling point and its specific heat can be taken to be the same as that of water. Steam is available at 203.6 kPa, and the condenser operates at 13.33 kPa. The feed enters the evaporator at 295 K. The total evaporation is to be 25,000 kg/h of water. Calculate the heat transfer required and the steam consumption in kilograms per hour if the overall coefficient is 2800 W/ $m^2 \cdot K$.

Solution. From steam tables, the water-vapor temperature at 13.33 kPa is 325 K and the steam temperature at 203.6 kPa is 394 K.

The latent heat of vaporization of water at 325 K is 2375 kJ/kg. So

$$\text{Evaporation} = 25{,}000 \text{ kg/h} = \frac{25{,}000}{3600} = 6.944 \text{ kg/s}$$

The feed and thick liquor are denoted by F and B, respectively. By solids balance,

$$F(0.1) = B(0.65)$$

By water balance,

$$F(0.9) = B(0.35) + 6.944$$

By solving the equations simultaneously,

$$F = 8.21 \text{ kg/s} \qquad B = 1.26 \text{ kg/s}$$

Taking boiling point in the evaporator as the datum temperature, the evaporator heat load is

$$Q = 6.944(2375) + 8.21(4.1868)(325 - 295)$$
$$= 17{,}523 \text{ kW or kJ/s}$$

The temperature difference is

$$\Delta T = 394 - 325 = 69 \text{ K}$$

The heat-transfer surface is

$$\frac{17{,}523}{69(2.8)} = 90.7 \text{ m}^2$$

The latent heat of steam (from steam tables) is 2198 kJ/kg, and the steam consumption is

$$\frac{(17{,}523 \text{ kJ/s})(3600 \text{ s/h})}{2198 \text{ kJ/kg}} = 28{,}700 \text{ kg/h}$$

Multieffect Evaporator. As stated before, multieffect evaporation is used to effect steam economy. In practice, the heating areas of evaporators in a multieffect evaporation system are kept equal to obtain economy in construction.

The heat-transfer rates in each effect are given by

$$q_1 = U_1 A \, \Delta T_1$$

$$q_2 = U_2 A \ \Delta T_2$$

$$q_3 = U_3 A \ \Delta T_3 \tag{4-3}$$

These are approximately equal, and hence

$$U_1 \ \Delta T_1 = U_2 \ \Delta T_2 = U_3 \ \Delta T_3 = \frac{q}{A} \tag{4-4}$$

which enables one to get the ΔT distribution in the effects for the first trial.

The multieffect-evaporator calculations are also done as in the case of a single-effect evaporator by solving a number of simultaneous material- and heat-balance equations. However, this procedure is tedious when three or more effects are involved. Therefore, another variation of the method is used. A temperature distribution is assumed, and instead of solving a set of simultaneous equations, a value of the evaporation from one of the effects is also assumed. Then the material- and heat-balance equations are solved one by one for each effect. If the total evaporation calculated does not equal the required evaporation, the value of the assumed evaporation is readjusted, and the trial is repeated. The accuracy of the assumed temperature differences is ascertained by the calculation of the heat-transfer surfaces which should be equal for all the effects. Example 4-3 illustrates the method.

Example 4-3. A triple-effect forced-circulation evaporator is to concentrate NaOH solution from 10 to 50%. Feed is at 100°F and 75,000 lb/h. The evaporator operates at 28-inHg vacuum in the last effect. The backward feed arrangement is to be used. The overall heat-transfer coefficients are to be: in I, 600; in II, 600; in III, 1000, all in Btu/h · ft^2 · °F. What are the steam consumption, economy, and heating surfaces needed if live steam at 15 psig is used? (Refer to Fig. 4-2.)

Solution. Given data are

$$U_1 = 600 \qquad U_2 = 600 \qquad U_3 = 1000$$

$$t_s = 250°\text{F} \qquad x_F = 0.1 \qquad x_1 = 0.5$$

$$t_F = 100°\text{F}$$

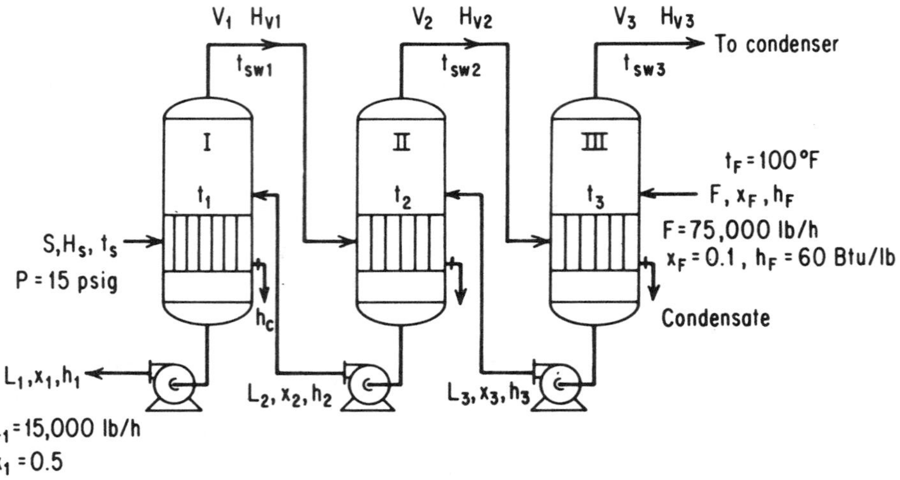

Figure 4-2. Backward-feed triple-effect evaporation system (Example 4-3).

From the enthalpy composition diagram[1b] for NaOH, $h_F = 60$ Btu/lb. From steam tables, $H_s = 1164.1$ Btu/lb and $h_c = 218.9$ Btu/lb. The saturation temperature for water in III is 100°F. Flows of the feed, thick liquor, and evaporation are

$$F = 75{,}000 \text{ lb/h} \qquad \text{Solids in feed} = 7500 \text{ lb/h}$$

$$L_1 = \frac{7500}{0.5} = 15{,}000 \text{ lb/h}$$

$$V = 75{,}000 - 15{,}000 = 60{,}000 \text{ lb/h}$$

$$\text{Total evaporation } V = V_1 + V_2 + V_3$$

Trial 1. To know the temperature drop available, the boiling-point elevations are required. To determine these, the concentrations in II and III must be known. For a preliminary estimate, one should divide the total evaporation in three effects based on a judicious guess, or to start with, one can assume equal evaporations in all the effects to calculate the concentrations and boiling-point elevations. In this example, assume equal evaporation in each effect. Therefore,

$$L_3 = 75{,}000 - 20{,}000 = 55{,}000 \qquad x_3 = 0.1\left(\frac{75{,}000}{55{,}000}\right) = 0.136$$

$$L_2 = 55{,}000 - 20{,}000 = 35{,}000 \qquad x_2 = 0.1\left(\frac{75{,}000}{35{,}000}\right) = 0.214$$

$$L_1 = 35{,}000 - 20{,}000 = 15{,}000 \qquad x_1 = 0.1\left(\frac{75{,}000}{15{,}000}\right) = 0.500$$

Note that one can assume unequal evaporations in the effects. The preliminary guess should be based on the considerations of the U values and the probable ΔT that may be available in each effect. To know the boiling points of the solutions and the boiling-point elevations, assume a ΔT of 25°F between the steam and the solution in the first effect and also assume equal ΔP per effect. Pressure in the third effect is known. Therefore, the saturation temperature of water and the boiling point of the solution in each effect can be obtained with the use of the steam tables and Duhring lines[2] for NaOH solutions as given in Table 4-1.

TABLE 4-1

Effect	Conc., wt%	Operating Pressure, psia	bp of Water, °F	bp of Solution, °F	bp Rise, °F
III	13.6	0.98*	100*	110	10
II	21.4	2.35†	131	148	17
I	50.0	3.72‡	150‡	225	75
				Total	102°F

*Based on the given 28-inHg vacuum in the third effect and $x_3 = 0.136$.

†Calculated on the assumption of equal ΔP per effect.

‡Based on the assumption of 25°F Δt between the steam and the solution in the first effect and $x_1 = 0.5$.

Total boiling-point elevation is 102°F, and therefore the temperature drop available is (250 – 100) – 102 = 48°F. Divide this into the ratio of the heat transfer coefficients using $\Delta T_1{:}\Delta T_2{:}\Delta T_3 = 1/U_1{:}1/U_2{:}1/U_3$ as follows:

$$\Delta T_1 = 18 \qquad \Delta T_2 = 18 \qquad \Delta T_3 = 12$$

Using steam tables, now prepare Table 4-2.

TABLE 4-2 Calculations of Enthalpies for First Trial

	Evaporator	Temp., °F	H_s or H_v,† Btu/lb	h_c, Btu/lb	λ_v, Btu/lb	Flow Rate, lb/h
Steam	I	250	1164.1	218.9	945.2	8
ΔT_1	I	18				
bp (t_1)	I	232	1163.5			V_1
bpr*	I	75				
t_{sv1}	to II	157	1129.0	125	1003.0	V_1
ΔT_2	II	18				
bp (t_2)	II	139	1122.2			V_2
bpr	II	17				
t_{sv2}	to III	122	1114.7	92.0	1023.7	V_2
ΔT_3	III	12				
bp (t_3)	III	110	1109.2			V_3
bpr	III	10				
t_{sv3}		100	1105.2	68.0	1037.2	V_3

*Boiling point rise.

†These are obtained by $H_v = H_{t_{sv}} + 0.46(\text{bpr})$ where the subscript *sv* denotes saturated steam.

Heat Balance Equations.
First effect:

$$945.2S + (75{,}000 - V_2 - V_3)90 = 1163.5\ V_1 + 15{,}000(249)$$

Second effect:

$$1003V_1 + (75{,}000 - V_3)65 = 1122.2\ V_2 + 90(75{,}000 - V_2 - V_3)$$

Third effect:

$$1024.7\ V_2 + 75{,}000(60) = 1109.2\ V_3 + 65(75{,}000 - V_3)$$

Since the last equation contains only two variables, it is convenient to assume V_3. Let $V_3 = 19{,}000$ lb/h. Then from the third-effect equation

$$V_2 = 19{,}728 \text{ lb/h}$$

From the second-effect equation

$$V_1 = 21{,}677 \text{ lb/h}$$

$$\text{Total V} = 19{,}000 + 19{,}728 + 21{,}677 = 60{,}405 \text{ lb/h} \doteq 60{,}000^*$$

Reassume

$$V_3 = \frac{19{,}000(60{,}000)}{60{,}405} \doteq 18{,}875 \text{ lb/h}$$

Then $V_2 = 19{,}600 \qquad V_1 \doteq 21{,}550 \text{ lb/h}$

$$V = V_1 + V_2 + V_3$$

$$= 18{,}875 + 19{,}600 + 21{,}550 = 60{,}025 \text{ lb/h} \doteq 60{,}000 \text{ lb/h}$$

From the first-effect equation,

$$S = 27{,}000 \text{ lb/h}$$

Values of the enthalpies for the liquid streams are given in Table 4-3.

TABLE 4-3

Stream	x_n	**Temp., °F**	h_ℓ**, Btu/lb**	**Flow Rate, lb/h**
F	0.1	100	60	75,000
L_1	0.5	232	249	15,000
L_2	0.214	139	90	$75{,}000 - V_2 - V_3$
L_3	0.136	112	65	$75{,}000 - V_3$

Now check the equality of the heat-transfer surfaces.

$$A_1 = \frac{945.2(27{,}000)}{600(18)} = 2363 \text{ ft}^2$$

*The calculated evaporation in this trial differs only by 0.7 percent from the required evaporation. Therefore, this is a good estimate and no further trial is actually necessary. A second trial, however, is shown here.

$$A_2 = \frac{1004(21{,}550)}{600(18)} = 2003 \text{ ft}^2$$

$$A_3 = \frac{1023.7(19{,}600)}{1000(12)} = 1674 \text{ ft}^2$$

$$\text{Average } A = 2013 \text{ ft}^2$$

The areas are differing too much. Therefore, the assumed ΔT distribution is not correct. Redistribute ΔT's as follows:

$$\Delta T_1 = \frac{2363}{2012.3}(18) \doteq 21$$

$$\Delta T_2 = \frac{2003}{2013.3}(18) \doteq 18$$

$$\Delta T_3 = \frac{1674}{2013.3}(12) \doteq 10$$

$$\text{Total } \Delta T = \overline{49°\text{F}}$$

Also readjust concentrations.

$$x_2 = \frac{7500}{75{,}000 - 18{,}875 - 19{,}600} = 0.205$$

$$x_1 = \frac{7500}{75{,}000 - 18{,}875} = 0.1336$$

The boiling-point elevation in the second effect will be slightly less than the assumed boiling-point elevation for the first trial.

Trial 2. The calculations completed in a similar manner in trial 1 give $V_1 = 21{,}430$ lb/h, $V_2 = 19{,}650$ lb/h, $V_3 = 18{,}950$ lb/h, and $S = 26{,}840$ lb/h.

Heat-transfer surfaces are now checked again as follows.

	% Deviation from Average *A*
$A_1 = \frac{26{,}840(945.2)}{600(21)} = 2013.4 \text{ ft}^2$	+ 0.26
$A_2 = \frac{1005.6(21{,}430)}{600(18)} = 1995.4 \text{ ft}^2$	− 0.63
$A_3 = \frac{1025.8(19{,}650)}{1000(10)} = 2015.7 \text{ ft}^2$	+ 0.38

The average area is 2008 ft^2. The maximum deviation from the average value is 0.63 percent. A more accurate trial will involve fractional ΔT's and this is not necessary since U values are not known more accurately. Hence, the area of each effect is 2008 ft^2, the steam consumption is 26,840 lb/h, and the steam economy is 60,000/26,840 = 2.24 lb/lb.

References

1. *Chemical Engineers' Handbook*, 5th ed., R. H. Perry (ed.), McGraw-Hill, New York, 1973: (*a*) pp. 10-32 and 10-33; (*b*) p. 3-204.
2. G. G. Brown et al., *Unit Operations*, Wiley, New York, 1956, p. 486.

Chapter

5

Distillation

Important topics in distillation that should be reviewed in detail are: vapor-liquid equilibria, calculation of dew and bubble points, and application of the McCabe-Thiele diagram for the determination of theoretical plates in solving problems connected with the binary systems.

Vapor Pressure

The vapor pressure of a component P_A is a measure of its volatility at any given temperature. It is essentially independent of pressure except at very high pressures. The vapor pressure of a substance is generally represented by a two-constant or three-constant Antoine equation of the type

$$\ln P = \begin{cases} A + \dfrac{B}{T} & \text{two-const. eq.} \quad (5\text{-}1) \\[2ex] A - \dfrac{B}{C + t} & \text{three-const. Antoine eq.} \quad (5\text{-}2) \end{cases}$$

where A, B, and C are constants and T and t are temperatures in K and °C, respectively. The constant C is generally given by

$$C = 239 - 0.19t_B \tag{5-3}$$

where t_B is the boiling point, °C.

Vapor Pressures of Immiscible Liquids

The total vapor pressure exerted by a mixture of immiscible liquids at a given temperature is the sum of the vapor pressures of the individual components at that temperature. The boiling point of a mixture of immiscible liquids is lower than that of any one of its components.

Relative Volatility α_{ij}

For a vapor phase in equilibrium with its liquid phase, the relative volatility of a component i with respect to component j is given by

$$\alpha_{ij} = \frac{y_i/x_i}{y_j/x_j} \tag{5-4}$$

where y is the mole fraction of a component in the vapor phase and x is the mole fraction of the same component in the liquid phase. For a binary system, $y_j = 1 - y_i$ and $x_j = 1 - x_i$; therefore

$$\alpha = \left(\frac{y_i}{1 - y_i}\right)\left(\frac{1 - x_i}{x_i}\right) \tag{5-5}$$

In general

$$\alpha = \left(\frac{y}{1 - y}\right)\left(\frac{1 - x}{x}\right) \tag{5-6}$$

from which

$$y = \frac{\alpha x}{1 + (\alpha - 1)x} \qquad x = \frac{y}{\alpha + y(1 - \alpha)} \tag{5-7}$$

For ideal mixtures, α is the ratio of the vapor pressures or

$$\alpha_{ij} = \frac{P_i}{P_j} \tag{5-8}$$

and for a binary system,

$$y_1 = \frac{P_1x_1}{P_1x_1 + P_2(1 - x_1)} = \frac{P_1x_1}{P_t} \qquad x_1 = \frac{P_t - P_2}{P_1 - P_2} \tag{5-9}$$

where P_1 and P_2 are vapor pressures of the components 1 and 2, respectively.

Fugacity Coefficient

The fugacity coefficient of a component in a gas mixture is defined by either

$$\nu = \frac{f}{p} \tag{5-10}$$

or

$$\left(\frac{\delta G}{\delta p}\right)_T = RT\ d\ \ln f \tag{5-10a}$$

where ν = fugacity coefficient
f = fugacity or actual pressure
p = partial pressure if the gas behaves ideally
T = temperature (absolute)
G = free energy

For nonideal gas mixtures,

$$\nu_i = \frac{f_i}{y_iP_t} \tag{5-11}$$

where P_t is the total pressure of the system. For nonideal liquid mixtures,

$$f_i = \gamma_i x_i P_i \tag{5-12}$$

$$\nu_i y_i P_t = \gamma_i x_i P_i \tag{5-13}$$

$$\alpha_{ij} = \frac{y_i/x_i}{y_j/x_j} = \frac{\gamma_i}{\gamma_j}\left(\frac{\nu_j}{\nu_i}\right)\left(\frac{P_i}{P_j}\right) \tag{5-14}$$

Van Laar Equations[1]

γ_1 and γ_2, the activity coefficients of the two components of a liquid mixture, are given by

$$T \ln \gamma_1 = \frac{B}{(1 + Ax_1/x_2)^2} \qquad T \ln \gamma_2 = \frac{AB}{(A + x_2/x_1)^2} \tag{5-15}$$

where A and B are constants. At the azeotropic composition,

$$x_i = y_i \qquad \text{and} \qquad \gamma_i = \frac{P_t}{P_i} \tag{5-16}$$

Also

$$\gamma_i = \frac{y_i P_t}{x_i P_i} \tag{5-17}$$

Knowing the azeotropic composition, one can calculate A and B in Eq. (5-15).

Margules[2] equations are given by

$$\ln \gamma_1 = x_2^2[A + 2x_1(B - A)] \quad \text{and} \quad \ln \gamma_2 = x_1^2[B + 2x_2(A - B)] \tag{5-18}$$

Dew Point

The dew point of a mixture is the temperature and pressure at which the first drop of the liquid is formed when a vapor mixture is cooled. This may be expressed in terms of the equilibrium relationship; thus

$$\sum \frac{y_i}{K_i} = \sum x_i = 1.0 \tag{5-19}$$

For ideal systems,

$$\sum x_i = \sum \frac{y_i}{P_i} P_t = 1.0 \tag{5-20}$$

The temperature at which the first drop forms is called the dew-point temperature, and the corresponding pressure is the dew-point pressure.

Bubble Point

The bubble point of a liquid mixture is the temperature and pressure at which the liquid begins to boil such that the vapor is in equilibrium with the liquid. This condition can be expressed by the relation

$$\Sigma K_i x_i = \Sigma y_i = 1.0 \tag{5-21}$$

For ideal systems,

$$\sum y_i = \sum \frac{P_i x_i}{P_t} = 1.0 \tag{5-22}$$

In the above equations, K_i is the equilibrium constant of a component defined by

$$K_i = \frac{y_i}{x_i} \tag{5-23}$$

Example 5-1. A mixture of 60 mol % n-hexane, 10 mol % n-heptane, and the remainder steam is cooled at constant pressure 1 atm from an initial temperature of 205°C. Hydrocarbons and water are immiscible. Calculate: (**a**) the temperature at which condensation first occurs, (**b**) the composition of the first liquid phase, (**c**) the temperature at which the second liquid phase appears, and (**d**) the composition of the second liquid phase as it appears.

Vapor pressures of n-hexane and n-heptane are given by

$$\ln P = A + \frac{B}{T}$$

where T is in K, P is in mmHg, and constants A and B are as given below.

	A	*B*
Hexane	17.7109	−3787
Heptane	17.9184	−4193

Solution. Check for the condensation of steam. For the vapor pressure of steam, refer to steam tables. The vapor pressure of steam at the point of condensation is

$$P_s = 0.3(760) = 228 \text{ mmHg} = 4.41 \text{ psi}$$

From steam tables at 4.41 psi,

$$t = 155 + \frac{4.41 - 4.203}{4.741 - 4.203}(5)$$

$$= 156.92°\text{F} = 69.4°\text{C} = 342.4 \text{ K}$$

This is lower than 205°C. Hence, the steam will condense when cooled to this temperature. Check for condensation of *n*-heptane and *n*-hexane. Vapor pressure of *n*-hexane at 342.4 K = 773.4 mmHg. Vapor pressure of *n*-heptane at 342.4 K = 290.80 mmHg. Assume Raoult's law applies.
Then

$$y_i P_t = x_i P_i \qquad x_i = \frac{y_i P_t}{P_i}$$

Hence, for hexane $x_{\text{hex}} = \dfrac{0.6(760)}{773.4} = 0.5896$

For heptane $x_{\text{hep}} = \dfrac{0.1(760)}{290.8} = 0.2613$

and $\Sigma x_i = 0.5896 + 0.2613 = 0.8509 < 1.0$

Readjust x_{hex} and x_{hep}:

$$x_{\text{hex}} = \frac{0.5896}{0.8509} = 0.693$$

$$x_{\text{hep}} = \frac{0.2613}{0.8509} = \underline{0.307}$$

$$\text{Total} = 1.000$$

Therefore, at this temperature no hydrocarbons condense.

a. Condensation first occurs at 156.92°F = 69.4°C.

b. Liquid phase is pure water. There is no hydrocarbon in the liquid.

c. Assuming t = 66°C = 339 K as the temperature for the condensation of the second phase (t = 150.8°F),

$$P_s = 3.718 + \frac{150.8 - 150}{155 - 150}(4.203 - 3.718)$$

$$= 3.7956 \text{ psi} = 196.24 \text{ mmHg}$$

$$y_{H_2O} = \frac{196.24}{760} = 0.2582$$

The vapor pressure of hexane is 692.15 mmHg, and the vapor pressure of heptane is 257.15 mmHg.

$$y_i \text{ (hydrocarbon)} = 1 - 0.2582 = 0.7418$$

$$\frac{y_{hex}}{y_{hep}} = \frac{6}{1} \qquad \text{and} \qquad y_{hex} + y_{hep} = 0.7418$$

$$y_{hep} = \frac{0.7418}{6 + 1} = 0.1060$$

$$y_{hex} = 0.7418 - 0.1060 = 0.6358$$

$$x_{hex} = \frac{0.6358(760)}{692.15} = 0.698$$

$$x_{hep} = \frac{0.106(760)}{257.15} = \underline{0.313}$$

$$\text{Total} = 1.011$$

Readjust x_{hex} and x_{hep}.

$$x_{hex} = \frac{0.698}{1.011} = 0.69$$

$$x_{hep} = \frac{0.313}{1.011} = \underline{0.31}$$

$$\text{Total} = 1.0$$

$$P_{hex} = \frac{0.6358(760)}{0.69} = 700.3 \text{ mmHg}$$

$$\ln P = A + \frac{B}{T}$$

$$\ln 700.3 = 17.7109 - \frac{3787}{T}$$

$$T = 339.36 \text{ K} = 66.25°\text{C} = 151.4°\text{F}$$

This is close to the assumed temperature.

d. Therefore, for $t = 151.4°\text{F}$,

$$x_{hex} = 0.69 \qquad x_{hep} = 0.31$$

Calculation of the Bubble-Point Temperature Using *K* Values

Example 5-2. A liquid-phase mixture of hydrocarbons has the following composition at 125 psia: $C_2 = 2.0$ mol %, $=C_3 = 29$ mol%, $C_3 = 29.0$ mol %, n-$C_4 = 40.0$ mol %. Calculate the bubble-point temperature of the mixture at 125 psia. *K* values[3] are given in Fig. 5-1.

Solution.

$$y_i = Kx_i$$

At the bubble point

$$\Sigma y_i = \Sigma K_i\, x_i = 1.0$$

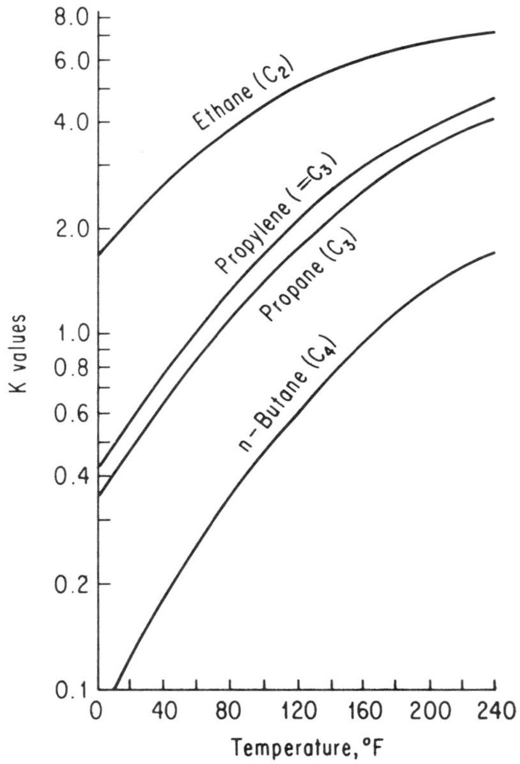

Figure 5-1. Approximate *K* values for light hydrocarbons.

If noncondensables are present at the bubble point,

$$\Sigma y_i = \Sigma K_i x_i = 1 - y_{nc} \qquad i \neq nc$$

The bubble point is to be calculated by trial and error. A temperature is assumed, K values at this temperature are obtained, and $\Sigma K_i x_i$ is calculated. y_{nc} is the mole fraction of the noncondensables. A useful relation to obtain the next temperature estimate is

$$t_{n+1} = t_n - 100\left(\sum K_i x_i - 1.0\right)$$

Assumed temperature $t_1 = 100°F$. See Table 5-1.

TABLE 5-1

		$t_1 = 100°F$		$t_2 = 86.6°F$		$t_3 = 86.5°F$	
Component	x_i	K_i	$y_i = K_i x_i$	K_i	$y_i = K_i x_i$	K_i	$y_i = K_i x_i$
C_2	0.02	4.3	0.086	3.90	0.078	3.89	0.0778
C_3	0.29	1.65	0.4785	1.43	0.4147	1.43	0.4148
$=C_3$	0.29	1.44	0.4176	1.22	0.3538	1.21	0.3509
n-C_4	0.40	0.46	0.1840	0.39	0.1548	0.39	0.1564
			1.1661		1.0013		0.9999

Therefore, the bubble-point temperature is 86.5°F. The vapor phase composition is

$$C_2 = 7.78 \text{ mol } \% \qquad C_3 = 41.48 \text{ mol } \%$$

$$=C_3 = 35.09 \text{ mol } \% \qquad n\text{-}C_4 = 15.64 \text{ mol } \%$$

Dew-Point Calculation Using *K* Values

Knowing the composition of a vapor phase, the composition of the liquid phase in equilibrium with this vapor can be obtained. At the dew point,

$$\sum \frac{y_i}{K_i} = \sum x_i = 1.0 \tag{5-24}$$

If nonvolatiles are present,

$$\sum\frac{y_i}{K_i} = 1.0 - x_{nv} \qquad i \neq nv \tag{5-25}$$

x_{nv} is the mole fraction of the nonvolatile component in the liquid phase.

To calculate the dew point, assume a temperature, get K values at the assumed temperature, and then calculate $\Sigma y_i/K_i$. If $\Sigma y_i/K_i \neq 1.0$, assume another temperature and repeat the calculation. A useful equation to obtain the next temperature for an estimate is

$$t_{n+1} = t_n + 100\left(\sum\frac{y_i}{K_i} - 1.0\right) \tag{5-26}$$

Example 5-3. A vapor phase mixture at 125 psia has the following composition: C_2 = 2 mol %, $=C_3$ = 29 mol %, C_3 = 29 mol %, and n-C_4 = 40 mol %. Calculate the dew-point temperature and the composition of the condensed phase at the dew-point temperature. K values can be obtained from Fig. 5-1.

Solution. Assume a temperature and estimate the composition. The results are shown in Table 5-2.

TABLE 5-2

		Assume t = 125°F		**Assume t = 119°F**		
Component	y_i	K_i	$x_i = y_i/K_i$	K_i	$x_i = y_i/K_i$	**mol %**
C_2	0.02	5.10	0.00392	4.6	0.00435	0.435
$=C_3$	0.29	2.18	0.13302	2.05	0.14146	14.146
C_3	0.29	1.85	0.15676	1.75	0.16571	16.571
n-C_4	0.40	0.625	0.6400	0.58	0.68966	68.966
		Σx_i =	0.9337	Σx_i =	1.00118	

The dew-point temperature is 119°F. The composition of condensed liquid is shown in Table 5-3.

TABLE 5-3

Component	mol %	
C_2	0.435	
$=C_3$	14.146	
C_3	16.571	
n-C_4	68.966	
	100.118	$\doteq$ 100

Boiling-Point Diagrams

The boiling-point diagram for a binary system is the representation of the temperature-composition relationship at a constant pressure. For example, Fig. 5-2*a* represents a boiling-point diagram of the hexane-octane system. In general, the boiling-point diagrams must be determined experimentally.

For systems which obey Raoult's law, it is possible to calculate the boiling-point diagram from the vapor pressures of the pure components. Raoult's law states that at a given temperature, the partial pressure of one component of a mixture is equal to its mole fraction in the liquid phase times its vapor pressure in the pure state at the same temperature or for a binary system

$$p_{\mathrm{A}} = x_{\mathrm{A}} P_{\mathrm{A}} \qquad \text{for component A} \tag{5-27}$$

$$p_{\mathrm{B}} = (1 - x_{\mathrm{A}}) P_{\mathrm{B}} \qquad \text{for component B}$$

If the vapor obeys Dalton's law of partial pressures, the mole fraction of the component A in the vapor phase in equilibrium with the liquid phase is given by

$$y_A = \frac{p_A}{P_t} = \frac{P_A x_A}{P_t} \tag{5-28}$$

Example 5-4. Vapor pressures of pure hexane and octane are given in Table 5-4. Construct the boiling-point and equilibrium diagrams for the system at 1 atm.

TABLE 5-4 Vapor Pressure Data for Hexane and Octane

	Vapor Pressure, mmHg	
Temperature, °F	**Hexane**	**Octane**
155.7	760	121
175.0	1025	173
200	1480	278
225	2130	434
250	3000	654
258.2	3420	760

Solution. Let x_A be the mole fraction of hexane in the liquid. Then $(1 - x_A)$ is the mole fraction of octane in the liquid. Let y_A and y_B be the mole fractions of the hexane and octane in the vapor, respectively. According to Raoult's law,

$$y_A = \frac{P_A x_A}{P_t} \qquad (a)$$

where P_A is vapor pressure of hexane. Also, since $y_A + y_B = 1$,

$$\frac{P_A x_A}{P_t} + \frac{P_B(1 - x_A)}{P_t} = 1$$

where P_B is the vapor pressure of octane and P_t is the total pressure from which

$$x_A = \frac{P_t - P_B}{P_A - P_B} \qquad (b)$$

First, at a given temperature, x_A is calculated using Eq. (b), and then y_A is calculated using Eq. (a). Thus at a temperature of 200°F,

$$x_A = \frac{760 - 278}{1480 - 278} = 0.401$$

$$y_A = \frac{1480}{760} x_A = \frac{1480}{760}(0.401) = 0.7809$$

Similar calculations are done at other temperatures with the results given in Table 5-5.

TABLE 5-5 Calculations of Equilibrium Compositions for *n*-Hexane–Octane System

Temperature, °F	Mole Fraction Hexane in Liquid, x_A	Mole Fraction Hexane in Vapor, y_A
155.7	1.0	1.000
175	0.689	0.929
200	0.401	0.781
225	0.1922	0.539
250	0.0452	0.1784
258.2	0.00	0.000

The boiling-point diagram is plotted in Fig. 5-2*a* and the equilibrium diagram in Fig. 5-2*b*.

Flash Vaporization

This consists of vaporizing a fraction of a liquid in such a manner that the vapor is in equilibrium with the undistilled liquid. For a multicomponent mixture, the following equations apply:

$$x_i = \frac{x_{Fi}/L}{K_i V/L + 1} \quad \text{or} \quad x_i = \frac{x_{Fi}/V}{K_i + L/V} \tag{5-29}$$

where L = moles of undistilled liquid
V = moles of vapor (distilled liquid)
x_{Fi} = mol fraction of component i in the feed
K_i = ratio of equilibrium vapor to equilibrium liquid $= y_i/x_i$
x_i = mol fraction of component i in undistilled liquid
y_i = mol fraction of component i in distilled vapor

For an ideal system,

$$K_i = \frac{P_i}{P_t} \qquad \text{(5-29}a\text{)}$$

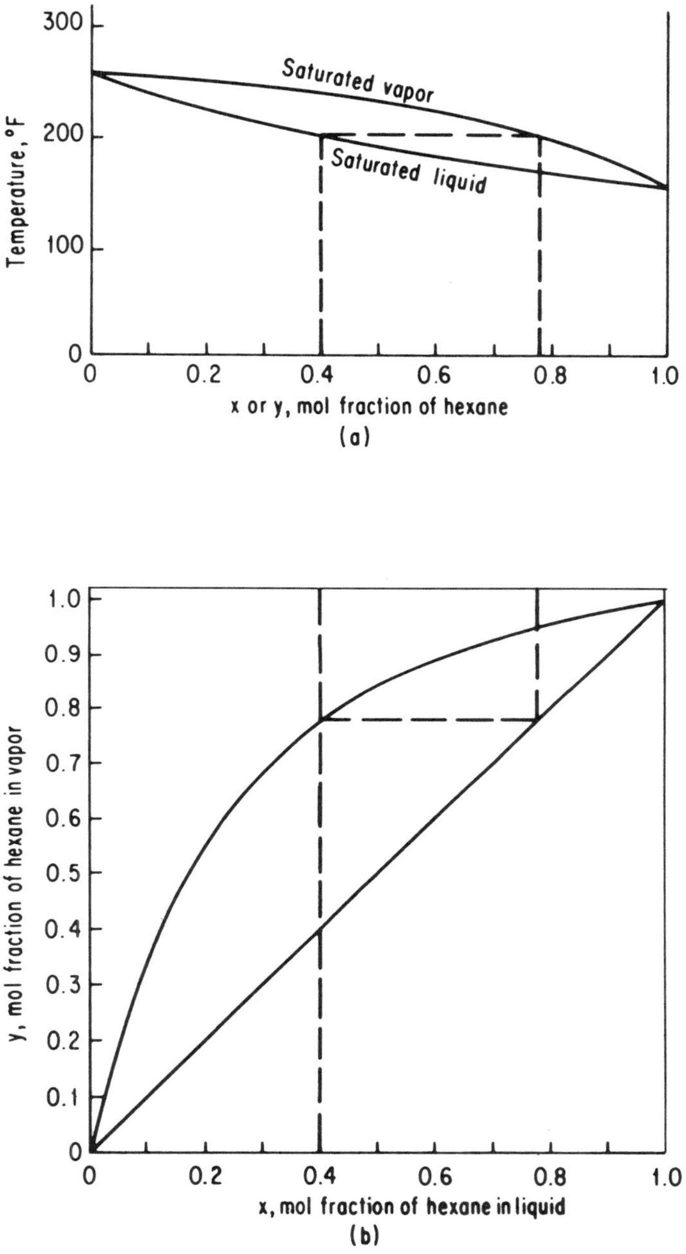

Figure 5-2. Hexane-octane system: (*a*) boiling-point diagram; (*b*) equilibrium diagram.

The solution of the flash-vaporization problem involves trial and error. Since $y_i = K_i x_i$,

$$y_i = \frac{Kx_{Fi}/V}{K_i + L/V} \tag{5-29b}$$

either $\Sigma x_i = 1$ or $\Sigma y_i = 1.0$.

If 1 mol of the liquid is considered and f is the fraction vaporized, then $(1 - f)$ is the liquid left behind and by material balance,

$$x_{Fi} = fy_i + (1 - f)x_i$$

and therefore

$$y_i = -\frac{1-f}{f}x_i + \frac{x_{Fi}}{f} \tag{5-30}$$

Equation (5-30) is easier to use in solving the flash-vaporization problems. In the case of a multicomponent system (more than two components), a trial-and-error solution is required. In the case of the binary systems, the fraction vaporized is given by

$$f = \frac{x_{Fi}(K_1 - K_2)/(1 - K_2) - 1}{K_1 - 1} \tag{5-31}$$

Example 5-5. A liquid mixture at 125 psia has the composition given in Example 5-3. Calculate the fraction vaporized and the compositions of the liquid and vapor phases if it is flashed isothermally at 100°F and 125 psia.

Solution. The dew point of the mixture is 119°F assuming the vapor composition is identical with the liquid composition given in the problem. The bubble point of the mixture is 86.5°F. The flash temperature is between the dew and bubble points.

Feed Composition.

$$y_i = \frac{x_{Fi}K_i/V}{K_i + L/V} \qquad \text{by Eq. (5-29}b\text{)}$$

Assume 1 mol of feed. Then fraction vaporized $= f = V$ and the fraction of feed remaining as liquid $= L = 1 - V$. Hence

$$\frac{L}{V} = \frac{1 - V}{V}$$

See Table 5-6.

TABLE 5-6 Calculation of y_i at Different Values of $f = V$

	x_F	K_{100}	$f = 0.5$	$f = 0.41$	$f = 0.45$
C_2	0.02	4.30	0.0325	0.0366	0.0346
$=C_3$	0.29	1.69	0.3693	0.3820	0.3740
C_3	0.29	1.44	0.3413	0.3525	0.3475
C_4	0.40	0.46	0.2520	0.2363	0.2431
	1.00		$\Sigma y_i =$ 0.9951	$\Sigma y_i =$ 1.0074	$\Sigma y_i =$ 0.9992 $\doteq 1$

From last trial, when $f = 0.45$, $\Sigma\ y_i \cong 1.0$. Hence the fraction vaporized is 0.45 mol/mol of feed, and the fraction of feed remaining as liquid is 0.55 mol/mol of feed. The liquid compositions are then calculated using the relation $x_i = y_i/K_i$ and are given in Table 5-7.

TABLE 5-7

Component	Vapor Composition, y_i	Liquid Composition, $x_i = y_i/K_i$
C_2	0.0346	0.00805
$=C_3$	0.3740	0.2213
C_3	0.3475	0.2430
C_4	0.2431	0.5285
	$\Sigma y_i =$ 0.9992	$\Sigma x_i =$ 1.00085

Differential Distillation

A liquid is charged to a still and slowly heated; the vapor formed is continuously condensed out. The Rayleigh equation relating final moles L_2 in the still to the initial charge L_1 moles of the liquid is

$$\ln \frac{L_1}{L_2} = \int_{x_{i2}}^{x_{i1}} \frac{dx_i}{y_i - x_i} \tag{5-32}$$

In theory, the vapor formed is in equilibrium with the liquid. The solution of this equation requires graphical integration. The integral is evaluated by plotting $1/(y_i - x_i)$ versus x_i and using Simpson's rule or other method to find the integral.

If the relative volatility α is constant, the solution of the equation is

$$\ln \frac{L_1}{L_2} = \frac{1}{\alpha - 1}\left(\ln \frac{x_{i1}}{x_{i2}} + \alpha \ln \frac{1 - x_{i2}}{1 - x_{i1}}\right) \tag{5-33}$$

For nonideal systems,

$$y = \frac{\gamma x P}{\nu P_t} = Kx \tag{5-34}$$

and

$$\ln \frac{L_1}{L_2} = \int_{x_{i2}}^{x_{i1}} \frac{dx_i}{x_i(K_i - 1)} = \int_{x_{i2}}^{x_{i1}} \frac{dx_i}{x_i(\gamma_i P_i/\nu_i P_t - 1)} \tag{5-35}$$

Binary Fractionation

Analysis of plate columns is based on the overall material and energy balances and phase equilibria. The important factors in the design calculations of the plate columns are: (1) number of plates required, theoretical and actual, for a desired separation; (2) heat input to the reboiler; (3) heat output from the condenser; (4) plate spacing and diameter of column; and (5) type and construction of plates.

Overall Material Balance—Binary Systems. Referring to Fig. 5-3, the total material balance gives

$$F = D + B \tag{5-36}$$

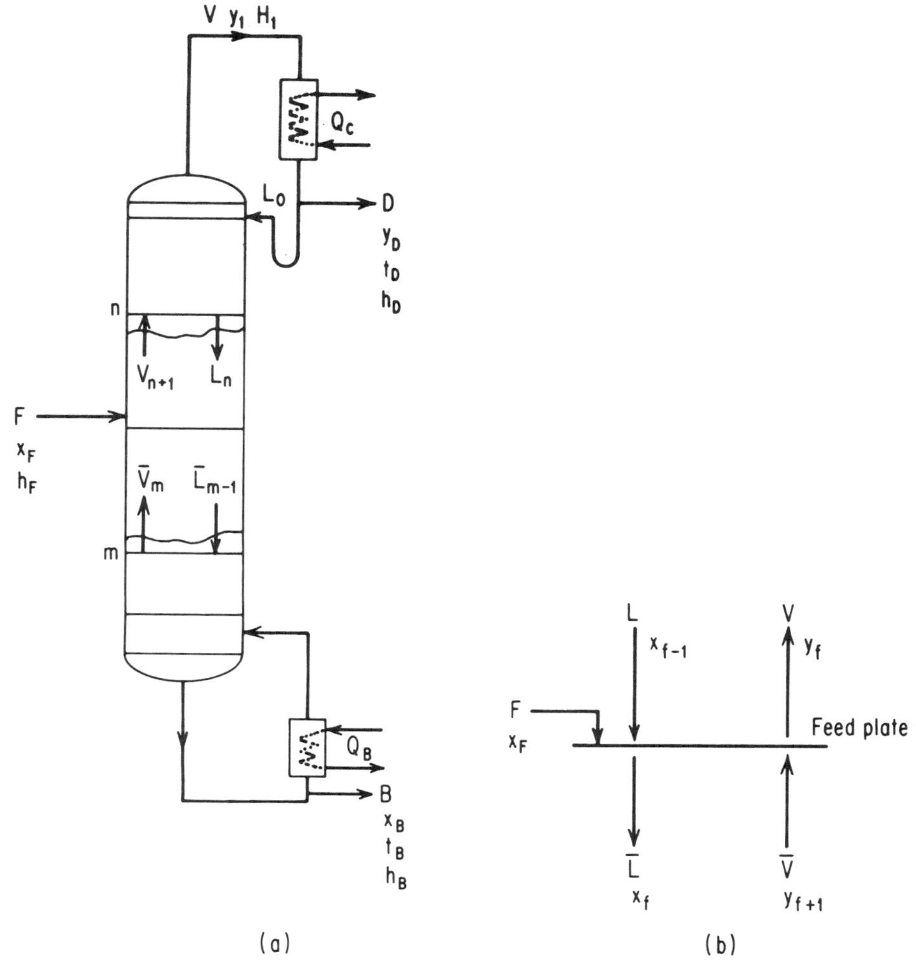

Figure 5-3. (*a*) Material and energy balances for a binary system around distillation column; (*b*) material balance over the feed plate.

and component A balance gives

$$Fx_F = Dx_D + Bx_B \tag{5-37}$$

where F = feed, lb · mol/h
D = distillate, lb · mol/h
B = bottoms product, lb · mol/h

From Eqs. (5-36) and (5-37), one obtains

$$\frac{D}{F} = \frac{x_F - x_B}{x_D - x_B} \quad \text{and} \quad \frac{B}{F} = \frac{x_D - x_F}{x_D - x_B} \tag{5-38}$$

The McCabe-Thiele solution assumes constant molar overflows. The operating line equations are different for the rectification and stripping sections. These equations are listed for several cases of column operation.

Reflux at Its Bubble Point, Total Condenser. The equation of the operating line is

$$y_{n+1} = \frac{L}{V}x_n + \frac{D}{V}x_D \tag{5-39}$$

where L is the liquid flow, lb · mol/h, from plate to plate in the rectifying section and V is the vapor flow, lb · mol/h, from plate to plate in the rectifying section. Also,

$$y_{n+1} = \frac{R}{R+1}x_n + \frac{x_D}{R+1} \tag{5-40}$$

where the external reflux ratio is

$$R = \frac{L_0}{D} = \frac{L}{D} \tag{5-41}$$

The slope of the operating line is

$$\frac{L}{V} = \frac{R}{R+1} \tag{5-42}$$

and the intercept of the line is $x_D/(R+1)$.

Stripping Section. The equation of the operating line is

$$y_{m+1} = \frac{\overline{L}}{\overline{V}}x_m - \frac{B}{\overline{V}}x_B \tag{5-43}$$

where $\overline{L}$ is the liquid overflow in stripping section, lb · mol/h, and $\overline{V}$ is the vapor flow in stripping section, lb · mol/h, or

$$y_{m+1} = \frac{\bar{L}}{\bar{L} - B} x_m - \frac{B}{\bar{L} - B} x_B \tag{5-44}$$

Using the material balance at the feed plate (Fig. 5.5),

$$F + L + \bar{V} = V + \bar{L} \tag{5-45}$$

$$\bar{L} = L + qF \qquad \text{or} \qquad V = \bar{V} + (1 - q)F$$

$$q = \frac{\bar{L} - L}{F}$$

$$= \frac{\text{heat to convert 1 mol of feed to saturated vapor}}{\text{latent heat of vaporizaton per mol of feed}} \tag{5-46}$$

The equation of the operating line (stripping section) in terms of q is

$$y_{m+1} = \frac{L + qF}{L + qF - B} x_m - \frac{Bx_B}{L + qF - B} \tag{5-47}$$

The equation of the q line is

$$y = \frac{q}{q - 1} x - \frac{x_F}{q - 1} \tag{5-48}$$

The method of the McCabe-Thiele graphical solution is shown in Fig. 5-4.

Values of *q*. The values of q are determined as follows.

Cold feed: $$q > 1 \qquad q = 1 + (C_p)_L \frac{t_b - t_F}{\lambda} \tag{5-49}$$

Feed at bp: $$q = 1 \tag{5-50}$$

Partially vaporized feed:

$$0 < q < 1 \qquad q = \frac{f_\ell \lambda}{\lambda} = f_\ell \tag{5-51}$$

where f_ℓ is the liquid fraction in the feed.

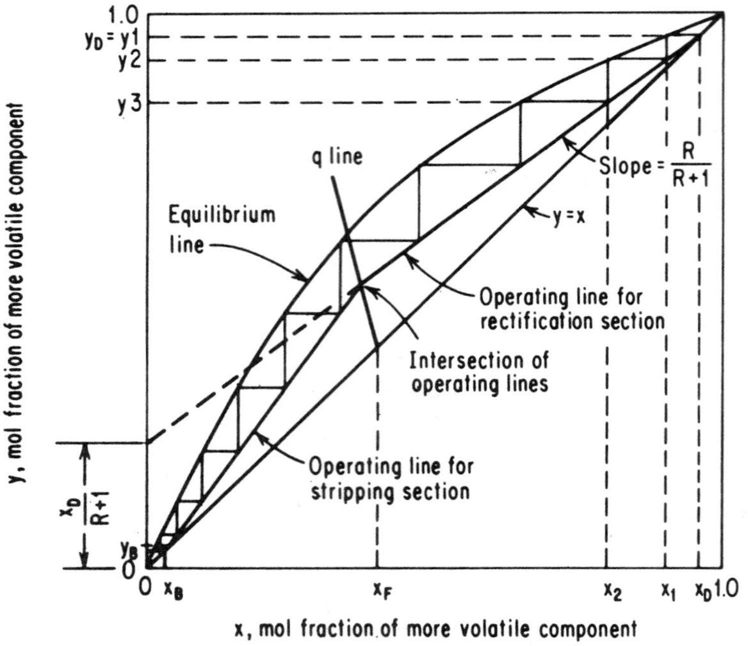

Figure 5-4. McCabe-Thiele solution for the number of theoretical plates for a binary system.

Feed at dew point (saturated vapor):

$$q = 0 \tag{5-52}$$

Feed, superheated vapor:

$$q < 0 \qquad q = \frac{C_{pv}(t_F - t_d)}{\lambda} \tag{5-53}$$

where t_d = dew-point temperature, °F
$(C_p)_L$ = specific heat of liquid feed, Btu/lb · mol · °F
C_{pv} = specific heat of superheated vapor feed, Btu/lb · mol · °F
t_b = boiling point, °F
t_F = feed temperature, °F
λ = latent heat, Btu/lb · mol of feed

Note that the q, or feed, line has a slope of ∞ (i.e., it is vertical) when $q = 1$ or the feed is at its boiling point. It is parallel to the x

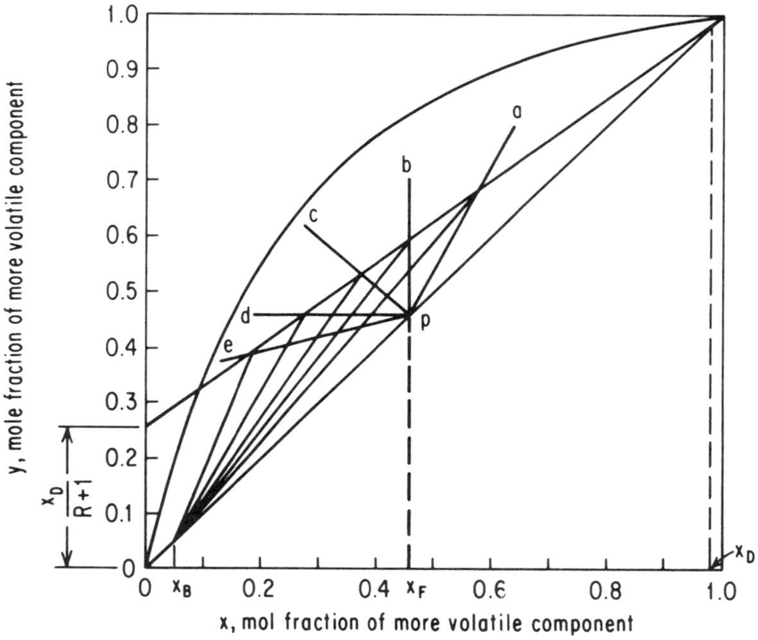

Figure 5-5. Effect of feed condition on q line: pa: q line when the feed is cold liquid. pb: q line when the feed is liquid at its boiling point. The line is vertical. pc: q line when the feed is a mixture of liquid and vapor (partially vaporized feed). pd: q line when the feed is a saturated vapor. The line is parallel to the x axis. pe: q line when the feed is a superheated vapor.

axis when the feed is a saturated vapor. The q line also passes through the intersection point of the 45° line and the vertical line through x_F. All the intersections of the operating lines fall on the q line. The effect of the feed condition on the q line is shown in Fig. 5-5.

Feed-Plate Location

The optimum location of the feed plate is represented by a triangle which has one corner on the rectification operating line and another on the stripping section operating line.

Overall Energy Balance

Assume heat loss to the surroundings of $Q_L = 0$. Energy balance around the whole column gives

$$Fh_F + Q_B = Dh_D + Bh_B + Q_c \tag{5-54}$$

Condenser duty is given by

$$Q_c = D(R + 1)(H_1 - h_D) \tag{5-55}$$

where h and H represent enthalpy, Btu/lb · mol. Subscripts B, D, F, and 1 indicate bottom, distillate, feed, and tray 1, respectively. If no subcooling of the condensate is done,

$$Q_c = D(R + 1)\lambda \tag{5-55a}$$

where λ is the latent heat of condensation of overhead vapor, Btu/lb · mol.

If t_F is taken as the datum temperature, the feed is 100% liquid, and t_D and t_B are distillate and bottoms temperatures, respectively,

$$\begin{aligned} h_F &= 0 \\ h_D &= C_{PD}(t_D - t_F) \\ h_B &= C_{PB}(t_B - t_F) \end{aligned} \tag{5-56}$$

where C_{PD} and C_{PB} are the molar specific heats of the distillate and bottoms, respectively. Hence

$$Q_B = DC_{PD}(t_D - t_F) + BC_{PB}(t_B - t_F) + Q_C \tag{5-57}$$

If heat loss Q_L is not negligible, the reboiler duty would be $Q_B + Q_L$.

Heating and Cooling Requirements

Steam Required at the Reboiler. Assuming radiation from the column to be small and negligible and the column to be operating adiabatically, the amount of the steam required at the reboiler is given by

$$S = \frac{\overline{V}\lambda}{\lambda_s} = \frac{Q_B}{\lambda_s} \qquad \text{lb/h} \tag{5-58}$$

where λ = latent heat of mixture, Btu/lb · mol
λ_s = latent heat of steam, Btu/lb
$\overline{V}$ = vapor flow in stripping section, lb · mol/h

Cooling Water Required at the Condenser. With no subcooling and total condensation, the cooling water required at the condenser is given by

$$\frac{V\lambda}{t_2 - t_1} \qquad \text{lb/h} \tag{5-59}$$

and with subcooling of the condensate, the cooling water needed is

$$\frac{V[\lambda + C_{PD}(t_b - t_R)]}{t_2 - t_1} \qquad \text{lb/h} \tag{5-60}$$

where t_b = bubble-point temperature of the overhead vapor, °F
t_R = actual temperature of condensed distillate, °F
C_{PD} = specific heat of the distillate, Btu/lb · mol · °F
V = vapor flow in the rectification section, lb · mol/h
= $(R + 1)D$
$t_2 - t_1$ = temperature rise of cooling water, °F
λ = latent heat of overhead vapor mixture, Btu/lb · mol

Minimum Reflux

When the equilibrium curve shows concavity downward, the minimum reflux ratio (Fig. 5-6) can be calculated by

$$R_m = \frac{x_D - y'}{y' - x'} \tag{5-61}$$

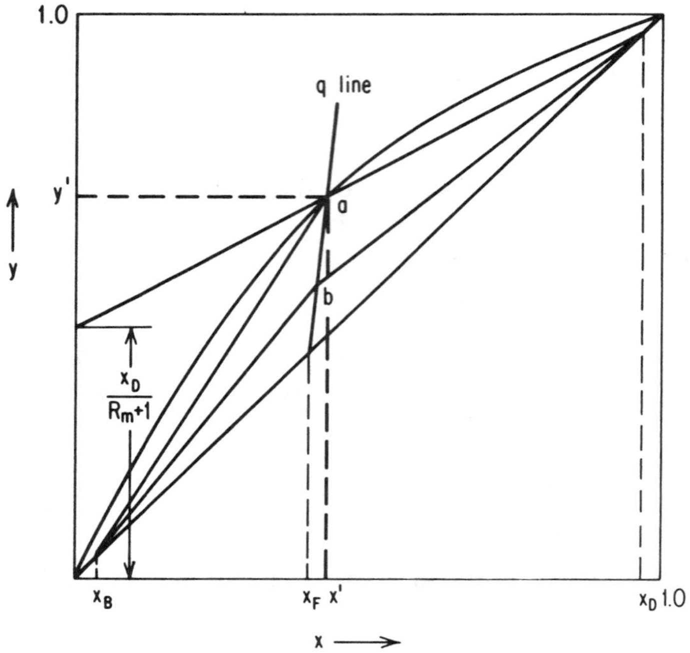

Figure 5-6. Minimum reflux ratio when the equilibrium curve is concave downward. (*Note:* $x' = x_F$ when the feed is at its boiling point.)

where R_m is the minimum reflux ratio. At the minimum reflux ratio, the number of plates required is infinite for a given separation. When $q = 1$, i.e., the feed is at the bubble point, the minimum reflux ratio for a binary system can be calculated by

$$R_m = \frac{1}{\alpha - 1}\left[\frac{x_D}{x_F} - \frac{\alpha(1 - x_D)}{1 - x_F}\right] \tag{5-62}$$

When $q = 0$, i.e., the feed is saturated vapor at the dew point, the minimum reflux ratio can also be calculated by

$$R_m = \frac{1}{\alpha - 1}\left(\frac{\alpha x_D}{y_F} - \frac{1 - x_D}{1 - y_F}\right) - 1 \tag{5-63}$$

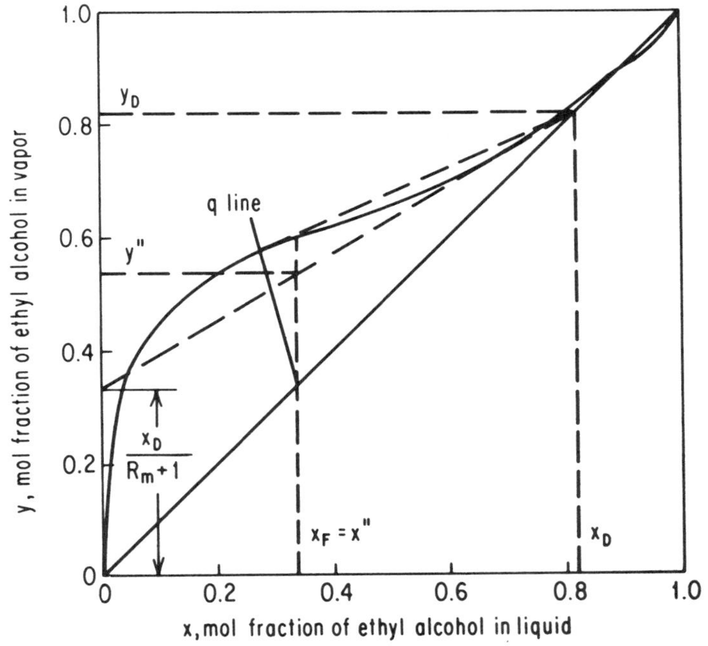

Figure 5-7. Minimum reflux ratio when the equilibrium curve is concave upward (ethyl alcohol–water system).

In these cases, α is estimated at the conditions represented by the intersection of the q line and the equilibrium curve. Equation (5-61) cannot be applied in all cases. If the equilibrium curve has an unusual curvature, the minimum reflux R_m is to be found by drawing an operating line through $y_D = x_D$ that is tangent to the equilibrium curve. The slope of this tangent line is to be used to find R_m (Fig. 5-7):

$$R_m = \frac{x_D - y''}{y'' - x''} \tag{5-64}$$

Alternatively, R_m can be calculated from the intercept of the operating line drawn through the point x_D, y_D and tangential to the equilibrium line. Note that R_m is independent of the q line in this case.

Cold Reflux

If the reflux is cooled below its bubble point, the amount of vapor from the top tray is less because of condensation caused by heating the reflux from t_R to t_b. In this case, internal reflux L_n is different from L_0, the reflux fed to the top tray, and is given by

$$L_n = L_0 + \frac{L_0 C_{PD}(t_b - t_R)}{\lambda_{av}} \tag{5-65}$$

Therefore, apparent reflux ratio is

$$R' = \frac{L_n}{D} = \frac{L}{V-L} = R\left[1 + \frac{C_{PD}(t_b - t_R)}{\lambda_{av}}\right] \tag{5-66}$$

where C_{PD} is the specific heat, Btu/lb · mol · °F, and λ_{av} is the average latent heat of overhead vapor mixture, Btu/lb · mol.

The equation of operating line in the rectification or enriching section is given by

$$y_{n+1} = \frac{R'}{R'+1}x_n + \frac{x_D}{R'+1} \tag{5-67}$$

Partial Condensation

A partial condenser may produce one of the following results: (1) if the time of contact of the liquid and vapor is adequate, equilibrium condensation occurs; (2) if the condensate is removed rapidly, differential condensation occurs; and (3) if cooling is rapid, no mass transfer occurs and the compositions of the vapor and liquid are the same. In case 1, the condenser rates as one theoretical plate. In the design of new equipment, it is safer to ignore the enrichment by the partial condensation and include an additional theoretical tray in the column itself, since it is difficult to ensure that equilibrium condensation will occur.

Total Reflux

At total reflux, the operating lines coincide with the 45° line. Then the number of plates is minimum, but there is no product (Fig. 5-8).

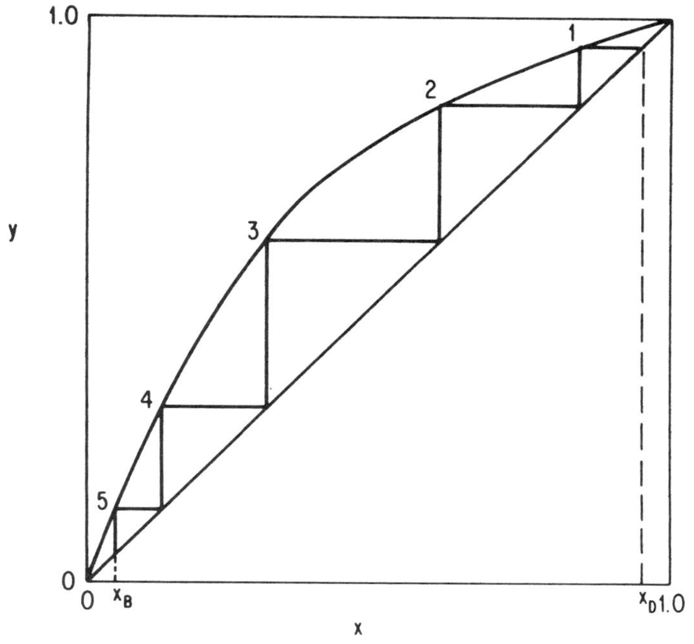

Figure 5-8. Theoretical plate construction for a column operating at total reflux. (*Note:* The operating line coincides with the 45° line.)

Multifeed Column

Equations for the operating lines and q lines are as follows (Fig. 5-9). For enriching section,

$$y_{n+1} = \frac{L}{V}x_n + \frac{D}{V}x_D = \frac{L}{L+D}x_n + \frac{D}{L+D}x_D = \frac{R}{R+1}x_n + \frac{1}{R+1}x_D \quad (5\text{-}68)$$

For the intermediate section (column section in between two feed points),

$$y_{p+1} = \frac{L''}{V''}x_p + Dx_D - Fx_{F1} \quad (5\text{-}69)$$

For stripping section,

$$y_{m+1} = \frac{\overline{L}}{\overline{\overline{V}}}x_m - \frac{\overline{B}}{\overline{\overline{V}}}x_B \quad (5\text{-}70)$$

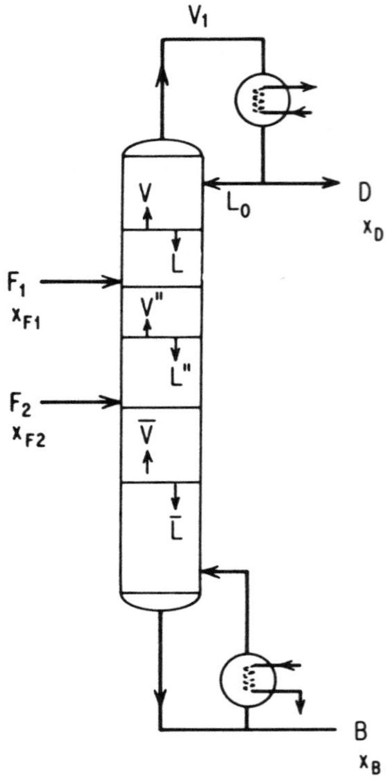

Figure 5-9. Multifeed column.

where $L'' = L + L_{F1}$ and $V'' = L'' + D - F_1$. Also, $\bar{L} = L'' + L_{F2}$ and $\bar{V} = L' - B$.

Equations of the q lines for feeds 1 and 2 are

$$y = \begin{cases} \dfrac{q_1}{q_1 - 1}x - \dfrac{x_{F1}}{q_2 - 1} & \text{feed 1} \quad (5\text{-}71) \\ \dfrac{q_2}{q_2 - 1}x - \dfrac{x_{F2}}{q_2 - 1} & \text{feed 2} \quad (5\text{-}72) \end{cases}$$

The above equations then become, for the rectification section,

$$y_{n+1} = \frac{R}{R + 1}x_n + \frac{x_D}{R + 1} \tag{5-73}$$

and, for the intermediate section, the q-line equation is given by

$$y_{p+1} = \frac{L + q_1F_1}{L + q_1F_1 + D - F_1}x_p + \frac{Dx_D - F_1x_{F1}}{L + q_1F_1 + D - F_1} \tag{5-74}$$

The bottom section operating line is given by

$$y_{m+1} = \frac{L + q_1F_1 + q_2F_2}{L + q_1F_1 + q_2F_2 + D - F_1 - F_2}x_m - \frac{Dx_D - F_1x_{F1} - F_2x_{F2}}{L + q_1F_1 + q_2F_2 + D - F_1 - F_2} \tag{5-75}$$

Side-Stream Withdrawal

When the side stream is in the enriching section, the operating line above side stream is given by (Fig. 5-10)

$$y_{n+1} = \frac{L}{V}x_n + \frac{Dx_D}{V} = \frac{R}{R + 1}x_n + \frac{D}{R + 1}x_D \tag{5-76}$$

For the section between the feed point and side stream, the operating-line equation is

$$y_{p+1} = \frac{L''}{V}x_p + \frac{Dx_D + L_sx_s}{V} \qquad \text{since } V'' = V \tag{5-77}$$

The locus of the intersection of the two operating lines is given by

$$(V'' - V)y = (L - L'')x - L_sx_s \tag{5-78}$$

Since $V'' = V$ and $V'' - V = 0$, the locus of the intersection of the operating line passes through the 45° line at x_s and has a slope of ∞; therefore it is a vertical line. Since $L_s = L - L''$ or $L'' = L - L_s$,

$$y_{p+1} = \frac{L - L_s}{V}x_p + \frac{Dx_D + L_sx_s}{V} \tag{5-79}$$

At the feed point, the equation of the q line is

$$y = \frac{q}{q - 1}x - \frac{x_F}{q - 1} \tag{5-80}$$

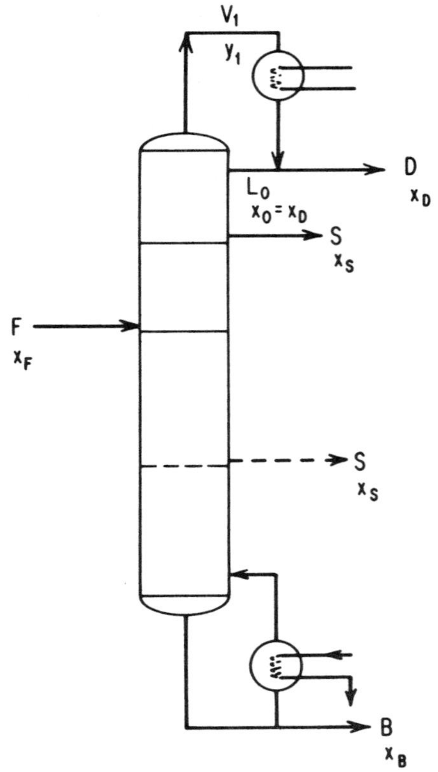

Figure 5-10. Distillation column with side-stream withdrawal.

The bottom-section operating-line equation is

$$y_{m+1} = \frac{(L - L_s) + q_F}{(L - L_s) + q_F - B} x_m - \frac{Bx_B}{(L - L_s) + q_F - B} \tag{5-81}$$

If the side stream is withdrawn from the bottom section, the following equations apply.

$$y_{m+1} = \frac{\bar{L}_m}{\bar{V}_{m+1}} x_m - \frac{L_s' x_s' + Bx_B}{\bar{V}_{m+1}} \qquad \text{above side stream} \tag{5-82}$$

$$\left.\begin{aligned} V''_{r+1} y_{r+1} &= L''_r x_n - Bx_B \\ y_{r+1} &= \frac{L''_r}{V''_{r+1}} x_r - \frac{Bx_B}{V''_{r+1}} \end{aligned}\right\} \qquad \text{below side stream}$$

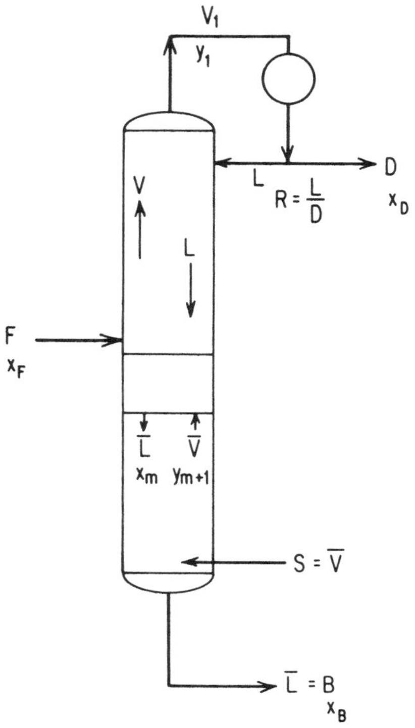

Figure 5-11. Stripping column with use of open steam.

where $L_r'' = L_m - L_s$ and $V_{r+1}'' = V$. In this case also, the side-draw line is vertical since $V_{r+1}'' = \overline{V}$.

Use of Open Steam

Open steam is useful only where water is one of the components and is removed as a bottom product (Fig. 5-11).

Case 1.

Saturated Steam at Tower Pressure. The overall balance is

$$F + S = D + B \tag{5-83}$$

where S is flow rate of steam used.

$$Fx_F = Dx_D + Bx_B \tag{5-84}$$

The operating line for the rectifying section is given by

$$y_n = \frac{L}{V}x_n + \frac{Dx_D}{V} = \frac{R}{R+1}x_n + \frac{x_n}{R+1} \tag{5-85}$$

q line:

$$y = \frac{q}{q-1}x - \frac{x_F}{q-1} \tag{5-86}$$

For saturated steam at tower pressure, the material balance for the lower section is given by

$$\overline{L}x_m + S(0) = \overline{V}y_{m+1} + Bx_B \tag{5-87}$$

and since $\overline{L} = B$ and $\overline{V} = S$,

$$\frac{\overline{L}}{\overline{V}} = \frac{y_m + 1}{x_m - x_B} \quad \text{or} \quad y_{m+1} = \frac{\overline{L}}{\overline{V}}x_m - \frac{\overline{L}}{\overline{V}}x_B \tag{5-87a}$$

When $x_m = x_B$, $y_{m+1} = 0$ at the bottom of the column; therefore the stripping-section operating line passes through $y = 0$, and $x = x_B$.

Case 2.

Superheated Steam. If steam is superheated, the liquid on the bottom tray vaporizes. The amount of vaporization is given by

$$\frac{S(\overline{H}_s - H_s)}{\lambda_m} \quad \text{lb}\cdot\text{mol/h} \tag{5-88}$$

where S = lb · mol/h of superheated steam
$\overline{H}_s$ = enthalpy of superheated steam, Btu/lb · mol
H_s = enthalpy of saturated steam at tower pressure, Btu/lb · mol
λ_m = molar latent heat of bottom liquid mixture at tower pressure, Btu/lb · mol

Therefore

$$\overline{V} = S + \frac{S(\overline{H}_s - H_s)}{\lambda_m} = S\left(1 + \frac{\overline{H}_s - H_s}{\lambda_m}\right) \tag{5-89}$$

and

$$\overline{L} = \overline{V} - S + B \tag{5-89a}$$

from which the internal $\dfrac{\overline{L}}{\overline{V}}$ is calculated.

Fenske Equation

The minimum number of plates at total reflux can be obtained graphically on the xy diagram between compositions x_D and x_B using the 45° line as the operating line for both the rectifying and stripping sections.

For ideal mixtures and constant relative volatility, the minimum number of plates $N_{\min}$ can be calculated by the Fenske equation as

$$N_{\min} = \frac{\log [x_D(1 - x_B)/x_B(1 - x_D)]}{\log \alpha_{AB}} - 1 \qquad (5\text{-}90)$$

If α changes moderately over the column,

$$\alpha_{\text{av}} = \sqrt{\alpha_{\text{top}}\alpha_{\text{bottom}}} \qquad (5\text{-}91)$$

Optimum Reflux Ratio

As the reflux ratio is increased from the theoretical minimum, the total cost of running a distillation system comprising the depreciation of the installed equipment and the operating cost of coolant, heating medium, and electricity first decreases, reaches the minimum value, and then increases. The reflux ratio at which the total cost is minimum is called the optimum reflux ratio. In practice, a reflux ratio equal to 1.2 to 2 times minimum reflux ratio is used. The total cost is not very sensitive in this range and a better operating flexibility is obtained.

Example 5-6. A fractionating column is to be designed to separate 10,000 lb/h of a mixture of 40 mol % benzene and 60 mol % toluene into an overhead product containing 97 mol % benzene and a bottoms product containing 98 mol % toluene. (**a**) Establish an overall material balance on the column; (**b**) calculate the minimum number of plates at total reflux graphically and by the use of the Fenske equation; and (**c**) using a reflux ratio of 3.7 mol per mol of distillate product, calculate the number of theoretical plates and the minimum reflux ratio, and indicate the position of the feed plate for the following cases: (**1**) if the feed is liquid at its boiling point, (**2**) if the feed is saturated vapor, (**3**) if the feed is liquid at 20°C (specific heat of feed = 0.44 Btu/lb · °F), (**4**) if the feed is a mixture of 67 mol % vapor and 33 mol % liquid.

Data: In all cases, assume total condenser is employed. Consider the reboiler as a theoretical plate. Molal latent heat of benzene = 13,248 Btu/lb · mol. Molal latent heat of toluene = 14,328 Btu/lb · mol. The equilibrium data for the benzene-toluene system are given below.

x	0	0.581	0.411	0.258	0.130	0.0107
y	0	0.777	0.632	0.456	0.263	0.039

Solution.

a. *Overall Material Balance*. For total feed, the weight of 1 mol of feed is

$$0.4(78) + 0.6(92) = 86.4 \text{ lb/lb} \cdot \text{mol}$$

so the total feed is

$$\frac{10{,}000}{86.4} = 115.74 \text{ lb} \cdot \text{mol/h}$$

Overall balance gives

$$F = B + D$$

or

$$115.74 = B + D \tag{1}$$

Benzene balance gives

$$0.4F = 0.02B + 0.97D$$

or

$$0.02B + 0.97D = 0.4(115.74) = 46.296$$

or

$$B + 48.5D = 2314.8 \tag{2}$$

Solving (1) and (2) simultaneously gives

$$D = 46.296 \text{ lb} \cdot \text{mol/h} \qquad B = 69.444 \text{ lb} \cdot \text{mol/h}$$

The benzene in the distillate product is

$$0.97(46.296) = 44.91 \text{ lb} \cdot \text{mol/h}$$

The toluene in the distillate product is

$$0.03(46.296) = 1.39 \text{ lb} \cdot \text{mol/h}$$

The benzene in the bottoms product is

$$0.02(69.44) = 1.39 \text{ lb} \cdot \text{mol/h}$$

The toluene in the bottoms product is

$$0.98(69.44) = 68.05 \text{ lb} \cdot \text{mol/h}$$

b. This part solves for the minimum number of theoretical plates at total reflux.

Graphical Method. Plot the equilibrium diagram. In this case the operating lines coincide with the 45° line. Starting from the distillate composition, step off the theoretical plates by the McCabe-Thiele method (Fig. 5-12). When this is done, the number of theoretical plates is found to be 8 which includes the reboiler. The minimum number of theoretical plates = 8 − 1 = 7 at total reflux.

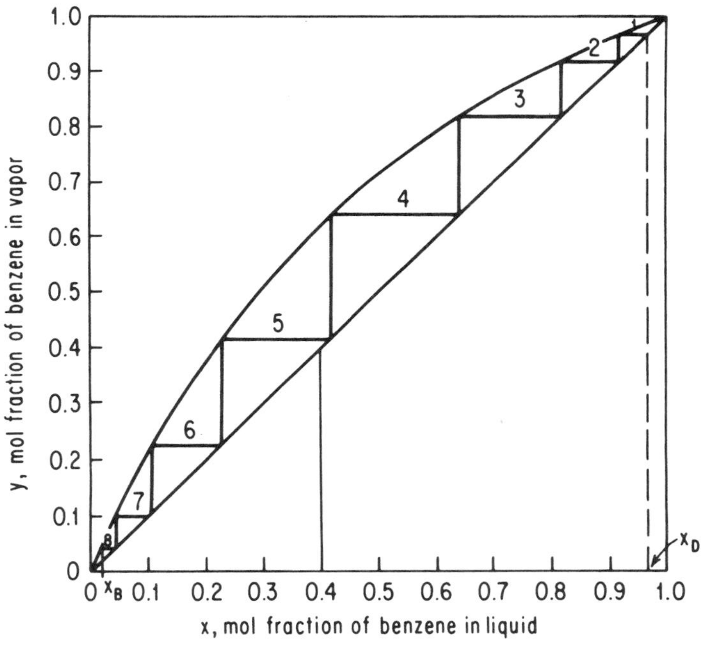

Figure 5-12. Construction of the minimum number of theoretical plates at total reflux (Example 5-6*b*).

Fenske Equation. The Fenske equation is

$$N_{\min} + 1 = \frac{\log\,[x_D(1 - x_B)/x_B(1 - x_D)]}{\log\,\alpha_{AB}}$$

$$x_D = 0.97 \qquad x_B = 0.02$$

For α_{top}, $y = 0.97$ and $x = 0.918$ (from equilibrium diagram), so

$$\alpha_{top} = \left(\frac{y}{1 - y}\right)_t \left(\frac{1 - x}{x}\right)_t = \frac{0.97}{0.03}\,\frac{1 - 0.918}{0.918} = 2.89$$

Similarly,

$$\alpha_{bottom} = \frac{0.041}{0.959}\,\frac{0.98}{0.02} = 2.095$$

$$\alpha_{av} = \sqrt{\alpha_{top}\alpha_{bottom}} = \sqrt{2.89(2.095)} = 2.461$$

$$N_m + 1 = \frac{\log\,[0.97(1 - 0.02)/0.02(1 - 0.97)]}{\log 2.461}$$

$$N_m = 8.2 - 1 = 7.2 \text{ theoretical plates}$$

c. *Calculations of Theoretical Stages for Different Conditions.*

1. The feed is liquid at its boiling point. The q line is vertical. Draw a vertical line through $x_F = 0.4$ meeting the equilibrium line at $y' = 0.63$. In this case, $x' = 0.40$.

$$\text{Intercept of rectification operating line} = \frac{0.97}{3.7 + 1} = 0.2064$$

Draw the operating line on the equilibrium diagram by joining x_D on the 45° line and the point $x = 0$, $y = 0.2064$. The operating line for the stripping section passes through the intersection point of the q line and the operating line for the rectification section and also through a point at which the vertical at $x_B = 0.02$ cuts the 45° line. When the theoretical stages are stepped off, it is found that the number of theoretical plates is 11.7 out of which one corresponds to the reboiler (Fig. 5-13). The number of theoretical plates in the column = 11.7 − 1 = 10.7. The minimum reflux ratio in this case is

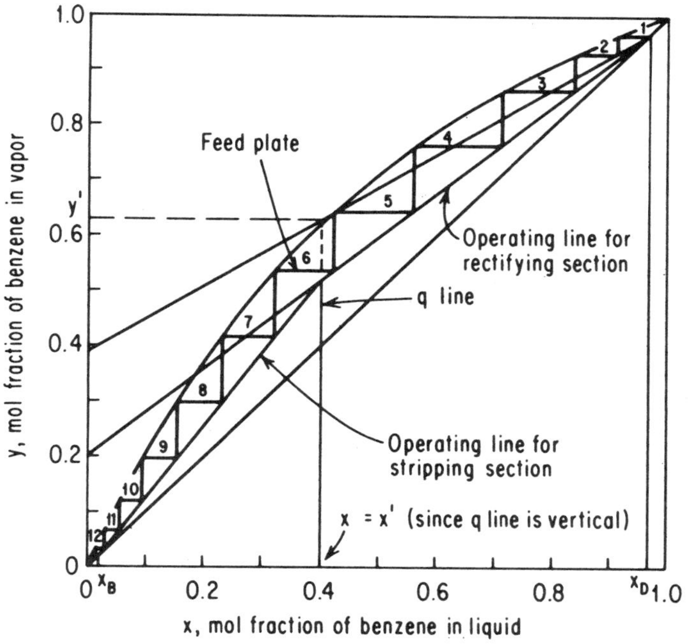

Figure 5-13. Determination of theoretical plates [Example 5-6*c*(1)]. Feed: liquid at its boiling point; *q* line vertical.

$$R_m = \frac{x_D - y'}{y' - x'} = \frac{0.97 - 0.63}{0.63 - 0.40} = 1.48$$

Alternatively, R_m can be calculated by drawing the straight line through $x_D = y_D = 0.97$ and passing through the intersection of the q and equilibrium lines and making an intercept of $x_D/(R + 1) = 0.39$; hence $R_m = 1.49$. The feed location is at the sixth theoretical plate.

2. In this case, the q line is horizontal. In Fig. 5-14, the operating lines are plotted in the same manner as in Case 1, except that the intersection point of the two operating lines now lies on the horizontal line.

The number of theoretical plates from Fig. 5-14 is 15.4 − 1 = 14.4. The feed location in this case is at the eighth theoretical plate. The minimum reflux ratio,

$$R_m = \frac{x_D - y'}{y' - x'} = \frac{0.97 - 0.4}{0.4 - 0.212} = 3.032$$

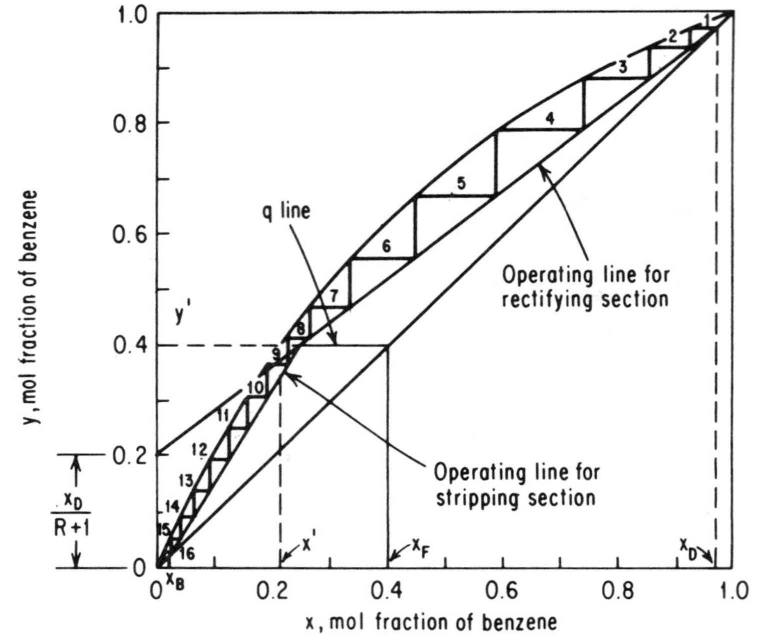

Figure 5-14. Determination of theoretical plates [Example 5-6*c*(2)]. Feed: saturated vapor; *q* line horizontal.

3. The equation of the *q* line is obtained in the following manner.

Molecular weight of feed = 86.4

Molar specific heat of feed = 86.4(0.44)
= 38.016 Btu/lb · mol · °F

Latent heat of feed = 0.4(13,248) + 0.6(14,328)
= 13,896 Btu/lb · mol

From the boiling-point diagram of benzene and toluene, the bubble point of feed is 95°C = 203°F. The feed temperature is 20°C = 68°F. From Eq. (5-49),

$$q = \frac{\lambda + C_p(t_b - t_F)}{\lambda}$$

$$= \frac{13{,}896 + 38.016(203 - 68)}{13{,}896} = 1.37$$

and the slope of the q line is

$$\frac{q}{q-1} = \frac{1.37}{1.37-1} = 3.703$$

The q line thus has a slope of 3.703. Draw the q line in the same manner as in Case 1, except that the q line in this case will have a slope of 3.703. The operating lines are drawn in Fig. 5-15 in the same manner as in Case 1. From Fig. 5-15 the number of theoretical plates including the reboiler is 10.5. The number of theoretical plates in the column is 10.5 − 1 = 9.5. The minimum reflux ratio in this case is

$$R_m = \frac{x_D - y'}{y' - x'} = \frac{0.97 - 0.695}{0.695 - 0.48} = 1.28$$

The feed plate is the fifth plate.

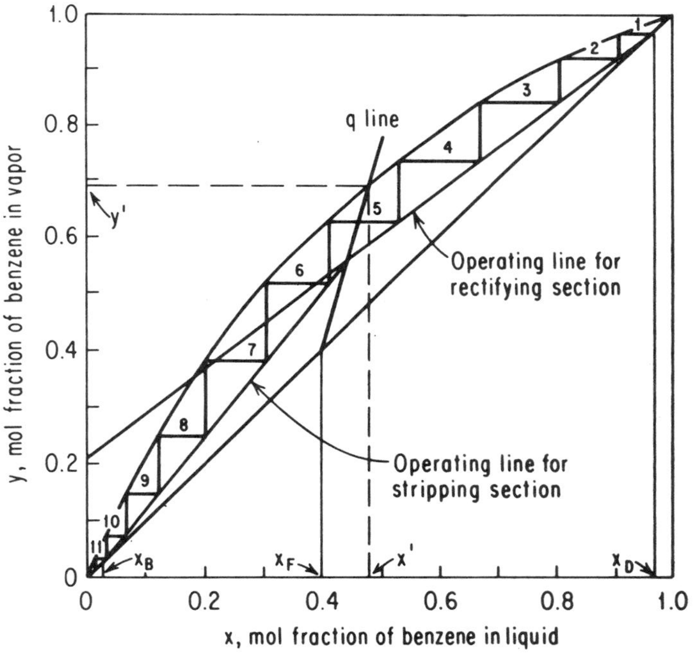

Figure 5-15. Determination of theoretical plates [Example 5-6*c*(3)].

4. In this case, the feed is a mixture of vapor and liquid, which means the feed is at its boiling point. The fraction of the feed to be vaporized is $q = 0.33$. Then, the equation of the q line is given by

$$y = \frac{0.33}{0.33 - 1}x - \frac{0.4}{0.33 - 1} = -0.493x + 0.597$$

The q line is drawn as before but with a slope of -0.493. The operating lines and the theoretical number of stages are stepped off in the usual manner as shown in Fig. 5-16. From the figure, the number of theoretical stages is $12 - 1 = 11$ stages in the column. The minimum reflux ratio is

$$R_m = \frac{x_D - y'}{y' - x'} = \frac{0.97 - 0.468}{0.468 - 0.265} = 2.473$$

The feed plate location is on the seventh plate.

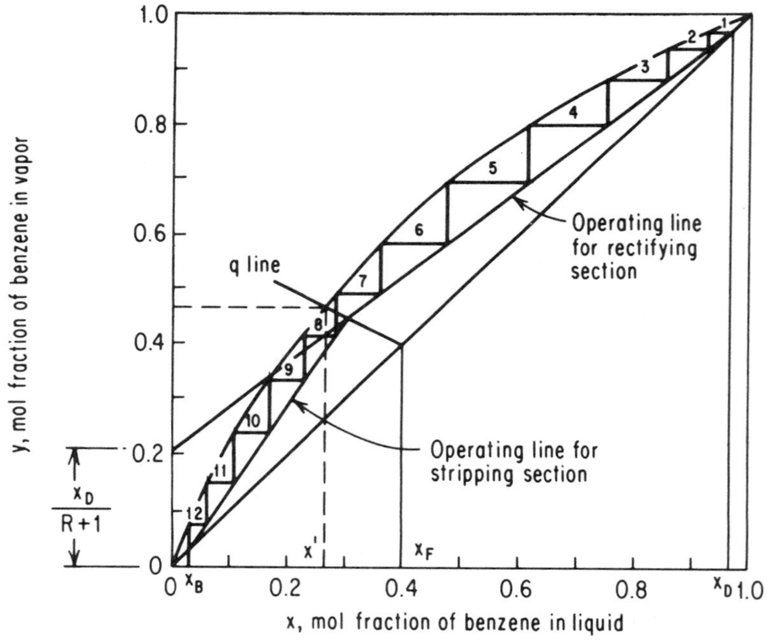

Figure 5-16. Determination of theoretical plates [Example 5-6*c*(4)].

Example 5-7. In Example 5-6**c3**: (**a**) Obtain the plate-to-plate liquid and vapor molar flows in each section of the column. (**b**) Obtain the condenser duty and number of gallons per minute of cooling water required if the water enters the condenser at 86°F and leaves at 106°F. The distillate leaves at the bubble point of 180°F, and the bottoms leaves at 229°F. (**c**) Obtain reboiler duty and amount of steam in pounds per hour if steam at 25 psia is used. Assume the heat loss from the column to be negligible. Assume also C_p of liquid independent of composition. (**d**) Establish the operating-line equation for the stripping section.

Solution.

a. *Molar Flows.*

$q = 1.37$ calculated in Example 5-6

$R = 3.7$ by the problem statement

$$V = L_0 + D = RD + D = D(R + 1) = 46.296(3.7 + 1)$$

$$= 217.6 \text{ lb}\cdot\text{mol/h}$$

$$L = L_0 = 3.7(46.296) = 171.3 \text{ lb}\cdot\text{mol/h}$$

$$\overline{L} = L + qF = 171.3 + 1.37(115.74) = 329.9 \text{ lb}\cdot\text{mol/h}$$

$$\overline{V} = V - (1 - q)F = 217.6 - (1 - 1.37)115.74$$

$$= 215.6 + 0.37(115.74) = 260.4 \text{ lb}\cdot\text{mol/h}$$

b. *Condenser Duty.*

$$Q_c = V\lambda$$

$$\lambda = 0.97(13{,}248) + 0.03(14{,}328) = 13{,}280 \text{ Btu/lb}\cdot\text{mol}$$

$$Q_c = 217.6(13{,}280) = 2{,}889{,}728 \text{ Btu/h}$$

$$\text{Flow of water} = \frac{2{,}889{,}728}{(106 - 86)500} = 288.98 \doteq 290 \text{ gpm}$$

c. C_p independent of composition. With this assumption the molar specific heat of the feed as calculated in Example 5-6**c3** is 38.02 Btu/lb · mol · °F. Using the feed temperature of 20°C (= 68°F) as the datum temperature and from Eqs. (5-56) and (5-57), one obtains Q, the heat required to heat distillate D to 180°F, as

$$Q = 46.296(38.02)(180 - 68) = 197{,}140 \text{ Btu/h}$$

Q_{SB} the heat required to heat bottoms product to 229°F, as

$$Q_{SB} = 69.444(38.02)(229 - 68) = 425{,}082 \text{ Btu/h}$$

Reboiler Duty.

$$Q_B = Q_c + Q_{SD} + Q_{SB} = 2{,}889{,}728 + 197{,}140 + 425{,}082$$
$$= 3{,}511{,}950 \text{ Btu/h}$$

Latent heat of steam $\lambda_s = 959.4$ Btu/lb

$$\text{Steam} = \frac{3{,}511{,}950}{959.4} = 3660 \text{ lb/h}$$

Alternatively, from Eq. (5-58),

$$\text{Steam required at the reboiler} = \frac{\overline{V}\lambda}{\lambda_s}$$

Now $\lambda = 0.98(14{,}328) + 0.02(13{,}248) = 14{,}306.4$ Btu/lb · mol

$$S = \frac{260.4(14{,}306)}{959.4} = 3883 \text{ lb/h}$$

d. *Operating-Line Equation for the Stripping Section.*

$$y_{m+1} = \frac{L + qF}{L + qF - B}x_m - \frac{Bx_B}{L + qF - B}$$

$$= \frac{329.9}{329.9 - 69.44}x_m - \frac{69.44(0.02)}{329.9 - 69.44}$$

or

$$y_{m+1} = 1.27x_m - 0.005332$$

Example 5-8. An acetone-water solution containing 20 mol % acetone is to be fractionated at 1 atm pressure. Ninety-eight percent of the acetone is to be received in the distillate at a concentration of 97 mol %. The feed is available at the bubble point at 80°F. The reflux will be returned at the bubble point. Open steam at 20 psig will be used at the base of the tower. Assume the heat loss from the tower to be negligible. Calculate: (**a**) the minimum reflux ratio, (**b**) a practical reflux ratio, (**c**) steam required per mole of feed, (**d**) the number of theoretical trays for a practical reflux ratio, and suggest (**e**) actual number of trays based on the efficiency value obtained from the O'Connell plot .[5]

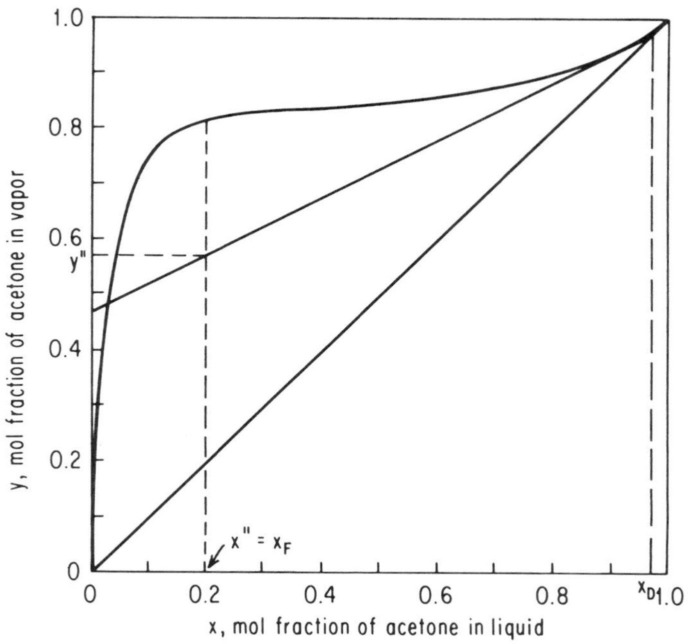

Figure 5-17. Minimum reflux ratio determination (Example 5-8).

Solution.

a. The equilibrium diagram is plotted in Fig. 5-17. Since the feed is liquid at the bubble point, the q line is vertical. At x_F, draw the vertical line at a tangent to the equilibrium curve from x_D. (See Fig. 5-17.) The minimum reflux ratio is

$$\frac{x_D - y''}{y'' - x_F} = \frac{0.97 - 0.573}{0.573 - 0.2} = \frac{0.397}{0.373} = 1.064$$

The minimum reflux ratio can more easily be found from the intercept as

$$R_m = \frac{x_D}{\text{intercept}} - 1 = \frac{0.97}{0.47} - 1 = 1.064$$

b. The optimum reflux ratio ranges from 1.2 to 2 times the minimum reflux ratio. Choose 2 times the minimum reflux ratio as the practical reflux ratio. The practical reflux ratio is 2.128. For the material balance, the basis is 100 lb · mol/h feed. From the statement of the problem,

$$D = \frac{100(0.2)(0.98)}{0.97} = 20.20 \text{ lb} \cdot \text{mol/h}$$

$$R = \text{actual reflux ratio} = 1.064(2.0) = 2.128$$

$$L_0 = L = RD = 2.128(20.20) = 43.0 \text{ lb} \cdot \text{mol/h}$$

$$V = L_0 + D = 43.0 + 20.20 = 63.2 \text{ lb} \cdot \text{mol/h}$$

Since the feed is at the bubble point, $q = 1$ and the q line is vertical.

$$\bar{L} = L + qF = 43 + 1(100) = 143 \text{ lb} \cdot \text{mol/h}$$

$$\bar{V} = V + F(q - 1) = 63.2 + 100(1 - 1) = 63.2 \text{ lb} \cdot \text{mol/h}$$

c. *Calculation of Steam Requirements.* For steam at 20 psig,

H_s = enthalpy of steam at 1 atm = 1150.5 Btu/lb

λ = 970.3 Btu/lb   neglecting contribution of acetone

$\bar{H}_s$ of steam at 20 psig = 1166.9 Btu/lb

$$\bar{V} = S\left[1 + \frac{1166.9 - 1150.3}{970.3}\left(\frac{18}{18}\right)\right]$$

$$S = \frac{63.2}{1.01711} = 62.1 \text{ lb} \cdot \text{mol/h}$$

Steam to tower = S = 62.1 lb · mol/h

Steam required per mole of feed = 0.621 lb · mol

d. *Balance on Bottom Plate.*

$$B = S + \bar{L} - \bar{V} = 62.1 + 143.0 - 63.2 = 141.9 \text{ lb} \cdot \text{mol/h}$$

Bottoms Composition.

Acetone in distillate = 0.97(20.2) = 19.594 lb · mol/h

Acetone in bottoms = 20 − 19.594 = 0.406 lb · mol/h

$$\text{Mol \% acetone in bottoms} = \frac{0.406}{141.9}(100) = 0.286 \text{ mol \%}$$

$$= 0.003 \text{ mol fraction}$$

Calculation of Theoretical Plates. The intercept of the operating line on the y axis is

$$\frac{x_D}{R+1} = \frac{0.97}{2.128+1} = 0.31$$

Draw this line on the equilibrium diagram in Fig. 5-18. It passes through $x_D = y_D = 0.97$ and its intercept on the y axis is 0.31.

The operating line in the stripping section passes through ($y = 0$, $x_B = 0.003$) and the point of intersection of the q line with the operating line of the rectification section. This is drawn in Fig. 5-18.

The number of the theoretical plates from the graph is 10 since there is no reboiler.

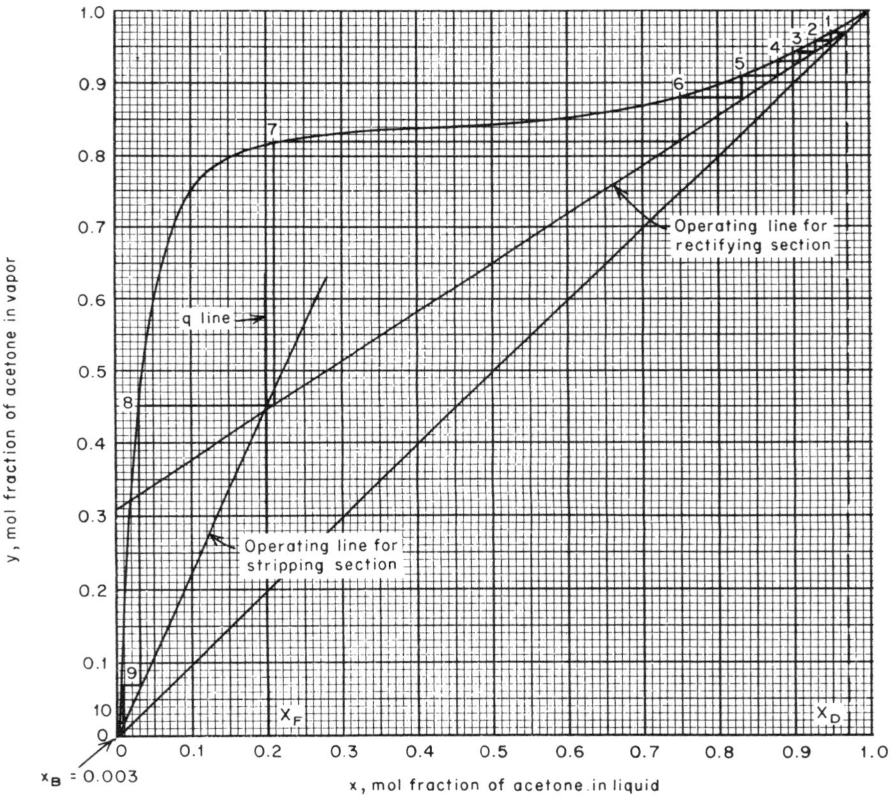

Figure 5-18. Graphical determination of theoretical stages (Example 5-8).

e. *Calculation of Actual Number of Plates*. From the composition of vapor at the top of the column and the equilibrium data,

$$\text{Column top temperature} = 135°\text{F}$$

$$\text{Column bottom temperature} = 212°\text{F}$$

$$\text{Average temperature} = \frac{212 + 135}{2} = 173.5°\text{F}$$

Average Viscosity.

Acetone: $\mu = 0.2 \text{ cP}$

H_2O: $\mu = 0.35 \text{ cP}$

Average: $\mu_m^{1/3} = 0.2(0.2)^{1/3} + (0.35)^{1/3}(0.8)$

$$\mu_m = 0.32 \text{ cP}$$

Average α.

$$\alpha_{\text{top}} = \frac{0.97/(1 - 0.97)}{0.96/(1 - 0.96)} = 1.35$$

$$\alpha_{\text{bottom}} = \frac{0.076/(1 - 0.076)}{0.003/(1 - 0.003)} = 27.33$$

$$\alpha_{\text{av}} = \sqrt{1.35(27.33)} = 6.07$$

$$\alpha_{\text{av}}\mu_1 = 6.07(0.32) = 1.94$$

From O'Connell plot,[5]

$$\text{Efficiency} \doteq 0.41$$

$$\text{Actual number} = \frac{10}{0.41} = 24.4 \doteq 25$$

The number of actual plates must be an integer.

Murphree Efficiency

The Murphree efficiency for the vapor phase is given by

$$E_M = \frac{y_n - y_{n-1}}{y_n^* - y_{n-1}}$$

where E_M = Murphree efficiency
y_n^* = composition of vapor in equilibrium with the liquid leaving stage n
y_n, y_{n-1} = actual nonequilibrium values for the vapor streams leaving nth and $(n-1)$th plate

y_n^* is to be read from the equilibrium curve. If the Murphree efficiency is assumed constant for all plates in the column, it is possible to use it to obtain a pseudoequilibrium curve which in turn can be used to obtain the actual number of stages for the column. For this, the pseudoequilibrium curve is drawn below the true equilibrium curve. To obtain a point on this curve, the vertical line is drawn between the true equilibrium curve and the operating line. The point of the pseudoequilibrium curve will be marked on this vertical line as indicated by the percent efficiency. For example, if the Murphree plate efficiency is 50 percent, the point will lie midway on the vertical.

Enthalpy Composition Diagram

When a constant molal overflow cannot be assumed, use can be made of the enthalpy composition or Ponchon-Savarit diagram to find out the number of theoretical plates.

For the preparation and application of the enthalpy composition diagram to determine the number of theoretical plates, the reader is referred to other textbooks.[4]

References

1. J. J. Van Laar, *Z. physik. Chem.*, vol. 185, 1929, p. 35.
2. M. Margules, *Sitzber. Akad. Wiss. Wian. Math. Naturw.* K1, II, 104, 1895, p. 1243.
3. C. S. Robinson and E. R. Gilliland, *Elements of Fractional Distillation*, McGraw-Hill, New York, 1950, pp. 42–43.
4. R. E. Treybal, *Mass Transfer Operations*, McGraw-Hill, New York, 1980, pp. 357–401.
5. *Chemical Engineers Handbook*, 5th ed., R. H. Perry (ed.), McGraw-Hill, New York, 1973, p. 18-14.

Chapter

6

Absorption

Absorption is an important unit operation used in industry. Problems involving absorption are frequently given in the P.E. examination. The candidates are required to become familiar with the equilibrium solubility relationships; absorption theory, including the height of transfer unit; the numbers of transfer units, mass-transfer coefficients, etc.; and the determination of the diameter and height of the absorption column.

Generally, the solution of problems on absorption requires establishing a material balance equation, an equation of the operating line, and an equation or graphical presentation of the equilibrium solubility relationship. Occasionally the solvent flow rate is not specified, and the candidate is advised to use a multiple of the minimum solvent flow rate. One must carefully examine the units of various parameters given in the problem statement and use the appropriate equation.

Equilibrium Solubility Curves

Equilibrium solubility relationships are available in many forms. Some of them may be converted into equations of the

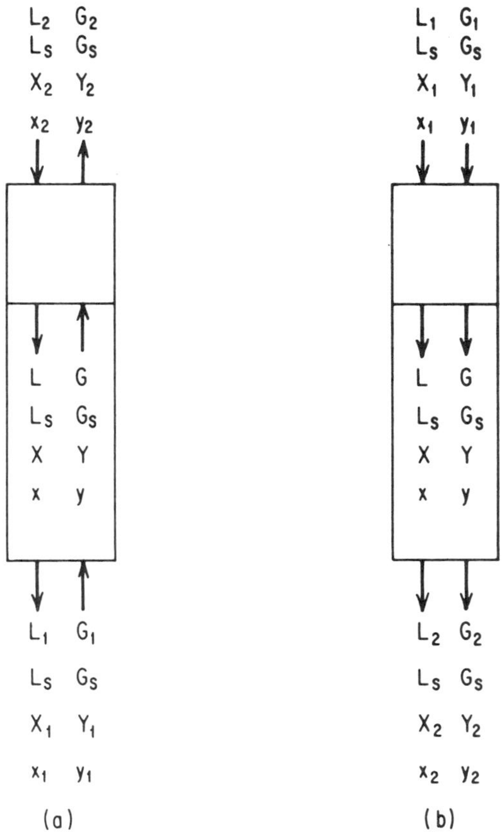

Figure 6-1. Absorption in packed column (*a*) countercurrent, (*b*) cocurrent.

type $Y^* = f(X)$ or $y^* = f(x)$ which can be readily used for the calculation of the transfer units by integration.

Figure 6-1 shows the operations of the countercurrent and cocurrent gas-absorption towers which may be any of the following types: packed, spray and bubble-cap, or sieve tray. The design of the packed absorption towers will be dealt with here. The solvent gas and solvent liquid flows are essentially unchanged in quantity as they pass through the tower, and it is convenient to express the material balances in terms of these. The equations of the material balance and operating line can be easily derived for various conditions of the absorption operation and are as follows:

1. Countercurrent Absorption (All Concentrations)

Material balance: $G_S(Y_1 - Y_2) = L_S(X_1 - X_2)$

Operating line: $G_S(Y - Y_2) = L_S(X - X_2)$ (6-1)

2. Countercurrent Absorption (Dilute Solution)

Material balance: $G_S(y_1 - y_2) = L_S(x_1 - x_2)$

Operating line: $G_S(y - y_2) = L_S(x - x_2)$ (6-2)

3. Cocurrent Absorption (All Concentrations)

Material balance: $G_S(Y_1 - Y_2) = L_S(X_2 - X_1)$

Operating line: $G_S(Y_1 - Y) = L_S(X - X_1)$ (6-3)

4. Cocurrent Absorption (Dilute Solutions)

Material balance: $G_S(y_1 - y_2) = L_S(x_2 - x_1)$

Operating line: $G_S(y_1 - y) = L_S(x - x_1)$ (6-4)

The gas stream at any point in the tower is the total gas flow G, which includes the diffusing solute of mole fraction y, partial pressure p, or mole ratio Y and a nondiffusing (insoluble) gas of flow rate G_S. Similarly, the liquid stream consists of total flow L containing the soluble gas of mole fraction x or mole ratio X and the essentially nonvolatile solvent of flow rate L_S. The auxiliary equations relating these quantities are

$$Y = \frac{y}{1 - y} = \frac{p}{P_t - p} \tag{6-5a}$$

$$y = \frac{Y}{1 + Y} = \frac{p}{P_t} \tag{6-5b}$$

$$X = \frac{x}{1 - x} \tag{6-5c}$$

$$x = \frac{X}{1 + X} \tag{6-5d}$$

$$G_S = G(1 - y) \tag{6-6a}$$

$$L_S = L(1 - x) \tag{6-6b}$$

where G_S = solvent gas flow rate, lb · mol/h · ft^2
L_S = solvent liquid flow rate, lb · mol/h · ft^2
L = total liquid flow rate, lb · mol/h · ft^2
Y = concentration of solute in the gas, lb · mol solute/lb · mol solvent gas
X = concentration of solute in liquid, lb · mol solute/lb · mol solvent liquid
y = concentration of solute in gas, mole fraction
x = concentration of solute in liquid, mole fraction
p = partial pressure of solute gas
P_t = total pressure
G = total gas flow rate, lb · mol/h · ft^2

In the auxiliary equations (6-5*a*) and (6-5*b*), p and P_t must have the same units.

Determination of Solvent Flow Rate

The following steps may be followed to determine the liquid solvent flow rates required to effect a given absorption:

1. The average gas flow rate is first calculated by

$$\text{Average gas flow rate} = \frac{G_{\text{in}} + G_{\text{out}}}{2} = G_{\text{av}}$$

where the subscripts refer to the inlet, outlet, and average flow rates. G_{in} is calculated from the given data and G_{out} is calculated as follows

$$G_{\text{out}} = G_{\text{in}} - \text{solute absorbed} + \text{solvent carried over as vapor}$$

$$\text{Solvent carried over} = \frac{p_s G_S}{P_t - p_s}$$

where p_s is the vapor pressure of the solvent at the operating temperature.

2. If the equilibrium solubility relationship is available in the form $y^* = mx$, estimate the solvent flow rate by

$$L_{av} = 1.6mG_{av}$$

where the star indicates the equilibrium composition and m is a constant.

3. If the equilibrium solubility relationship is available in the form $Y^* = f(X)$, then proceed as follows. (Refer to Fig. 6-2*a* and *b*.)
   a. Plot the equilibrium solubility line Y versus X.
   b. Locate the point A (X_2, Y_2). Note that for a pure solvent, $X_2 = 0$ and Y_2 is obtained from Y_1 and the specified percent recovery.
   c. Locate Y_1 on the ordinate and draw a horizontal line CB through Y_1 intersecting the equilibrium curve at B.
   d. Join AB without crossing the equilibrium curve, as in Fig. 6-2*b*.

If line AB tends to cross the equilibrium curve as in Fig. 6-2*a*, draw AB' tangential to the equilibrium curve. The slope of line AB or AB' is L_S/G_S. Minimum solvent flow rate is the slope of AB

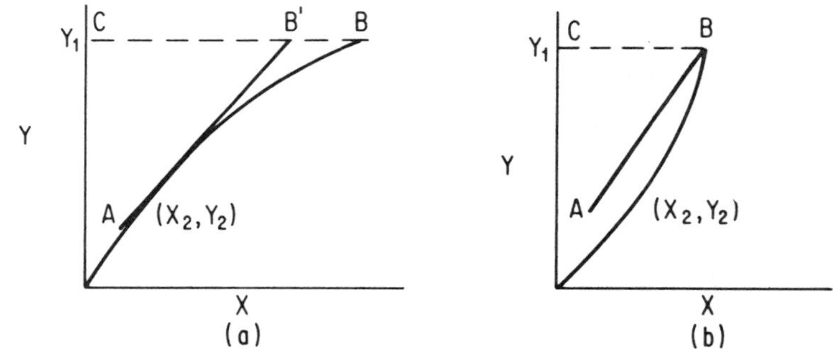

Figure 6-2. Estimation of minimum solvent flow: (*a*) equilibrium curve convex toward operating line; (*b*) equilibrium curve concave toward operating line.

or AB' times G_S. The operating solvent flow rate is 1.5 times the minimum solvent flow rate.

The multiplier 1.5 in the above relation is not a constant factor. It is determined by an economic analysis.

Note: In the above symbols, L, L_S, G, or G_S may be in lb · mol/h · ft^2 or lb · mol/h.

The method of graphical determination of the minimum solvent flow rate is illustrated in Example 6-3 and Fig. 6-5*a*.

Determination of Packing Height *Z*

There are three most commonly used equations for determining the height of a packed column. These are:

1. Principal resistance is in the gas phase. (This may be used when neither the gas nor the liquid phase resistance dominates.)

$$Z = N_{OG}H_{OG} \qquad \text{ft} \tag{6-7}$$

2. Principal resistance is in the liquid phase. (This may be used when neither gas nor liquid phase resistance dominates.)

$$Z = N_{OL}H_{OL} \qquad \text{ft} \tag{6-8}$$

3. Resistance may be either in the gas phase or liquid phase or both phases.

$$Z = N_P(\text{HETP}) \qquad \text{ft} \tag{6-9}$$

where HETP is the height equivalent of a theoretical plate.

In the above equations, the number of transfer units, N_{OG} and N_{OL}, and the number of theoretical stages, N_P, can be determined as follows.

Determination of N_{OG}

1. For countercurrent absorption, and for any concentration of solute, the number of transfer units is given by

$$N_{OG} = \int_{y_2}^{y_1} \frac{(1-y)_{lm}\,dy}{(1-y)(y-y^*)} = \int_{y_2}^{y_1} k\,dy \tag{6-10}$$

where

$$(1-y)_{lm} = \frac{(1-y^*)-(1-y)}{\ln[(1-y^*)/(1-y)]} \tag{6-11}$$

and

$$k = \frac{(1-y)_{lm}}{(1-y)(y-y^*)} \tag{6-12}$$

N_{OG} is determined by graphical integration. A value of x is assumed between the known or specified limits (i.e., x_1 to x_2) and the corresponding value of y^* is obtained from the equilibrium relationship which is expressed as $y^* = f(x)$. For the same value of x, the corresponding value of y is obtained from the material-balance equation. From the values of y and y^* thus obtained, the value of k can be determined from Eq. (6-12). A plot of k versus y is made. The area under the curve gives N_{OG}. The equation can be simplified by replacing the logarithmic average by the arithmetic average as follows:

$$N_{OG} = \int_{y_2}^{y_1} \frac{dy}{y-y^*} + \frac{1}{2}\ln\frac{1-y_2}{1-y_1} = \int_{Y_2}^{Y_1} \frac{dY}{Y-Y^*} - \frac{1}{2}\ln\frac{1+Y_1}{1+Y_2} \tag{6-13}$$

2. In some countercurrent absorptions involving dilute solutions, both the equilibrium and operating lines may be straight. Then the equilibrium line is of the form $y^* = mx + c$ where m and c are constants. In these cases, the number of transfer units is given by

$$N_{OG} = \frac{y_1 - y_2}{(y-y^*)_{lm}} \tag{6-14}$$

where

$$(y-y^*)_{lm} = \frac{(y-y^*)_{\text{bottom}} - (y-y^*)_{\text{top}}}{\ln[(y-y^*)_{\text{bottom}}/(y-y^*)_{\text{top}}]} \tag{6-15}$$

In the above equation, the subscripts "bottom" and "top" denote the conditions at the bottom and top of packing.

3. In some countercurrent absorptions in dilute solutions, the equilibrium and operating lines are straight but not parallel. If the equilibrium line follows Henry's law, $y^* = mx$, and N_{OG} is given by

$$N_{OG} = \frac{1}{1 - mG/L} \ln \left[\frac{(1 - mG/L)(y_1 - mx_2)}{y_2 - mx_2} + \frac{mG}{L} \right] \qquad (6\text{-}16)$$

4. If the conditions are the same as in (3) above and if the equilibrium and operating lines are straight and parallel, i.e., $y^* = mx = (L/G)x$, N_{OG} is given by

$$N_{OG} = \frac{y_1 - y_2}{y_2 - mx_2} \qquad (6\text{-}17)$$

5. For the countercurrent absorption when dissolved solute reacts with the solvent so that there is no equilibrium pressure or concentration of the dissolved solute, i.e., $m = 0$ (e.g., absorption of CO_2 in a caustic solution), N_{OG} is given by

$$N_{OG} = \ln \left[\frac{\ln (1 - y_1)}{\ln (1 - y_2)} \right] \qquad (6\text{-}18)$$

6. When the conditions are the same as in (5) but the solution is dilute ($y < 0.05$), N_{OG} can be calculated by the relation

$$N_{OG} = \ln \frac{y_1}{y_2} \qquad (6\text{-}19)$$

7. For cocurrent absorption, when other conditions are the same as in (3) above (see Fig. 6-1*b* for an explanation of the symbols), N_{OG} is given by

$$N_{OG} = \frac{1}{1 + mG/L} \ln \frac{y_1 - mx_1}{y_2 - mx_2} \qquad (6\text{-}20)$$

N_{OG} can also be estimated by the graphical method.

Determination of N_{OL}. For countercurrent absorption, and for any concentration of the solute, N_{OL} is given by

$$N_{OL} = \int_{x_2}^{x_1} \frac{(1-x)_{lm}\,dx}{(1-x)(x^*-x)} = \int_{x_2}^{x_1} k\,dx \tag{6-21}$$

where

$$k = \frac{(1-x)_{lm}}{(1-x)(x^*-x)} \qquad \text{and} \qquad (1-x)_{lm} = \frac{(1-x)-(1-x^*)}{\ln[(1-x)/(1-x^*)]}$$

N_{OL} is determined by a graphical integration in the same manner as the determination of N_{OG}. A value of y is assumed between the known or specified limits (i.e., y_1 and y_2), and the corresponding value of x^* is obtained from the equilibrium line expressed in the form $y = f(x^*)$. For the same value of y, the corresponding value of x is obtained from the material-balance equation. Hence, the value of k is determined. The equation can be simplified by replacing the logarithmic average by the arithmetic average as follows:

$$N_{OL} = \int_{x_2}^{x_1} \frac{dx}{x^*-x} + \frac{1}{2}\ln\frac{1-x_2}{1-x_1} = \int_{X_2}^{X_1} \frac{dX}{X^*-X} - \frac{1}{2}\ln\frac{1+X_1}{1+X_2} \tag{6-22}$$

If the equilibrium relationship is linear, the above equation may be used to estimate N_{OL} without resorting to the graphical integration.

Determination of N_P.

1. Analytical Method. This is applicable when the operating line is straight, the solution is dilute, and Henry's law is applicable, i.e., $y^* = mx$, the number of theoretical plates is then given by

$$N_P = \frac{\ln[(1 - mG/L)(y_1 - mx_2)/(y_2 - mx_2) + mG/L]}{\ln(L/mG)} \tag{6-23}$$

2. Graphical Method. (See the following pages.)

Relation between N_{OG} and N_P. N_{OG} and N_P are related by the following equation.

$$N_{OG} = N_P \frac{\ln A}{A - 1} \tag{6-24}$$

where $A = mG/L$.

Determination of Height of Transfer Units H_{OG} and H_{OL}. These are experimentally determined and have the dimension of length. The analytical expression for H_{OG} is

$$H_{OG} = \frac{G}{K_y a(1 - y)_{lm}} = \frac{G}{K_G a P_t (1 - y)_{lm}} \tag{6-25}$$

where

$$(1 - y)_{lm} = \frac{(1 - y^*) - (1 - y)}{\ln [(1 - y^*)/(1 - y)]}$$

If the soluble gas concentration is low, $(1 - y)_{lm}$ may be assumed equal to unity.

In the above equation $K_G a$ and $K_y a$ are the overall mass-transfer coefficients. These overall coefficients may be calculated from the individual gas and liquid-film mass-transfer coefficients and the equilibrium solubility relationship. However, care must be exercised to identify the units of these film coefficients and the units of the parameters in the equilibrium relationship.

The analytical expression for H_{OL} is

$$H_{OL} = \frac{L}{K_x a(1 - x)_{lm}} = \frac{L}{K_L a C(1 - x)_{lm}} = \frac{L M_L}{K_L a \rho_L (1 - x)_{lm}} \tag{6-26}$$

where

$$(1 - x)_{lm} = \frac{(1 - x) - (1 - x^*)}{\ln [(1 - x)/(1 - x^*)]}$$

and M_L = molecular weight of the liquid solvent.

Estimation of K_Ga for a System from the K_Ga of Another System

If K_Ga is known for one system, the K_Ga of another system can be determined, if the diffusivities for the two systems are known, by using

$$K_Ga\ (\text{unknown}) = K_Ga\ (\text{known})\left[\frac{D_v\ (\text{unknown})}{D_v\ (\text{known})}\right]^{0.56}$$

In the preceding pages, K_Ga, K_La, K_xa, K_ya, k_Ga, k_La, k_xa, and k_ya indicate the mass-transfer coefficients. The capital letters indicate the overall mass-transfer coefficients; the small letters indicate the individual mass-transfer coefficients. The units of these mass-transfer coefficients are denoted by their subscripts as

$$K_Ga,\ k_Ga = \text{lb}\cdot\text{mol/h}\cdot\text{ft}^3\cdot\text{atm}$$

$$K_La,\ k_La = \text{lb}\cdot\text{mol/h}\cdot\text{ft}^3\cdot(\text{lb}\cdot\text{mol/ft}^3)$$

$$K_xa,\ k_xa,\ K_ya,\ k_ya = \text{lb}\cdot\text{mol/h}\cdot\text{ft}^3\cdot\text{mol fraction}$$

where M_L = molecular weight of liquid
ρ_L = density of liquid, lb/ft^3
C = concentration of solute in liquid, $\text{lb}\cdot\text{mol/ft}^3$
D_v = diffusivity, ft^2/s

Graphical Determination of N_{OG}

The following steps (see Fig. 6-3) are used: (1) Plot the equilibrium and operating line Y versus X; (2) draw the vertical lines, such as AB and DF, and draw a line bisecting these vertical lines; (3) starting from X_2, Y_2, draw triangle CMN such that $CO = OM$. This completes one unit of N_{OG}. Repeat constructing triangles like this until point $A(X_1, Y_1)$ is covered. For practical purposes, the last triangle covering (X_1, Y_1) may be taken as one unit of N_{OG}.

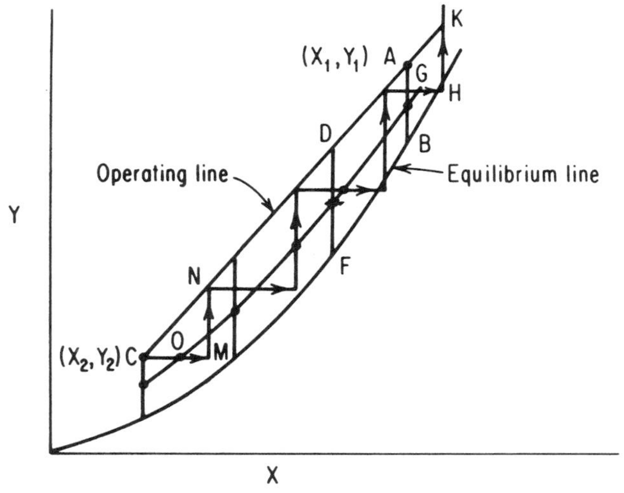

Figure 6-3. Graphical determination of N_{OG}.

The total number of triangles to cover the end composition is N_{OG}. Note that the triangles of N_{OG} may not touch or may cross the equilibrium curve.

Determination of Diameter of Column

The steps to be followed to estimate the diameter of an absorption column are:

1. Draw a sketch as in Fig. 6-1*a* or *b*, and show the gas and liquid flows and the compositions.
2. If not given, select a suitable type of packing. For operating pressure $\geqslant$ 100 mmHg, the ring- or saddle-type packings may be selected. For the lower operating pressures, the systematic packings may be considered.
3. If not otherwise dictated, select the packing size[2] from Table 6-1.

TABLE 6-1 Selection of Packing Size

Process Condition		Packing Size
Gas flow rate <	500 ft^3/min (0.236 m^3/s)	$\frac{3}{4}$in (19 mm)
Gas flow rate =	500–2000 ft^3/min (0.236–0.944 m^3/s)	1in (25 mm)
Gas flow rate > or vacuum operation	2000 ft^3/min (0.944 m^3/s)	$1\frac{1}{2}$–2in (38–50 mm) or bigger

4. Select the unit pressure drop, inH_2O/ft (or Pa/m), of the packing from Table 6-2.

TABLE 6-2 Selection of Unit Pressure Drop* for Nonfoaming Systems

Type of Operation	Unit Pressure Drop	
	inH_2O/ft of packing height	Pa/m of packing height
Atmospheric and high pressure	0.25 to 1.0 maximum 1.5	204 to 816 maximum 1224
Vacuum operation	0.05 to 0.2 maximum 1.0	41 to 164 maximum 816

*The pressure drop constraints given here are general guidelines and are not absolute recommendations. For vacuum operation, for example, a range of 0.1 to 1.0 inH_2O/ ft of packing has also been recommended[4] depending upon the system. Foaming limits the capacity and changes the ΔP characteristics of packing. In such cases, the design value of the unit pressure drop needs to be modified. A consideration of the effect of foaming is beyond the scope of this text.

5. Calculate the numerical value of $(L/G)\sqrt{\rho_G/\rho_L}$. If L is not known, for estimation purposes one may choose a value such that $(L/G)\sqrt{\rho_G/\rho_L}$ lies between 0.02 and 1. From $(L/G)\sqrt{\rho_G/\rho_L}$ and the chosen unit pressure drop, find the ordinate from Fig. 6-4 and calculate the design mass velocity.

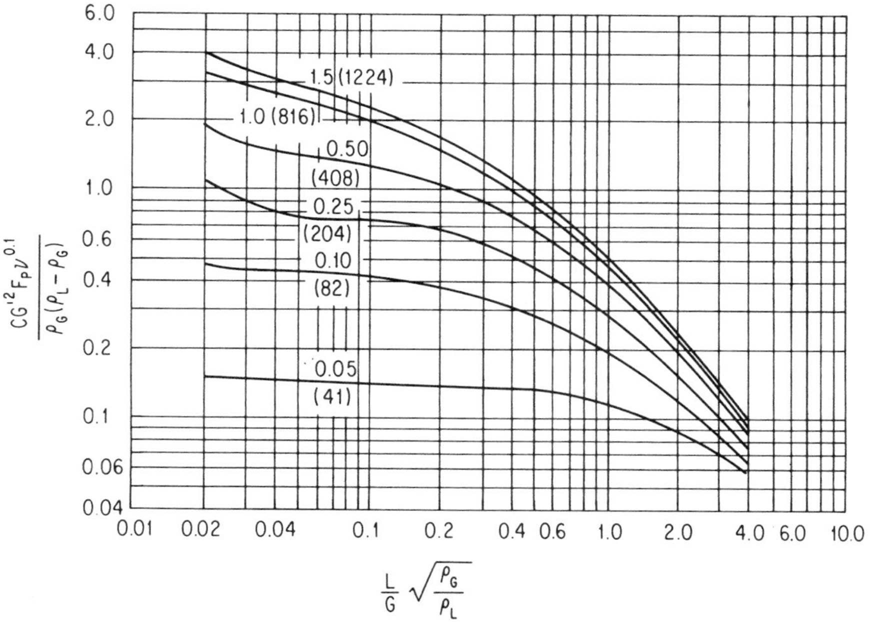

Figure 6-4. Generalized pressure-drop correlation. Parameter of curves is pressure drop in inches of water per foot of packing. Figures in parentheses are pascals per meter of packing height. In the British units, the symbols in the ordinate are G' (gas rate) = lb/ft$^2 \cdot$s, $C = 1$, ν (viscosity of liquid in centistokes) = cP/sp. gr., ρ_L, ρ_G (liquid and gas densities) = lb/ft^3. In SI units, $G' = \text{kg/m}^2 \cdot \text{s}$, $C = 42.84$, $\nu = \text{m}^2/\text{s}$, ρ_L and $\rho_G = \text{kg/m}^3$. The symbols in the abscissa must have consistent units. (*By permission, Norton Co., Akron, Ohio.*)

$$G' = \left[\frac{(\text{ordinate})(\rho_G)(\rho_L - \rho_G)}{(C)(F_p)(\nu^{0.1})}\right]^{0.5}$$

where $G' = \text{lb/s} \cdot \text{ft}^2$ or $\text{kg/s} \cdot \text{m}^2$ and consistent units are to be used.

6. The column diameter is to be calculated by

$$\text{Column diameter} = \begin{cases} 0.0188 \left(\dfrac{G_T}{G'}\right)^{0.5} \text{ ft} & G_T = \text{gas flow rate, lb/h} \\ 1.128 \left(\dfrac{G_T}{G'}\right)^{0.5} \text{ m} & G_T = \text{gas flow rate, kg/s} \end{cases}$$

Table 6-1A Properties of Various Packings

Packing factor (F_p), specific surface (ft^2/ft^3), and % free gas space of wet and dump-packed packings. Numbers without any bracket indicates F_p, numbers within ( ) indicates specific surface and numbers within [ ] indicates % free gas space. (Ref. 1,3,5)

		Nominal Packing Size, inch or #								
Type of* Packing	Material	1/4	3/8	1/2	5/8	3/4	1	1-1/2	2	3 or 3 1/2
Super Intalalox Saddles	Ceramic	-	-	-	-	-	60	-	30	-
Super Intalalox Saddles	Plastic	-	-	-	-	-	40	-	28	18
Intalox Saddles	Ceramic	725 (300) [75]	330	200 (190) [78]	-	145 (102) [77]	92 (78) [77]	52 (59) [80]	40 (36) [79]	22 (28) [80]
Intalox Snow Flake	Plastic							13 (28)	13 (28)	13 (28)
IMTP	metal				51		41	24	18	12
Hy-Pak rings	metal	-	-	-	-	-	45	29	26	16
Pall rings	Plastic	-	-	-	95 (104) [87]	-	55 (63) [90]	40 (39) [91]	26 (31) [92]	17

Type of Packing*	Material	1/4	3/8	1/2	5/8	3/4	1	1-1/2	2	3 or 3 1/2
Pall rings	Metal	-	-	-	81 (104) [93]	-	56 (63) [94]	40 (39) [95]	27 (31) [96]	18
Raschig rings	Ceramic	b,x 1600 (217) [62]	b,x 1000	c 580 (112) [64]	c 380	c 255 (74) [72]	d 179 (58) [74]	e 93 (37) [73]	f 65 (28) [74]	g,x 37 (19) [75]
Raschig rings [a]	Metal	x 700	x 390	x 300 (122) [85]	170 (103) [87]	155 (81) [89]	x 115	-	-	-
Raschig rings [b]	Metal	-	-	410 (111) [73]	300	220 (75) [80]	144 (56) [86]	83 (39) [90]	57 (29) [92]	x 32 (20) [95]
Berl Saddles	Ceramic	x 900 (274) [60]	-	x 240 (142) [62]	-	170 (87) [66]	110 (76) [68]	65 (46) [71]	x 45 (32) [72]	-
Tellerettes	Plastic	-	-	-	-	-	36 (55) [87]	-	h 11 (28) [95]	- 16 (30) [92]

Notes: a=1/32" wall c=3/32" wall e=3/16" wall g=3/8" wall h= type K
b=1/16" wall d=1/8" wall f= 1/4" wall x= Extrapolated

* All packings except Tellerettes are products of NORTON Company, Akron ,OH. Tellerettes is a product of Ceilcote, Berea, OH.

7. This step involves checking the packing size and wetting rate. Maximum allowable packing sizes are given in Table 6-3.

TABLE 6-3 Maximum Allowable Packing Sizes

Packing Type	Maximum Allowable Packing Size
Raschig and partition ring	1/30th of column diameter
Saddle type	1/15th of column diameter
Pall-type rings	1/10th of column diameter

The *wetting rate* is obtained by dividing the liquid flow in gpm or m^3/s by the cross section in ft^2 (or m^2). This rate should be greater than or equal to the minimum wetting rate which can be calculated by using the relationships given in Table 6-4.

TABLE 6-4 Minimum Wetting Rate[2]

Packing Size	Minimum Wetting Rate* gpm/ft²	Minimum Wetting Rate* m³/s·m²
<3 in (75 mm)	$0.106A_p$	$2.2 \times 10^{-5}A_p$
≥3 in (75 mm)	$0.164A_p$	$3.4 \times 10^{-5}A_p$

*A_p is the specific surface, ft^2/ft^3 or m^2/m^3. The specific surfaces of the packings are given elsewhere.[3d]

The *minimum wetting rate* thus obtained is not the absolute minimum below which flow conditions are poor and above which the packing becomes thoroughly wet. It is desirable that the column operates above the minimum wetting rate estimated by the criteria given in Table 6-4. If the actual wetting rate is below the minimum wetting rate, one of the following should be used: higher packing size, an increased pressure drop, a combination of both, a systematic packing, or recirculation of liquid. The turbulence created by a lower packing size would improve the mass transfer and may compensate for the loss of wetting.

Note: Distillation in a packed column can be run at a wetting rate lower than what is predicted by the above formulas without adversely affecting the packing efficiency.

Example 6-1. A gas stream containing 10% by volume CO_2 at a pressure of 16 inH_2O, a temperature of 176°F, and a flow rate of 70,000 lb/h is to be countercurrently scrubbed with 195,000 lb/h of 8% (by weight) caustic solution at 176°F. The entering gas has the following composition (by volume): $CO_2 = 10\%$, $N_2 = 80\%$, and $O_2 = 10\%$. The pressure at the top of the column is 0.1 psig. A pressure drop of 0.38 inH_2O per foot of packing will be used. Determine the diameter of the column for 70% recovery of CO_2. Choose a suitable plastic packing. The caustic solution has a viscosity of 0.9 cP and a specific gravity of 1.06 at the operating temperature. Check the wetting of the packing.

Solution.

Step 1. Establish the gas and liquid flows and estimate $(L/G)\sqrt{\rho_G/\rho_L}$. At the bottom of the column, the gas flow is 70,000 lb/h.

$$\text{Average molecular weight of gas} = \Sigma M_i x_i$$

$$= 44(0.1) + 28(0.8) + 32(0.1)$$

$$= 30$$

$$CO_2 \text{ in the gas stream} = \frac{70{,}000}{30}(0.1) = 233.3 \text{ lb}\cdot\text{mol/h}$$

For 70 percent recovery,

$$CO_2 \text{ absorbed} = 233.3(0.7)(44) = 7185 \text{ lb/h}$$

$$\text{Inert gas in feed} = \frac{70{,}000}{30}(1 - 0.1) = 2100 \text{ lb}\cdot\text{mol/h}$$

$$\text{NaOH conc. in feed solvent} = \frac{\frac{8}{40}}{\frac{8}{40} + \frac{92}{18}}$$

$$= 0.0377 \text{ mol fraction}$$

The vapor pressure of water at 176°F is 6.87 psia. The water vapor pressure of solution is 6.87(1 − 0.0377) = 6.61 psia. Since the pressure at the top of the column is 0.1 psig, assuming saturation by water vapor, water vapor in inert gas at the top of the column is

$$\frac{6.61}{14.8 - 6.61} = 0.807 \text{ lb}\cdot\text{mol/lb}\cdot\text{mol inert gas}$$

$$\text{Gas remaining after } CO_2 \text{ absorption} = 2100 + 233.3(0.3)$$
$$= 2170 \text{ lb}\cdot\text{mol/h}$$

$$\text{Water vapor carried by the gas} = 2170(0.807) = 1751.2 \text{ lb}\cdot\text{mol/h}$$

Top gas composition is listed in Table 6-5.

TABLE 6-5

Component	mol/h		Mole Fraction	
CO_2	233.3(0.3) =	70	0.018	
Water vapor		1751.2	0.447	
N_2	2100(8/9) =	1866.66	0.476	
O_2	2100(1/9) =	233.33	0.060	
Total		3921.19	1.001	≐ 1.00

The molecular weight of gas at top of the column is

$$0.018(44) + 18(0.447) + 28(0.476) + 32(0.06) = 24.09$$

The gas density at the top is

$$\frac{PM_W}{RT} = \frac{14.8(24.09)}{10.72(460 + 176)} = 0.052 \text{ lb/ft}^3$$

The gas flow at the top is

$$3921.19(24.09) = 94{,}461.5 \text{ lb/h}$$

$$\frac{L}{G}\sqrt{\frac{\rho_G}{\rho_L}} \text{ at top} = \frac{195{,}000}{94{,}461.5}\sqrt{\frac{0.052}{1.06(62.4)}} = 0.058$$

The liquid flow at the bottom of the column equals caustic flow in plus gas flow in minus gas flow out:

$$195{,}000 + 70{,}000 - 94{,}462 = 170{,}538 \text{ lb/h}$$

The gas density at the bottom of the column is

$$\frac{15.28(30)}{10.72(460 + 176)} = 0.067 \text{ lb/ft}^3$$

$$\frac{L}{G}\sqrt{\frac{\rho_G}{\rho_L}} \text{ at bottom of column} = \frac{170{,}538}{70{,}000}\sqrt{\frac{0.067}{1.06(62.4)}} = 0.078$$

$$\frac{L}{G}\sqrt{\frac{\rho_G}{\rho_L}}\ \text{average} = \frac{0.058 + 0.078}{2} = 0.07$$

Step 2. Select a $1\frac{1}{2}$-in plastic pall ring as packing,[3c] since

$$\text{Gas flow} = \frac{70{,}000}{0.067(60)} = 17{,}413\ \text{ft}^3/\text{min} > 2000\ \text{ft}^3/\text{min}$$

$$F_p = 32$$

Step 3. The specified pressure drop is 0.38 inH_2O per foot of packing.
Step 4. From Fig. 6-4 the ordinate is 1.0 at 0.38 inH_2O per foot. Then

$$G' = \left[\frac{(\text{ordinate})(\rho_G)(\rho_L - \rho_G)}{CF_p(\nu)^{0.1}}\right]^{1/2} = \left[\frac{1.0(0.063)(66.14 - 0.063)}{32(0.9/1.06)^{0.1}}\right]^{1/2}$$

$$= 0.3636\ \text{lb/ft}^2\cdot\text{s}$$

$$G_T = \text{average vapor flow rate} = \frac{94{,}462 + 70{,}000}{2} = 82{,}231\ \text{lb/h}$$

$$\text{Column diameter} = 0.0188\left(\frac{G_T}{G'}\right)^{1/2} = 0.0188\left(\frac{82{,}231}{0.3636}\right)^{1/2}$$

$$= 8.94\ \text{ft} \qquad \text{say 9 ft}$$

Packing size is well below $\frac{1}{10}$ of the column diameter.

$$\text{Minimum wetting rate} \doteq 0.1A_p$$

$$= 0.1(39) = 3.9\ \text{gpm/ft}^2$$

$$\text{Actual wetting rate} = \frac{187{,}269(4)}{\pi(9)^2(1.06)(500)} = 5.6\ \text{gpm/ft}^2$$

The selected packing will have sufficient wetting.

Example 6-2. Estimate the packing height for the problem in Example 6-1. The overall mass-transfer coefficient K_Ga may be obtained from Fig. 18-86 of Perry.[3a] K_Ga is proportional to $L^{0.28}$. The temperature correction factor for K_Ga may be obtained from the following data:

At 52°C and $L = 3000$, $K_Ga = 18$ lb · mol/h · ft^3 · atm.
At 55°C and $L = 3000$, $K_Ga = 20$ lb · mol/h · ft^3 · atm.
L is liquid mass velocity in lb/h · ft^2.

Solution.

Step 1. Estimate the percent conversion of NaOH to Na_2CO_3. From Example 6-1, CO_2 absorbed = 7185.6 lb/h and

$$2NaOH + CO_2 \longrightarrow Na_2CO_3 + H_2O$$

$$44 \text{ lb } CO_2 \longrightarrow 2(40) \text{ lb NaOH}$$

$$7185.6 \text{ lb } CO_2 = \frac{2(40)(7185.6)}{44} = 13{,}064.7 \text{ lb NaOH}$$

$$\text{NaOH in caustic} = 195{,}000(0.08) = 15{,}600 \text{ lb/h}$$

$$\text{Conversion} = \frac{13{,}064.7}{15{,}600} = 83.7 \text{ percent}$$

Step 2. Estimation of K_Ga. It is to be noted that K_Ga for CO_2 absorption depends on the temperature, percent conversion, caustic normality, and superficial liquid mass velocity; it is independent of gas mass velocity.

$$K_Ga \text{ of the system} = C_1C_2C_3(K_Ga)_R$$

where $(K_Ga)_R = K_Ga$ at the reference point
C_1 = correction for normality and conversion
C_2 = correction for temperature
C_3 = correction for superficial mass velocity

a. Estimate K_Ga at the top of the tower. From Fig. 18-86 of Ref. 3*a*, the K_Ga value for the $1\frac{1}{2}$-in plastic pall ring may be obtained as follows:

$$(K_Ga)_R = 2.2 \text{ lb}\cdot\text{mol/h}\cdot\text{ft}^3\cdot\text{atm}$$

at the reference point. Conditions at the reference point are:

Caustic solution	1 *N* NaOH
Conversion	25 percent
Temperature	75°F
Superficial liquid mass velocity	5000 lb/h · ft^2

In this problem, normality is the number of gram moles in 1000 ml of solution. The basis is 100 g solution.

$$\text{Volume} = (100\text{ g})\left(\frac{1}{1.06\text{ g/ml}}\right) = 94.34\text{ ml}$$

$$\text{Number of g}\cdot\text{mol} = \frac{8}{40} = 0.2$$

$$\text{Normality} = \frac{0.2(1000)}{94.34} = 2.1$$

The conversion at top is 0 percent and the temperature is 176°F. At 0 conversion and 176°F (80°C) the correction factor for normality and conversion C_1 is 1.5 (from Fig. 18-86 of Ref. 3*a*). The correction for the temperature from Fig. 18-92 of Ref. 3*b* is found as follows:

$$\text{Superficial liquid mass velocity} = \frac{195{,}000}{\frac{1}{4}\pi(9^2)}$$
$$= 3065\text{ lb/h}\cdot\text{ft}^2$$

At 52°C and $L' = 3000\text{ lb/h}\cdot\text{ft}^2$, $K_Ga = 18$. At 55°C and $L' = 3000\text{ lb/h}\cdot\text{ft}^2$, $K_Ga = 20$. Rate of increase of K_Ga is

$$\frac{20-18}{55-52} = 0.67/°\text{C}$$

C_2 is the correction for temperature (176°F = 80°C):

$$C_2 = \frac{18 + 0.67(80-52)}{18} = 2.04$$

C_3 is the correction for superficial mass velocity:

$$C_3 = \left(\frac{3065}{5000}\right)^{0.28} = 0.87$$

Hence, at the top of the tower

$$K_Ga = (2.2)(1.5)(2.04)(0.87)$$
$$= 5.86\text{ lb}\cdot\text{mol/h}\cdot\text{ft}^3\cdot\text{atm}$$

b. K_Ga at bottom of the tower. Correction factor for 2.1 N and 83.7 percent conversion is 0.33. At the bottom

$$K_Ga = 2.2(0.33)(2.04)(0.87)$$
$$= 1.29 \text{ lb} \cdot \text{mol/h} \cdot \text{ft}^3 \cdot \text{atm}$$

Taking the logarithmic mean,

$$K_Ga = \frac{5.86 - 1.29}{\ln (5.86/1.29)} = 3.02 \text{ lb} \cdot \text{mol/h} \cdot \text{ft}^3 \cdot \text{atm}$$

Step 3. Estimation of G, y_1, y_2, and $(1 - y)_{lm}$ follows.
At bottom of tower,

$$\text{Total gas flow} = \frac{70{,}000}{30} = 2333.33 \text{ lb} \cdot \text{mol/h}$$

$$y_1 = 0.1 \frac{\text{lb} \cdot \text{mol solute}}{\text{lb} \cdot \text{mol gas}}$$

$$(1 - y)_{lm} = -\frac{y_1}{\ln (1 - y_1)} = -\frac{0.1}{\ln 0.9} = 0.949$$

At top of tower, the inert gas in feed is

$$2333.33(1 - 0.1) \doteq 2100 \text{ lb} \cdot \text{mol/h}$$

NaOH concentration in feed solvent is

$$\frac{\frac{8}{40}}{\frac{8}{40} + \frac{92}{18}} = 0.0377$$

The vapor pressure of water at 176°F is 6.87 psia. The vapor pressure of the solution is 6.87(1 − 0.0377) = 6.61 psia. The water vapor at top of the tower is

$$\frac{6.61}{14.8 - 6.61} = 0.807 \text{ lb} \cdot \text{mol/lb} \cdot \text{mol inert gas}$$

The inert gas in feed is 2100 lb · mol/h, and the gas remaining after absorption of CO_2 is

$$2100 + 2333.33(0.1)(0.3) = 2170 \text{ lb} \cdot \text{mol/h}$$

The water vapor carried by the gas is

$$0.807(2170) = 1751.20 \text{ lb} \cdot \text{mol/h}$$

The total gas flow at top of tower is

$$2170 + 1751.2 = 3921.2 \text{ lb} \cdot \text{mol/h}$$

Average gas flow is

$$\tfrac{1}{2}(3921.2 + 2333.33) = 3127.27 \text{ lb} \cdot \text{mol/h}$$

$$G = \frac{3127.27}{\frac{1}{4}\pi(9)^2} = 49.16 \text{ lb} \cdot \text{mol/h} \cdot \text{ft}^2$$

The amount of CO_2 in the exit gas is

$$2333.33(0.1)(0.3) = 70 \text{ lb} \cdot \text{mol/h}$$

$$y_2 = \frac{70}{3921.2} = 0.0179$$

$$(1 - y)_{lm} = -\frac{0.0179}{\ln(1 - 0.0179)} = 0.991$$

$$\text{Average } (1 - y)_{lm} = \tfrac{1}{2}(0.949 + 0.991) = 0.97$$

$$N_{OG} = \ln \frac{\ln(1 - y_1)}{\ln(1 - y_2)} = \ln \frac{\ln(1 - 0.1)}{\ln(1 - 0.0179)} = 1.764$$

$$H_{OG} = \frac{G}{K_G a P_t (1 - y)_{lm}} = \frac{49.16}{3.02(1)(0.97)} = 16.78 \text{ ft}$$

$$\text{Total packed height} = N_{OG} H_{OG} = 1.764(16.78) = 29.6 \qquad \text{say 30 ft}$$

Example 6-3. The vent gas from a reactor containing nitrogen and a solute gas of molecular weight of 80 is to be freed of the solute gas to meet the environmental standard by countercurrent absorption with a solvent in a packed column. The overhead gas from the packed column is vented to the atmosphere. Available data are as follows. Vent gas flow is 0.2 m^3/s at 0.11 MPa and 300 K. The concentration of the solute gas in the inert from the reactor is 4.5% by volume.

The concentration of the solute gas in the vent from the absorption column should not be more than 0.2% by volume. The equilibrium curve of the solute gas in the absorbing solvent is shown in Fig. 6-5*a*. Pure solvent of molecular weight 240 and viscosity 0.9 mPa · s is used, and the design solvent flow rate will be 1.5 times the minimum. The

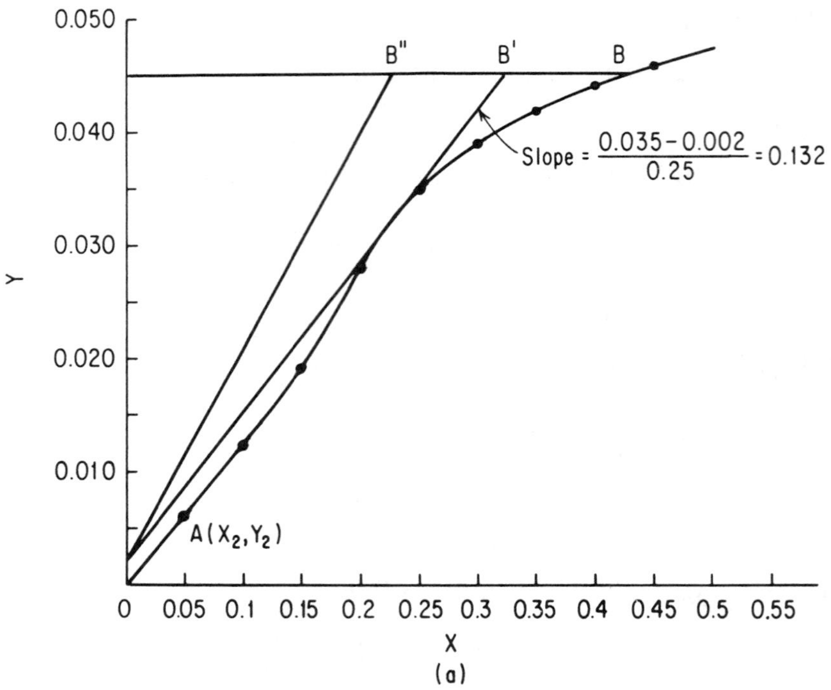

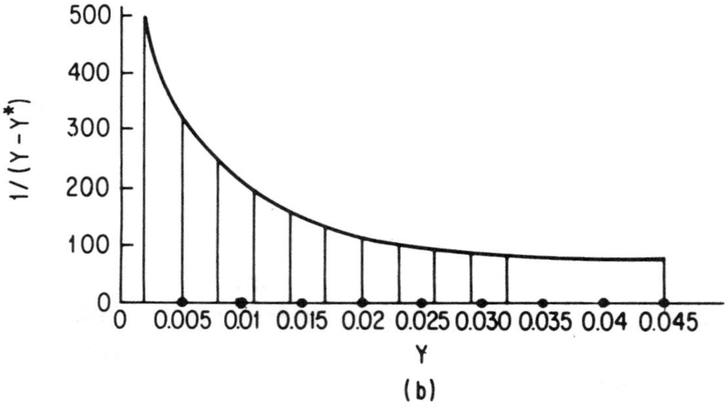

Figure 6-5. (*a*) Minimum and design solvent flow rates (Example 6-4); (*b*) integration for Example 6-4.

packing is 25 mm polypropylene Pall ring. The specific gravity of the solvent is 1.1. The overall mass-transfer coefficient is given by

$$K_G a = 25\ G'^{0.7} L'^{0.07} \qquad \text{kg/s} \cdot \text{m}^3 \cdot \text{MPa}$$

where $G' = \text{kg/s} \cdot \text{m}^2$ and $L' = \text{kg/s} \cdot \text{m}^2$. The diameter of the column is 387 mm and has a packing height of 3 m.

Determine whether the column is suitable for the service.

Solution.

Step 1. First determine whether the column meets the pressure-drop requirement. Base calculations on the inlet conditions. The molecular weight of the inlet gas is

$$M_w = 0.045(80) + 0.955(28) = 30.34 \text{ kg/kg} \cdot \text{mol}$$

$$\text{Gas density} = \frac{PM_w}{RT} = \frac{0.11 \text{ MPa} \times 10^6(30.34 \text{ kg/kg} \cdot \text{mol})}{(8314\text{Pa} \cdot \text{m}^3/\text{kg} \cdot \text{mol} \cdot \text{K})300 \text{ K}}$$

$$= 1.338 \text{ kg/m}^3$$

$$\text{Liquid density} = 1.1 \times 10^3 \text{ kg/m}^3$$

$$\text{Gas flow rate} = (0.2 \text{ m}^3/\text{s})(1.338 \text{ kg/m}^3) = 0.2676 \text{ kg/s}$$

$$= 0.00882 \text{ kg} \cdot \text{mol/s}$$

$$\text{Gas mass velocity} = \frac{0.2676}{0.785(0.387)^2} = 2.276 \text{ kg/m}^2 \cdot \text{s}$$

Following the procedure outlined in connection with Fig. 6-2*a*, the solvent flow rate is determined as follows. Since a pure solvent is used, $X_2 = 0$ and

$$y_1 = 0.045 \qquad Y_1 = \frac{0.045}{1 - 0.045} = 0.0471$$

$$G_2 = 0.00882(1 - 0.045) = 0.00842 \text{ kg} \cdot \text{mol/s}$$

$$y_2 = 0.002 \text{ maximum} \qquad \text{or} \qquad Y_2 = \frac{y_2}{1 - y_2} = \frac{0.002}{1 - 0.002} \doteq 0.002$$

(X_2, Y_2) is located in Fig. 6-5*a*.

$$\text{Minimum solvent flow rate} = (\text{slope of the line } AB)G_S$$

$$= 0.132(0.00842) = 0.001111 \text{ kg} \cdot \text{mol/s}$$

$$= (0.001111 \text{ kg} \cdot \text{mol/s})(240 \text{ kg/kg} \cdot \text{mol})$$

$$= 0.27 \text{ kg/s}$$

$$\text{Actual solvent flow rate} = 0.27(1.5) = 0.405 \text{ kg/s}$$

$$\text{Solvent mass velocity} = \frac{0.405}{0.785(0.387)^2} = 3.445 \text{ kg/m}^2 \cdot \text{s}$$

$$\text{Liquid viscosity} = 0.9 \text{ mPa} \cdot \text{s} = 0.9 \times 10^{-3} \text{ (kg/m} \cdot \text{s}^2\text{)(s)}$$

$$= 9 \times 10^{-4} \text{ kg/m} \cdot \text{s}$$

$$\text{Kinematic liquid viscosity} = \frac{9 \times 10^{-4} \text{ kg/m} \cdot \text{s}}{1.1 \times 10^3 \text{ kg/m}^3} = 8.182 \times 10^{-7} \text{ m}^2\text{/s}$$

Referring to Fig. 6-4,

$$\frac{L}{G}\left(\frac{\rho_G}{\rho_L}\right)^{0.5} = \frac{3.445}{2.276}\left(\frac{1.338}{1.1 \times 10^3}\right)^{0.5} = 0.053$$

$$\frac{CG'^2 F_p \nu^{0.1}}{\rho_G(\rho_L - \rho_G)} = \frac{42.84(2.276)^2(52)(8.182)^{0.1} \times (10^{-7})^{0.1}}{1.338(1100 - 1.338)} = 1.93$$

Locating 0.053 (abscissa) and 1.93 (ordinate) on Fig. 6-4, the pressure drop is read as

$$\Delta P = 612 \text{ Pa/m of packing}$$

$$\text{Total pressure drop} = (612 \text{ Pa/m})(3 \text{ m}) = 1836 \text{ Pa} = 0.001836 \text{ MPa}$$

The allowable pressure drop including the line loss is 0.01 MPa. Hence, the column would work as far as the pressure drop and flooding are concerned.

Step 2. Determine the number of transfer units using Eq. (6-13).

$$N_{OG} = \int_{Y_2}^{Y_1} \frac{dY}{Y - Y^*} - \frac{1}{2} \ln \frac{1 + Y_1}{1 + Y_2}$$

The first half of the right-hand side requires graphical integration. Locate the operating line with a slope of 0.132(1.5) = 0.198 and draw the operating line AB'' (Fig. 6-5*a*). From the operating line, it may be seen that X varies from $X_2 = 0$ to $X = 0.228$ mol/mol. So between $X_2 = 0$ and $X_1 = 0.228$, assume values of X, read the values of Y from the operating line and Y^* from the equilibrium line, and tabulate as in Table 6-6. Plot $1/(Y - Y^*)$ versus Y to find the integral of the above equation (Fig. 6-5*b*). The two right-hand columns of the table will be used to estimate H_{OG}. The area under the curve, partly by Simpson's rule (Fig. 6-5*b*), is 6.162.

$$N_{OG} = 6.162 - \frac{1}{2} \ln \frac{1 + 0.0471}{1 + 0.002} = 6.14$$

TABLE 6-6 Calculations for Evaluation of Integral $\int_{Y_2}^{Y_1} \frac{dY}{Y - Y^*}$

X	Y	Y^*	$1/(Y - Y^*)$	$\frac{1}{2}(Y + Y^*)$	$\frac{1}{2}(Y + Y^*)_{\text{average}}$
$X_2 = 0$	0.0020	0	500	0.001	
0.05	0.01140	0.006	185.2	0.0087	
0.1	0.0210	0.012	111.1	0.0165	
0.15	0.0304	0.019	87.7	0.0247	0.0206
0.2	0.0398	0.028	84.7	0.0339	
$X_1 = 0.228$	0.045	0.0324	79.4	0.0387	

Step 3. Determine the height of transfer unit from Eq. (6-25):

$$H_{OG} = \frac{G}{K_G a P_t (1 - y)_{lm}}$$

$$(1 - y)_{lm} \doteq (1 - Y)_{lm} = \frac{(1 - Y) + (1 - Y^*)}{2} = 1 - \frac{Y + Y^*}{2}$$

$$= 1 - \frac{0.0206}{2} = 0.990$$

$$K_G a = 25\ G'^{0.7} L'^{0.07}$$

$$G' = 2.276\ \text{kg/m}^2 \cdot \text{s}$$

$$L = 3.445\ \text{kg/m}^2 \cdot \text{s}$$

$$K_G a = 25(2.276)^{0.7}(3.443)^{0.07} = 48.48\ \text{kg/m}^3 \cdot \text{s} \cdot \text{MPa}$$

$$H_{OG} = \frac{G'}{K_G a P_t (1 - y)_{lm}} = \frac{2.276}{48.48(0.11)(0.99)} = 0.431\ \text{m}$$

Height of packing required = $N_{OG} H_{OG} = 6.14(0.431) = 2.65$ m

Since the available packed height is 3 m, the column would be adequate for the service specified in the problem.

Example 6-4. A gas stream containing 0.05 mol fraction of the solute gas is to be countercurrently scrubbed in a packed column to reduce the solute content to 0.0001 mol fraction. Estimate the packing height. Use these data: gas mass velocity = 10 lb · mol/h · ft^2; liquid mass velocity =

400 lb · mol/h · ft^2; overall mass-transfer coefficient based on overall driving force in liquid phase = 120 lb · mol/ft^3 · h · mol fraction. The equilibrium solubility relationship is given by $y = 20x$ where y and x are mole fractions of the solute in the vapor phase and liquid phase, respectively. The solvent entering the column is pure.

Solution. Since the principal resistance is in the liquid phase,

$$N_{OL} = \int_{x_2}^{x_1} \frac{dx}{x^* - x} + \frac{1}{2} \ln \frac{1 - x_2}{1 - x_1}$$

Equilibrium line: $y = 20x^*$ or $x^* = 0.05y$

Operating line: $G(y - y_2) = L(x - x_2)$

or

$$y = \frac{L}{G}(x - x_2) + y_2$$

$$= \frac{L}{G}x + y_2 \qquad \text{since } x_2 = 0$$

$$= \frac{400}{10}x + 0.0001$$

$$= 40x + 0.0001$$

$$x^* = 0.05y = 0.05(40x + 0.0001) = 2x + 0.000005$$

$$x^* - x = x + 0.000005$$

Overall material balance:

$$G(y_1 - y_2) = L(x_1 - x_2) = 10(0.05 - 0.0001) = 400x_1$$

$$x_1 = 0.00125$$

$$N_{OL} = \int_{x_2=0}^{x_1=0.00125} \frac{dx}{x + 0.000005} + \frac{1}{2} \ln \frac{1 - x_2}{1 - x_1}$$

$$= \ln \frac{x_1 + 0.000005}{x_2 + 0.000005} + \frac{1}{2} \ln \frac{1 - x_2}{1 - x_1}$$

$$= 5.33$$

where $x_1 = 0.00125$ and $x_2 = 0$.

$$H_{OL} = \frac{L}{K_x a(1 - x)_{lm}} \doteq \frac{L}{K_x a} = \frac{400}{120} = 3.33$$

$$\text{Height of packing} = 5.53(3.33) = 18.4 \text{ ft}$$

Example 6-5. The capacity of a column is to be increased by 55 percent by replacing the existing 1-in metal Raschig rings ($F_P = 115$) with some other packing without significantly changing the percent flooding. The end compositions, L/G ratio, pressure, temperature, etc., remain unchanged. Select the packing.

Solution. The new packing factor required is

$$F_{P2} = \left(\frac{G_1}{G_2}\right)^2 F_{P1} = \left(\frac{G_1}{1.55G_1}\right)^2 (115) = 47.9$$

From Table 18-5 of Ref. 3, select 1-in metallic Pall rings.

Example 6-6. During the unloading of a tank car into a storage tank, air containing 0.02 mol fraction of a water soluble gas comes out of the storage tank. This air is to be scrubbed with water in a countercurrent packed column to reduce the concentration of the gas to 0.0001 mol fraction. The following data are available: gas flow rate = 1000 scfm/ft^2 tower cross section; pure water rate = 1500 lb/h · ft^2 tower cross section; equilibrium relationship $y^* = 1.8x$. y^* and x are mole fractions of the solute in vapor and liquid phase, respectively. $K_y a$ = 2 lb · mol/ft^3 · h · mol fraction. Determine the packing height.

Solution. By Eq. (6-16),

$$N_{OG} = \frac{1}{1 - mG/L} \ln\left[\frac{(1 - mG/L)(y_1 - mx_2)}{y_2 - mx_2} + \frac{mG}{L}\right]$$

$$m = 1.8 \qquad G = \frac{1000}{359} = 2.79 \text{ lb}\cdot\text{mol/h}\cdot\text{ft}^2$$

$$y_1 = 0.02 \qquad y_2 = 0.0001 \qquad x_2 = 0$$

$$L = \frac{1500}{18} = 83.33 \text{ lb}\cdot\text{mol/h}\cdot\text{ft}^2$$

$$\frac{mG}{L} = \frac{1.8(2.79)}{83.33} = 0.0603 \qquad 1 - \frac{mG}{L} = 1 - 0.0603 = 0.9397$$

$$N_{OG} = \frac{1}{0.9397} \ln\left[\frac{0.9397(0.02)}{0.0001} + 0.0603\right]$$

$$= 1.0642(5.236) = 5.572$$

$$H_{OG} = \frac{G}{K_y a} = \frac{2.79}{2} = 1.395$$

$$\text{Height} = N_{OG}H_{OG} = 5.572(1.395) = 7.77 \text{ ft}$$

Example 6-7. A gas mixture containing 0.01 mol fraction of a solute is to be scrubbed at a pressure of 1 atm in a packed column. The gas flow rate is 200 lb · mol/h and the pure solvent flow rate is 200 lb · mol/h. The tower is 5 ft in diameter and 10 ft high. The overall height of the transfer unit for the operating condition and packing used is given by

$$H_{OG} = 5\frac{G^{0.1}}{L^{0.4}}$$

where G and L are gas and liquid flows, lb/h · ft^2. The molecular weights of the gas and liquid are 29 and 18, respectively. The equilibrium relationship is given by $p = x$, where p is the partial pressure of the solute in atmosphere and x is mol fraction of the solute in the solution.

Estimate: (**a**) Percent recovery for countercurrent absorption, (**b**) percent recovery for cocurrent absorption.

Solution.

Step 1. Estimate H_{OG}, $H_{OG} = 5G^{0.1}/L^{0.4}$.

$$G = \frac{200(29)}{\frac{1}{4}\pi(5^2)} = 295.39 \text{ lb/h} \cdot \text{ft}^2$$

$$L = \frac{200(18)}{\frac{1}{4}\pi(5)^2} = 183.35 \text{ lb/h} \cdot \text{ft}^2$$

$$H_{OG} = 5(295.39)^{0.1}(183.35)^{-0.4} = 1.098$$

Step 2. Estimate N_{OG}.

$$N_{OG} = \frac{\text{height}}{H_{OG}} = \frac{10}{1.098} = 9.1$$

Step 3. Estimate N_{OG} using analytical expression. Express the equilibrium line in the form $y^* = mx$.

Equilibrium line: $p = x$

$$y^* = \frac{p}{P_t} = \frac{x}{P_t} = x \qquad \text{since } P_t = 1 \text{ atm}$$

Slope of equilibrium line = m = 1. So the operating and equilibrium lines are parallel.

a. *Countercurrent Flow.*

$$N_{OG} = \frac{y_1 - y_2}{y_2 - mx_2} = \frac{y_1 - y_2}{y_2} \qquad \text{since } x_2 = 0$$

$$= \frac{y_1}{y_2} - 1$$

From Step 2,

$$\frac{y_1}{y_2} - 1 = 9.1 \qquad y_2 = 0.099y_1$$

$$\text{Recovery} = \frac{G(y_1 - y_2)}{Gy_1}100 = \frac{0.01 - 0.01(0.099)}{0.01}100$$

$$= 90.1 \text{ percent}$$

b. *Cocurrent Flow.*

$$N_{OG} = \frac{1}{1 + mG/L} \ln \frac{y_1 - mx_1}{y_2 - mx_2} = 9.1$$

$$m = \frac{L}{G} = 1 \qquad y_1 = 0.01 \qquad x_1 = 0$$

$$\frac{1}{1 + 1} \ln \frac{0.01}{y_2 - x_2} = 9.1$$

$$\frac{0.01}{y_2 - x_2} = e^{2(9.1)}$$

$$y_2 = x_2$$

Equation of the operating line:

$$x_2 = x_1 + \frac{G}{L}(y_1 - y_2) = y_1 - y_2 = 0.01 - y_2$$

From these two equations,

$$y_2 = 0.01 - y_2 \qquad y_2 = 0.005$$

$$\text{Recovery} = \frac{G(y_1 - y_2)}{Gy_1}100 = \frac{0.01 - 0.005}{0.01}100 = 50 \text{ percent}$$

References

1. J. S. Eckert, *Chem. Eng.*, vol. 82, April 14, 1975, pp. 70–76.

2. G. A. Morris and J. Jackson, *Absorption Towers*, Butterworth, London, 1953.

3. *Chemical Engineers' Handbook*, 5th ed., R. H. Perry (ed.), McGraw-Hill, New York, 1973: (*a*) p. 18-45; (*b*) p. 18-47, (*c*) p. 18-22. (*d*) p. 18-24.

4. M. J. Dolan, Norton Co., Akron, Ohio, private communication, Sept. 28, 1982.

5. Ralf F. Strigle, Jr., Random Packings and Packed Towers, Design and Applications, Gulf Publishing Company, Houston.

Chapter

7

Leaching

Leaching calculations can be done by (1) stage-to-stage algebraic calculations using the material balance and equilibrium relationship and (2) graphical methods such as the use of the right triangular, Ponchon-Savarit, or McCabe-Thiele diagram. The multistage countercurrent leaching is the most important and should be reviewed in detail.

Equilibrium Relationship in Leaching

The assumptions generally made to simplify the leaching calculations are as follows: (1) The system consists of three components, a solute, a solvent and an inert solid; (2) the inert solid (solute-free) is insoluble in the solvent, and the flow rate of the inerts from stage to stage is constant; and (3) in the absence of adsorption of the solute by the inerts, an equilibrium is attained on complete solution of the solute in the solvent. The nonequilibrium condition is taken into consideration by an efficiency factor.

If the conditions of equilibrium are met, the concentration of the solution leaving a stage is the same as the concentration of the solution adhering to the inerts. The equilibrium relationship is, therefore, $x_e = y_e$.

Representation of Equilibrium Data

The published data on a particular system are generally given in terms of the amount of the solution retained by the inerts and the extract composition. These can be used to construct the triangular or Ponchon-Savarit diagram. Example 7-1 illustrates the method of the calculation of the equilibrium data for the triangular diagram.

Example 7-1. The experimental data on the retention of oil by livers are given in the first two columns of Table 7-1. Construct the right triangular diagram for the system.

Solution. For smoothing out, the data can be plotted as in Fig. 7-1 using the overflow and underflow compositions in the table. The mass

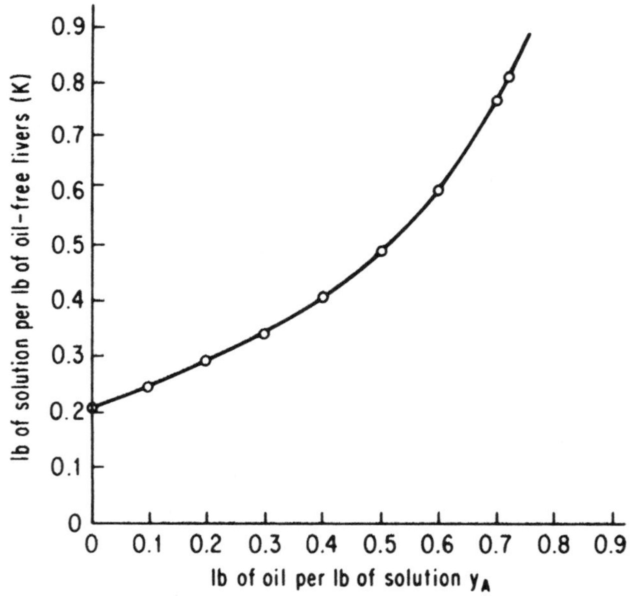

Figure 7-1. Retention of solution by livers as a function of solution concentration.

TABLE 7-1 **Retention of Oil by Livers**[1]

Experimental Data		Calculated Data		
		Underflow Compositions		
lb of Liver Oil in 1 lb of solution (Overflow Composition) y_A	lb of Solution Retained by 1 lb of Oil-Free Livers K	$x_A = y_A K/(K+1)$	$x_S = (1-y_A)K/(K+1)$	$x_I = 1/(K+1) = 1-(x_A+x_S)$
0.00	0.205	0	0.170	0.830
0.10	0.242	0.020	0.175	0.805
0.20	0.286	0.044	0.178	0.778
0.30	0.339	0.076	0.177	0.747
0.40	0.405	0.115	0.173	0.712
0.50	0.489	0.164	0.164	0.672
0.60	0.600	0.225	0.150	0.625
0.65	0.672	0.261	0.141	0.598
0.70	0.765	0.3034	0.130	0.567
0.72	0.810	0.3222	0.125	0.553

fractions of the solute, the solvent, and the inerts are calculated from these data by the following relations:

$$x_A = \frac{y_A K}{K+1} \qquad x_S = \frac{(1-y_A)K}{K+1} \qquad x_I = \frac{1}{K+1} = 1 - (x_A + x_S)$$

where x_A = mass fraction of the solute
x_S = mass fraction of the solvent
x_I = mass fraction of the inert solid
K = lb of the solution retained by 1 lb of oil-free livers
A, S, I = solute, solvent, and inert solid, respectively

The calculated mass fractions are given in the last three columns of Table 7-1. A right-triangular diagram is prepared as in Fig. 7-2.

Example 7-2. Prepare a Ponchon-Savarit diagram for the system in Example 7-1.

Solution. To prepare the Ponchon-Savarit diagram, the data given in Table 7-2 are first calculated. The Y coordinate is the ratio of the inerts to the solution retained by the inerts; the X coordinate is the solute fraction in the solution, or $y_A = x_A$.

TABLE 7-2 Calculation of Coordinates for the Ponchon-Savarit Diagram

$x_A = \frac{A}{A+S} = y_A$	$Y_I = \frac{I}{A+S}$
0.00	4.88
0.10	4.13
0.20	3.50
0.30	2.95
0.40	2.47
0.50	2.05
0.60	1.67
0.65	1.49
0.70	1.31
0.72	1.24

Note that $Y_I = 1/K$, where K is obtained from Table 7-1. The data are plotted in Fig. 7-3.

In the right-triangular diagram, the tie lines pass through 0 ($x_I = 1.0$), and in the Ponchon-Savarit diagram they are vertical for the ideal systems where $x_e = y_e$. The equilibrium data can be represented on an equilateral triangle also. However, because of its inconvenience, the right-triangular or Ponchon-Savarit diagram, is preferred.

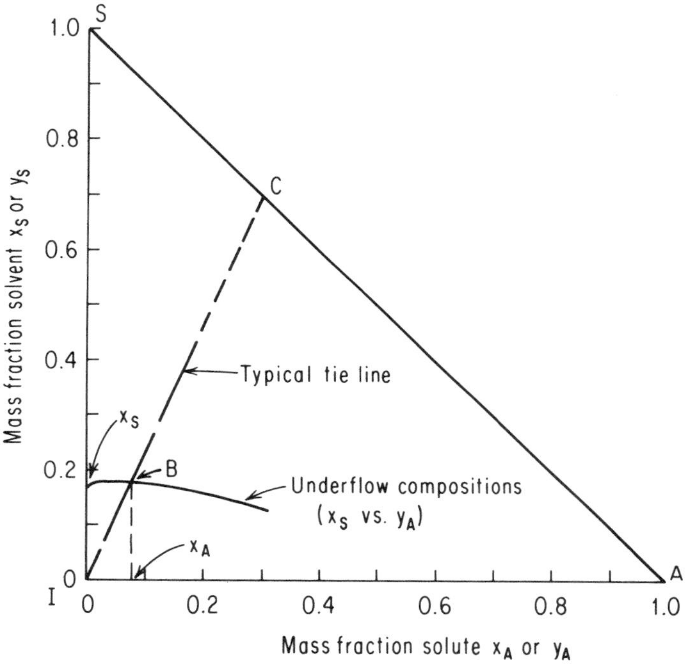

Figure 7-2. Oil extraction data on right-triangular diagram (Example 7-1). (*Note:* As an example, to read the composition of the underflow represented by point B, lines are drawn perpendicular to X and Y axes to give $x_S = 0.176$ and $x_A = 0.076$; then $x_I = 1 - 0.176 - 0.076 = 0.748$. To determine the composition of the overflow in equilibrium with the underflow represented by point B, the tie line is drawn through point B to intersect the hypotenuse SA in C. The point C gives the required overflow composition which is: $y_A = 0.3$, $y_S = 0.7$.)

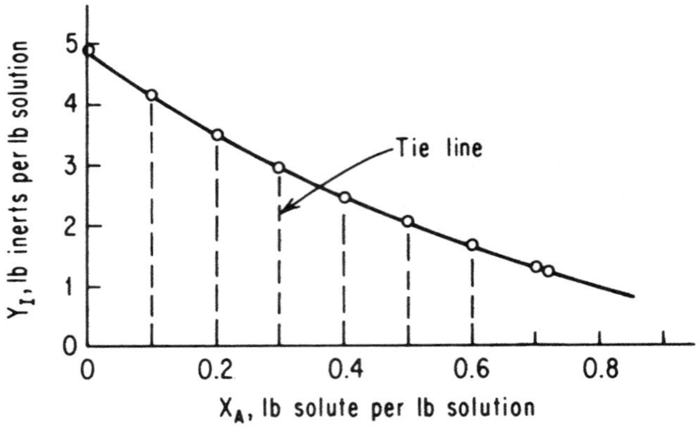

Figure 7-3. Oil extraction data on Ponchon-Savarit diagram (Example 7-2). $X_A = A/(A+S)$; $Y_I = I/(A+S)$.

Calculation of Equilibrium Stages

Consider a countercurrent leaching battery as shown in Fig. 7-4. The notation used is as follows:

V = overflow solution (no inerts in overflow)

L = underflow solution exclusive of inerts

R = total underflow including the solution and inerts

x = solute concentration in the solution retained by the inerts, mass fraction

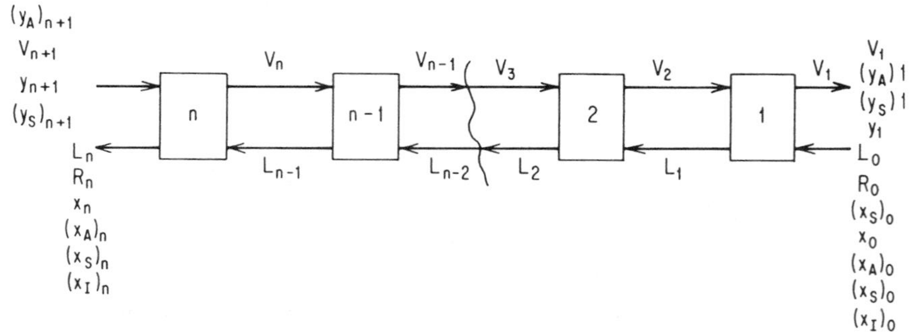

Figure 7-4. Multistage countercurrent extraction system.

y = solute concentration in the extract phase, mass fraction

y_S = mass fraction of the solvent in the extract solution

y_A = mass fraction of the solute in the extract solution

Since no inerts are present in the extract phase, $y = y_A$. The number of equilibrium stages can be calculated by the algebraic method making stage-to-stage calculations, using mathematical equations in some cases, or using graphical methods. The inerts are assumed to be constant from stage to stage and insoluble in the solvent. Also, no inerts are present in the extract (usual assumption) or overflow solution.

Material Balance Equations

In terms of the overflow and underflow solutions (exclusive of inerts), the overall material balance can be written as

$$V_{n+1} + L_0 = V_1 + L_n \tag{7-1}$$

and the solute balance as

$$V_{n+1}y_{n+1} + L_0x_0 = V_1y_1 + L_nx_n \tag{7-2}$$

In establishing the above equations, the inerts are assumed constant from stage to stage. In terms of the total overflow and underflow compositions, the overall material balance is

$$V_{n+1} + R_0 = V_1 + R_n \tag{7-3}$$

$$V_{n+1}(y_A)_{n+1} + R_0(x_A)_0 = V_1(y_A)_1 + R_n(x_A)_n \tag{7-4}$$

and the operating line equation is given by

$$y_{n+1} = \frac{1}{1 + (V_1 - L_0)/L_n}x_n + \frac{V_1y_1 - L_0x_0}{L_n + L_1 - L_0} \tag{7-5}$$

When the solution retained by the inerts is constant, both the underflow L_n and overflow V_n are constant and the equation of the operating line is a straight line. Since the equilibrium line is also straight, the number of the stages is given by

$$N = \frac{\log [(y_{n+1} - y_n^*)/(y_1 - y_1^*)]}{\log [(y_{n+1} - y_1)/(y_n^* - y_1^*)]} \tag{7-6}$$

Since in leaching, $y_1^* = x_0$ and $y_n^* = x_n$, Eq. (7-6) becomes

$$N = \frac{\log [(y_{n+1} - x_n)/(y_1 - x_0)]}{\log [(y_{n+1} - y_1)/(x_n - x_0)]} \tag{7-7}$$

Equation (7-7) cannot be used for the entire extraction battery if L_0 differs from $L_1, L_2, \ldots, L_n$ (underflows within the system). In this case, the compositions of all the streams entering and leaving the first stage are separately calculated by a material balance, and then Eq. (7-7) is applied to the remaining cascade.

The example below illustrates the method of solution of the problems when the solution retention by the inerts is constant.

Example 7-3. A countercurrent extraction system is to treat 100 tons/h of sliced sugar beets with fresh water as solvent. Analysis of the beets is as follows: water 48%, sugar 12%, with balance pulp. If 97% sugar is to be recovered and the extract phase leaving the system is to contain 15% sugar, determine the number of cells required if each ton of the dry pulp retains 3 tons of solution.

Solution. Basis of calculation: 100 tons of fresh sliced beets.

Sugar in beets	12 tons
Water in beets	48 tons
Pulp in beets	40 tons

Since 97 percent of the sugar is to be recovered, the sugar in the final extracted solution is

$$0.97(12) = 11.64 \text{ tons}$$

Water in the final extract solution = (0.85/0.15)(11.64) = 65.96 tons, and

$$V_1 = 11.64 + 65.96 = 77.6 \text{ tons}$$

Each ton of pulp retains 3 tons of solution. Hence the solution underflow = 3(40) = 120 tons.

$$L_1 = L_2 \ldots L_n = 120 \text{ tons} \qquad \text{but } L_0 = 60 \text{ tons}$$

(Note that inert dry pulp is excluded.) Therefore, first complete the calculations on the first stage. By overall balance,

$$L_0 + V_{n+1} = V_1 + L_n \qquad \text{or} \qquad 60 + V_{n+1} = 77.6 + 120$$

$$V_{n+1} = 137.6 \text{ tons}$$

For material balance over first stage,

$$L_1 = 120 \text{ tons solution} \qquad x_1 = y_1 = 0.15$$

$$L_1 + V_1 = L_0 + V_2$$

or

$$120 + 77.6 = 60 + V_2$$

$$V_2 = 137.6 \text{ tons}$$

By solute balance over first stage,

$$137.6(y_2) + 12 = 11.64 + 120(0.15)$$

$$y_2 = 0.1282$$

Now, since for the remaining $(N - 1)$ stages both the underflow and overflow solutions are constant, Eq. (7-11) can be applied as follows.

$$N - 1 = \frac{\log [(y_{n+1} - x_n)/(y_2 - x_1)]}{\log [(y_{n+1} - y_2)/(x_n - x_1)]}$$

Note the replacement of y_2 and x_1 for y_1 and x_0, respectively, because the first stage is dealt with separately.

The substitution of the values in the above equation gives

$$N - 1 = \frac{\log \left[\left(0 - \frac{0.36}{120} \right) \Big/ (0.1282 - 0.15) \right]}{\log \left[(0 - 0.1282) \Big/ \left(\frac{0.36}{120} - 0.15 \right) \right]}$$

$$N - 1 = 14.5 \qquad \text{and} \qquad N = 15.5 \text{ cells}$$

This is a fractional number of stages and in actual practice 16 cells will be used. This would give a slightly better recovery than is assumed in this example.

When the amount of the solvent retained by the inerts is constant, the number of the stages is given by the equation

$$N = \frac{\log[(y'_{n+1} - x'_n)/(y'_1 - x'_0)]}{\log[(y'_{n+1} - y'_1)/(x'_n - x'_0)]} \tag{7-8}$$

where x' is the mass ratio of the solute to solvent in the underflow and y' is the mass ratio of the solute to solvent in the overflow.

The above equation also needs the calculation of the first stage separately, since the solvent content of the fresh solids to the first stage is different from that in the other stages.

Example 7-4. In Example 7-3 determine the number of cells if each ton of dry pulp retains 3.5 tons of water. (This is left as an exercise to the reader.)

Answer. $N = 19$ stages.

When solution retention data by the inerts are available, the number of stages for a given extraction can be obtained by stage-to-stage calculations. An example is given next.

Example 7-5. Oil is to be extracted from meal with the use of benzene as solvent. The unit is to treat 1000 lb of meal (oil-free solid)/h. The feed meal contains 400 lb oil and 25 lb benzene. The wash solution contains 10 lb oil dissolved in 655 lb benzene. The discharged solids are to contain 60 lb unextracted oil. Solution retention data[2] are given below.

Concentration lb of oil/lb solution	0	0.1	0.2	0.3	0.4	0.5	0.6	0.7
Solution retained lb/lb of solid	0.5	0.505	0.515	0.53	0.55	0.571	0.595	0.620

Assuming countercurrent extraction, compute by stagewise calculations the following: (**a**) the composition of the strong solution, (**b**) the weight of solution leaving with the extracted meal, (**c**) the amount of strong solution, (**d**) the number of stages required.

Solution.

Step 1. To simplify interpolation, plot the solution retention data vs. the mass fraction of the solute in the solution. This is done in Fig. 7-5 (curve *A*).

Step 2. Establish overall material balance. Assume there is no inert meal in the overflow. V_1 and L_n are not known. However, the composition and amount of L_n can be established from the solution retention data. Since L_n has an equilibrium composition, the solvent in L_n is found from the solution retention data as follows.

y_A in solution	0	0.1	0.2	0.3	0.4	0.5	0.6	0.7
lb oil/lb inerts	0	0.0505	0.103	0.159	0.22	0.2855	0.357	0.434

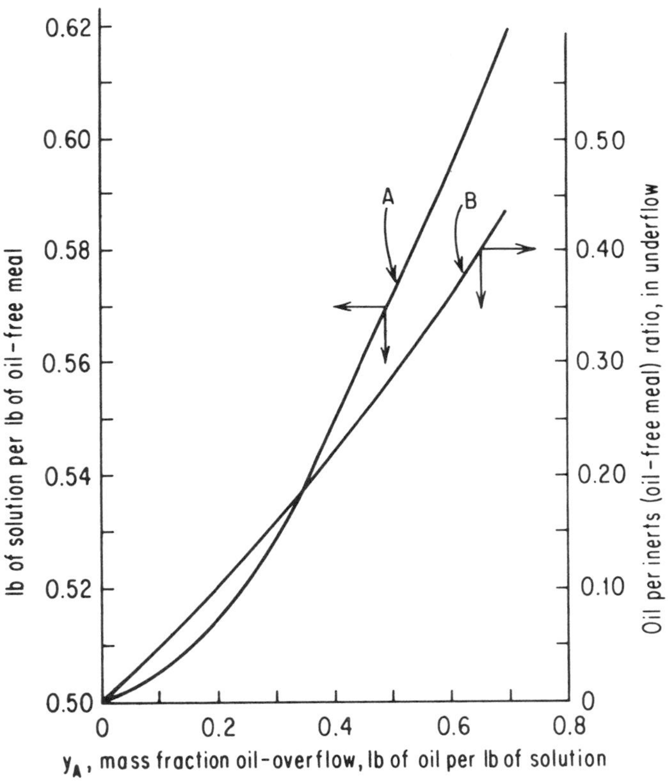

Figure 7-5. Solution retention data and oil/oil-free meal ratios as a function of solution concentration (Example 7-5).

These data are plotted in Fig. 7-5 as the curve B. The ratio of oil to inerts in $L_n = \frac{60}{1000} = 0.06$. From the curve B, for the oil/inerts ratio of 0.06, the solution concentration $(y_A)_n = 0.12$. Therefore, the solvent amount in the underflow L_n is

$$\frac{(1 - 0.12)(60)}{0.12} = 440 \text{ lb/h}.$$

Therefore $L_n = 440 + 60 = 500$ lb/h (inert solid free).

Step 3. Establish overall material balance.

$$L_0 + V_{n+1} = V_1 + L_n$$

$$V_{n+1} = 655 + 10 = 665 \text{ lb/h}$$

$$L_0 = 400 + 25 = 425 \text{ lb/h}$$

$$L_n = 500 \text{ lb/h}$$

Therefore

$$V_1 = 425 + 665 - 500 = 590 \text{ lb/h}$$

The amount of oil in the strong solution is

$$400 + 10 - 60 = 350 \text{ lb/h}$$

The oil concentration of the strong solution is

$$(y_A)_1 = \frac{350}{590} = 0.5932$$

Step 4. Material balance over first stage:

$$L_0 + V_2 = V_1 + L_1$$

$$L_0 = 425 \text{ lb/h}$$

$$V_1 = 590 \text{ lb/h}$$

From curve A of Fig. 7-5, corresponding to $y_A = 0.5932$, ordinate = 0.593 and $L_1 = 0.593(1000) = 593$ lb/h. Hence

$$V_2 = 590 + 593 - 425 = 758 \text{ lb/h}$$

$$x_1 = (y_A)_1 = 0.5932$$

Solute balance on first stage:

$$L_0 x_0 + V_2(y_A)_2 = L_1 x_1 + V_1(y_A)_1$$

or

$$400 + 758(y_A)_2 = 593(0.5932) + 590(0.5932)$$

Hence

$$(y_A)_2 = 0.398$$

$$\text{Solute in } V_2 = 0.398(758) = 301.7 \text{ lb/h}$$

Step 5. Material balances on stages 2 to n: The calculations are continued for second stage onward in a similar manner to that used in calculating material balance on the first stage. For example, the concentration of V_2 enables us to obtain L_2 from Fig. 7-6 and also $(y_A)_2 = x_2$. These values enable us to establish the overall material and solute balances on stage 2 and so on. A summary of the results is given in Table 7-3.

Answers. (**a**) Mass fraction oil in strong solution = 0.5932, (**b**) amount of solution leaving with extracted meal ≐ 500 lb/h, (**c**) amount of strong solution = 590 lb/h, (**d**) number of stages = 4.

Note: Calculations as shown in Table 7-3 indicate that fewer than four stages are required.

Use of Triangular Diagram

The compositions in terms of the mass fractions are useful in the calculation of the number of the equilibrium stages with the use of a triangular diagram. The material-balance equations are given as

Total material balance: $R_0 + V_{n+1} = R_n + V_1$ (7-9)

Note here that R_0 is the total underflow including the inerts which move from stage to stage unchanged. The solute balance gives

$$R_0(x_A)_0 + V_{n+1}(y_A)_{n+1} = R_n(x_A)_n + V_1(y_A)_1 \quad (7\text{-}10)$$

By a rearrangement of the material balances on the interme-

TABLE 7-3 Summary of Stage-to-Stage Calculations of Example 7-5

	Solution Underflow Leaving Stage $n = L_n$					Solution Entering Stage $n = V_{n+1}$			
	Quantities lb/h			Composition L_n		Quantities, lb/h		Composition V_{n+1}	
	Solution	Solution and Inert	Oil	Solvent		Oil		Solvent	Mass Fraction Oil
n	L_n	R_n^*	$L_n x_n$	$L_n(x_S)_n$	$(x_A)_n$†	V_{n+1}	$V_{n+1}(y_A)_{n+1}$	$V_{n+1}(y_S)_{n+1}$	$(y_A)_{n+1}$
0	425	1425	400	25.0	0.9411	590	350.0	240	0.5932
1	593	1593	351.8	241.2	0.5932	758	301.8	456.2	0.3982
2	550	1550	219.0	331.0	0.3982	715	169.0	546.0	0.2364
3	520	1520	122.9	397.1	0.2364	685	72.9	612.1	0.1064
4	505.5	1505.5	53.8	451.7	0.1064	670.5	3.8	666.7	0.0057
			60		←(Given)→	665	10	655	0.015

*Includes 1000 lb of oil-free meal. $R_n = 1000 + L_n$.

†Meal-free basis, or composition of L_n (not R_n).

diate stages $m - 1$, m, $m + 1$, etc., to express differences in the flows of the streams between the stages, the net flows towards the nth stage are obtained. The net total flow towards the nth stage is given by

$$R_{m-1} - V_m = R_m - V_{m+1} = R_{m+1} - V_{m+2} \tag{7-11}$$

and the net flow of the solute towards the nth stage is given by

$$R_{m-1}(x_A)_{m-1} - V_m(y_A)_m = R_m(x_A)_m - V_{m+1}(y_A)_{m+1} \tag{7-12}$$

Example 7-6. Oil is to be extracted from halibut livers by continuous countercurrent multistage extraction with ether. The solution retention data are given in Table 7-1. Halibut livers contain 0.3 mass fraction oil. The fresh solids charge is 100 lb/h, and 95 percent of oil is to be recovered.

Assuming the fresh solvent to be pure, calculate: (**a**) the composition of the solids discharged from the last stage; (**b**) minimum solvent-to-livers ratio; (**c**) using 1.6 times the minimum ratio, calculate the composition of the solution leaving the first stage, (**d**) the number of equilibrium stages; and (**e**) the number of actual stages required if stage efficiency is 73.3 percent.

Solution.

Step 1. Data (Table 7-4) of Example 7-1 are plotted on the triangular diagram (Fig. 7-6).

TABLE 7-4 Composition of Feed Solids

Mass fraction oil	$(x_A)_0 = 0.3$
Oil in fresh solids	0.3(100) = 30 lb
Inerts in fresh solids	0.7(100) = 70 lb
Oil in discharged solids	3 lb
Mass fraction of oil in discharged solids (excluding solvent)	3/73 = 0.0411

Locate this point on $\overline{IA}$ and call it B. Join $\overline{By_{n+1}}$ to intersect EF, the

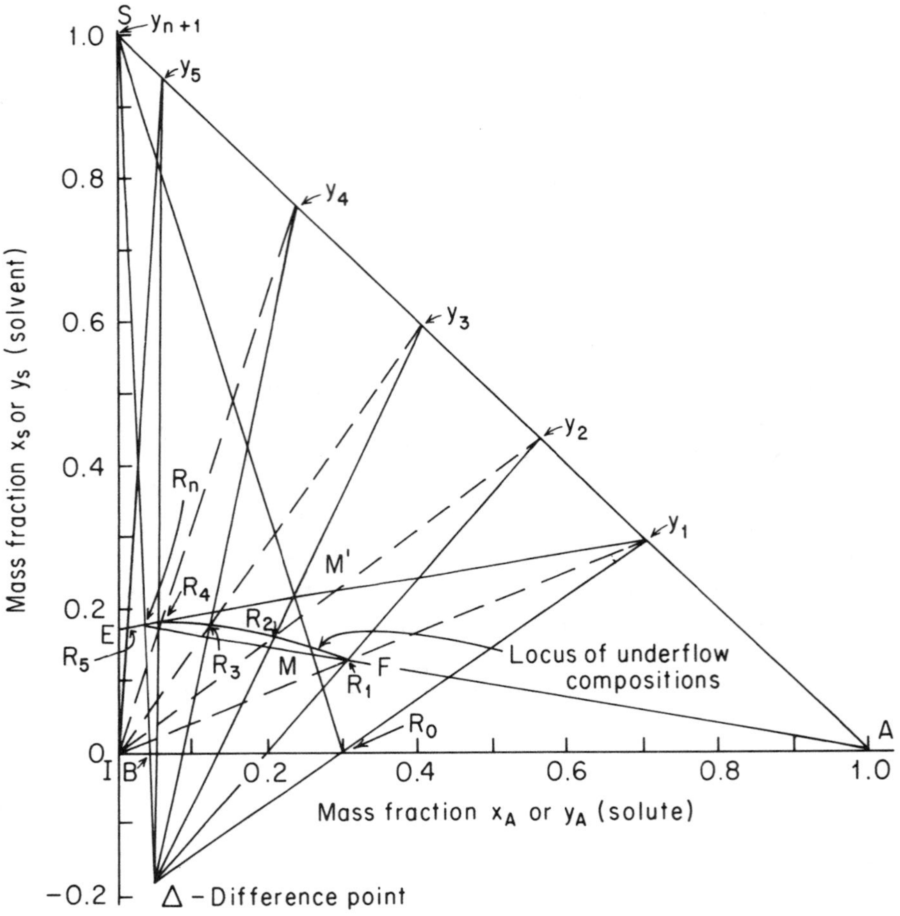

Figure 7-6. Graphical solution for the number of theoretical stages (Example 7-5).

locus of the underflow in R_n. The composition of stream R_n is

$$\text{Oil} = 0.033 \qquad \text{ether} = 0.179 \qquad \text{inerts} = 0.788$$

These are read directly from the graph.

Step 2. Calculation of minimum solvent/solids ratio:
Join R_n and A and $\overline{y_{n+1}R_0}$

$$\text{Minimum}\ \frac{\text{solvent}}{\text{solids}} = \frac{V_{n+1}}{R_0} = \frac{\overline{(x_A)_0M}}{\overline{My_{n+1}}} = \frac{1.9}{11.4} = 0.17$$

Step 3. Actual solvent/solids ratio = 0.170(1.6) = 0.272.
Step 4. Location of addition point:

$$\text{Total solvent required} = 0.272(100) = 27.2 \text{ lb}/100 \text{ lb solids}$$

$$\text{Solvent} + \text{solids} = 127.2 \text{ lb}$$

$$(x_A)_{M'} = \frac{30}{127.2} = 0.2358$$

$$(x_S)_{M'} = \frac{27.2}{127.2} = 0.2138$$

$$(x_I)_{M'} = \frac{70}{127.2} = 0.5504$$

Locate this point on $\overline{y_{n+1}R_0}$ as M'.

Step 5. Join R_n and M' and extend the line to meet the hypotenuse in y_1 [that is $(y_A)_1$ also]. The composition of the solution leaving stage 1 (from graph) is

$$(y_A)_1 = 0.705 \qquad (y_S)_1 = 0.295 \qquad (y_I)_1 = 0$$

Step 6. Join $\overline{y_1(x_A)_0}$ and extend to intersect $\overline{y_{n+1}R_n}$. The intersection point is the difference point. Coordinates of differences point:

$$(x_A)_\Delta = +0.05 \qquad (x_S)_\Delta = -0.18 \qquad (x_I)_\Delta = 1 - (x_A)_\Delta - (x_S)_\Delta = +1.13$$

Join the tie line $\overline{y_1 I}$ to intersect the underflow locus EF in R_1. Join $\overline{\Delta R_1}$ meeting the hypotenuse in y_2. Join $\overline{y_2 I}$ to intersect the locus of the underflow compositions, EF in R_2. Join $\overline{\Delta R_2}$ meeting the hypotenuse in y_3. Proceed in the above manner to obtain other stages till R_n is passed. From the graph, about 4.4 equilibrium stages are required.

Step 7.

$$\text{Actual stages} = \frac{\text{theoretical stages}}{\text{efficiency}} = \frac{4.4}{0.733} \doteq 6$$

Step 8. The compositions of the streams can be read from the graph directly for any stage.

The use of the Ponchon-Savarit method to solve multistage extraction problems is illustrated by Example 7-7.

Example 7-7. Oil is to be extracted from halibut livers by means of ether. The solution retention data are given in Table 7-2. Halibut livers contain 0.257 mass fraction oil. If 95 percent of the oil is to be extracted and the strong solution from the system is to contain 0.7 mass fraction oil, determine

1. Quantity and composition of the discharged solids
2. Pounds of ether (oil free) required to treat 1000 lb charge
3. Number of ideal stages required
4. Number of actual stages if the stage efficiency is 70 percent

Solution.

Step 1. Overall material balance: A, the oil in feed solids, is 257 lb; I, the inerts in feed solids, is 743 lb. The coordinates of points V_{n+1} and R_0 are

Point R_0: $$X = \frac{257}{257 + 0} = 1.0 \qquad Y = \frac{743}{257} = 2.891$$

Point V_{n+1}: $$X = 0 \qquad Y = 0$$

Plot these points on the Ponchon-Savarit diagram (Fig. 7-7).

Step 2. The amount of oil in strong solution is

$$0.95(257) = 244.15 \text{ lb}$$

The amount of solvent in strong solution is

$$\frac{0.3}{0.7}(244.15) = 104.64 \text{ lb}$$

The Ponchon-Savarit coordinates of this point are

$$X = \frac{244.15}{104.64 + 244.15} = 0.7 \qquad Y = 0$$

Plot this point V_1 on the diagram of Fig. 7-7.

Step 3. Location of point R_n: The ratio of Y/X for point R_n is

$$\frac{\text{Inerts}}{\text{Solute in } R_n} = \frac{743}{12.85} = 57.82$$

Draw a line through the origin with a slope Y/X to cut the underflow line in R_n (since R_n lies on the underflow curve). From the graph, $X = 0.075$, $Y = 4.34$, and

$$\frac{A}{S + A} = 0.075 \qquad \text{and} \qquad \frac{I}{S + A} = 4.34$$

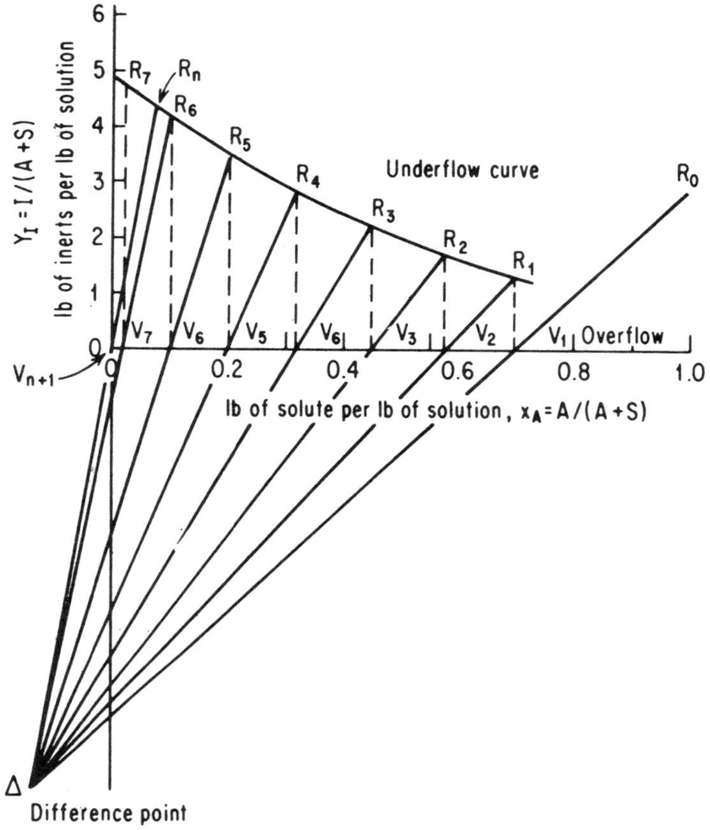

Figure 7-7. Theoretical stages by Ponchon-Savarit diagram (Example 7-6).

or $$S + A = \frac{743}{4.34} = 171.2 \text{ lb}$$

but $$A \text{ in } R_n = 12.85 \text{ lb}$$

$$S \text{ in } R_n = 171.2 - 12.85 = 158.35 \text{ lb}$$

Therefore $$R_n = 12.85 + 158.35 + 743 = 914.2 \text{ lb}$$

Step 4. Overall solvent balance gives

$$\frac{\text{Solvent}}{1000 \text{ lb solids}} = \frac{158.35 + 104.65}{1000} = 263 \text{ lb}/1000 \text{ lb}$$

or solvent = 0.263 lb per lb of fresh solids. The composition of discharged solids is

Mass fraction solute: $\dfrac{12.85}{914.2} = 0.014$

Mass fraction solvent: $\dfrac{158.35}{914.2} = 0.173$

Mass fraction inerts: $\dfrac{743}{914.2} = 0.813$

Step 5. Step the stages as follows. Extend $\overline{R_0V_1}$ and $\overline{R_nV_{n+1}}$ to find the difference point Δ. From V_1 draw a perpendicular (tie line) to intersect the underflow curve in R_1. Join $\overline{\Delta R_1}$ to intersect the overflow line in V_2. Draw a perpendicular from V_2 to cut the underflow curve in R_2. Proceed in this manner to obtain the remaining stages till a tie line falls on the other side of R_n. In Fig. 7-7, the desired separation requires between 6 and 7 stages, or roughly 6.3 stages.

Step 6. Actual number of stages equals

$$\frac{\text{Number of equilibrium stages}}{\text{Efficiency}} = \frac{6.3}{0.7} = 9$$

Caution: In using the Ponchon-Savarit diagram, note the difference in Y and X scales when plotting the line of slope Y/X as in Step 3 in the above example.

References

1. G. G. Brown, et. al., *Unit Operations*, Wiley, New York, 1956, p. 284.
2. W. L. McCabe and J. C. Smith, *Unit Operations of Chemical Engineering*, 3d ed., McGraw-Hill, New York, 1976, p. 617.

Chapter

8

Liquid-Liquid Extraction

Liquid-liquid extraction is an operation in which a solute dissolved in one liquid phase is transferred to a second liquid phase. In the treatment that follows in this chapter, the solvent is assumed to be insoluble in the solution from which solute is to be extracted.

Ternary Systems

As in leaching, equilibrium in a three-component system can be represented on either (1) equilateral-triangle, (2) right-triangle, or (3) Ponchon-Savarit diagrams. As in leaching, the right-triangle and Ponchon-Savarit diagrams are more convenient.

Methods of the construction of the equilibrium diagrams are illustrated by examples later. Compositions may be expressed in terms of the mass fractions or mole fractions.

Extract and Raffinate Solvents

The solvent which is used to recover the solute from another solution is termed the *extract solvent*. The raffinate solvent is the solvent from which the desired solute is to be separated.

Conjugate Phases

Two phases which exist in equilibrium such that their compositions are independent of the total two-phase mixture are called *conjugate solutions*.

Conjugate Systems

The equilibrium compositions of conjugate phases are plotted in Fig. 8-1. A line such as *AA′* connecting two conjugate phases is called a *tie line*. At the point *P*, called the *plait point*, the tie line disappears. At this point, the two conjugate phases are mutually soluble. The solubility curve *EPF* is obtained experimentally. Instead of showing many tie lines, a conjugate line or curve is drawn. This can be used to draw any tie line.

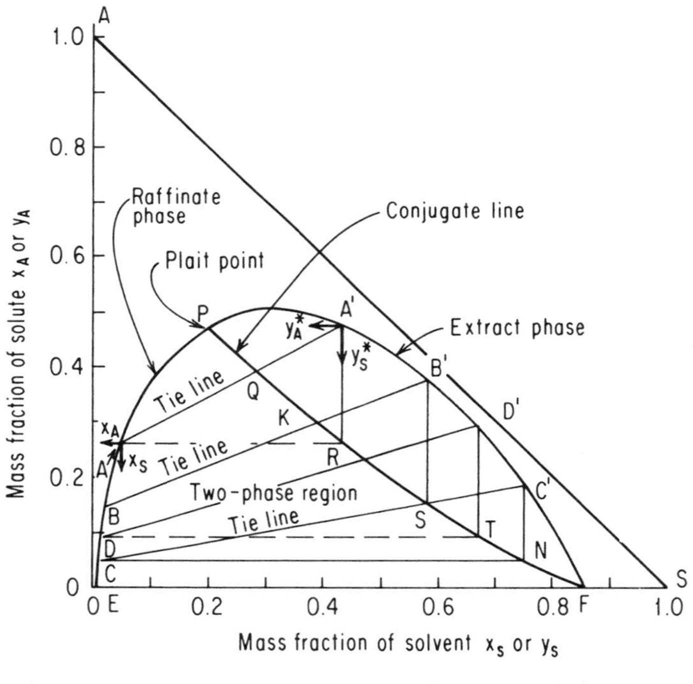

Figure 8-1. Construction of phase-equilibrium diagram on right-triangular diagram.

Construction of Conjugate Line

From the experimental data, the compositions of a few conjugate phases, such as AA', BB', and CC', are first plotted. Then straight lines parallel to two sides of the triangle are drawn from each composition of a conjugate phase to intersect each other as at Q, R, and S. A curve through points P, Q, R, S, and N is drawn, and this is the conjugate line. Now, to draw a desired tie line, a reverse procedure is followed; e.g., if the tie line for the conjugate phase D' is required, a straight line $D'T$ is drawn parallel to $0A$ to cut the conjugate curve at T. From T, a straight line TD is drawn, parallel to $0S$ to cut the solubility curve in D. Then DD' is the required tie line.

Extract and Raffinate Phases

The liquid phase, which is rich in extracted solute, is termed the extract phase; the raffinate phase is the liquid phase which is lean in solute to be extracted.

The solutions lying within the boundaries of the solubility curve will separate into two phases which are in equilibrium. The mass ratio of the two phases resulting in a mixture, e.g., as represented by a point K in the immiscibility region, is given by the ratio of opposite line segments; thus

$$\frac{\text{Mass of phase } E}{\text{Mass of phase } R} = \frac{KB}{KB'}$$

In the above equation, E represents the extract phase and R represents the raffinate phase. If A and S are the solute to be recovered and the solvent, respectively, the separation is increased as the plait point approaches zero and the line increases in slope.

Selectivity of Solvent

A solvent is said to be selective for a given solute when the mass fraction of the solute in the extract layer is greater than that in the raffinate layer. Selectivity of the solvent S for a solute component A is defined as follows.

$$\text{Selectivity} = \frac{y_A/y_B}{x_A/x_B} = \frac{y_A x_B}{x_A y_B} \tag{8-1}$$

where y = mass fraction of the component in the extract phase
x = mass fraction in the raffinate phase
A = desired solute
B = raffinate solvent

A selectivity diagram can be prepared from the conjugate-phase equilibrium data or from the binodal solubility curve, as shown in Fig. 8-2.

Ponchon-Savarit Diagram for Solvent Extraction

This diagram is especially useful for calculations involving the

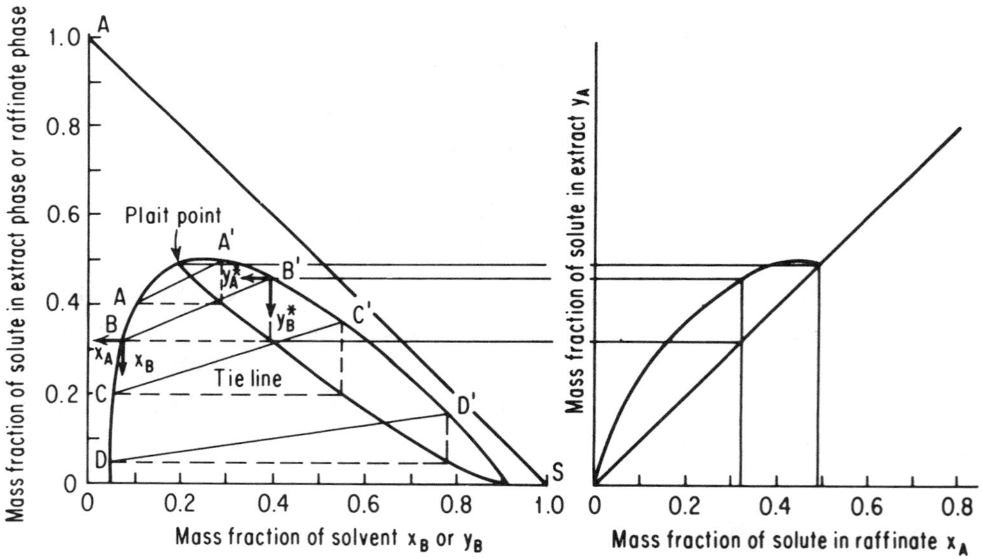

Figure 8-2. Construction of selectivity diagram from phase-equilibrium diagram.

continuous countercurrent extraction with reflux. The X and Y coordinates as in leaching are given by

$$X_A \text{ or } Y_A = \frac{\text{mass of solute component}}{\text{mass of solute + mass of raffinate solvent}}$$

$$X_S \text{ or } Y_S = \frac{\text{mass of extract solvent}}{\text{mass of solute + mass of raffinate solvent}}$$

Notice that the extract solvent replaces the inerts in leaching in the above definitions.

Single-Stage Extraction

Single-stage extraction may be conducted batchwise or continuously. The streams entering or leaving a stage are shown in Fig. 8-3*a*. An overall material balance over the stage gives

$$E_0 + R_0 = E_1 + R_1 = M_1 \tag{8-2}$$

The solute balance gives

$$E_0(y_A)_0 + R_0(x_A)_0 = E_1(y_A)_1 + R_1(x_A)_1 \tag{8-2a}$$

The point M_1 is located on the line joining E_0 and R_0 by the mixture rule or more easily by a calculation using the relations

$$R_0(x_A)_0 + E_0(y_A)_0 = M_1(x_A)_{M_1} \qquad \text{solute} \tag{8-2b}$$

$$R_0(x_S)_0 + E_0(y_S)_0 = M_1(x_S)_{M_1} \qquad \text{solvent} \tag{8-2c}$$

Thus

$$(x_A)_{M_1} = \frac{R_0(x_A)_0 + E_0(y_A)_0}{M_1} \tag{8-2d}$$

and

$$(x_S)_{M_1} = \frac{R_0(x_S)_0 + E_0(y_S)_0}{M_1} \tag{8-2e}$$

The point M_1 lies on line R_0E_0, and the tie line passing through M_1 (Fig. 8-3*b*) will give the compositions of the two phases R_1 and E_1 into which the mixture separates.

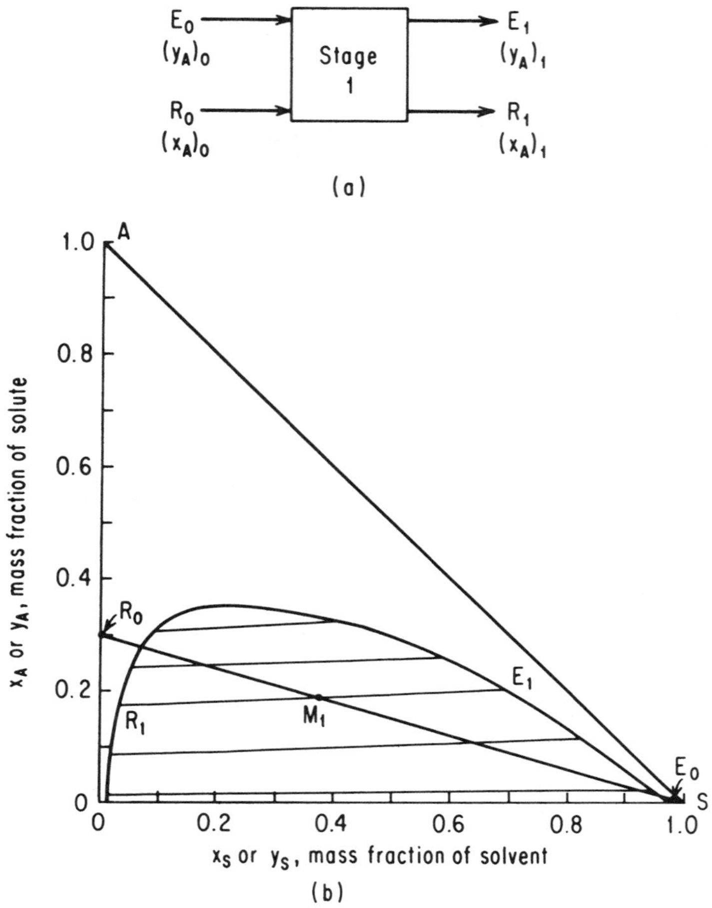

Figure 8-3. Single-stage extraction: (*a*) material balance; (*b*) representation on right-triangular diagram.

Also

$$\frac{\text{Mass of } E_1}{\text{Mass of } R_1} = \frac{\overline{R_1M_1}}{\overline{M_1E_1}} \tag{8-2f}$$

If E_0 is pure solvent, the point E_0 will coincide with apex S.

Stagewise Extraction

For the multistage extraction with fresh solvent in each stage, the equilibrium relationship is

$$(y_A)_n = f(x_A)_n$$

when $$(y_A)_n = K(x_A)_n \qquad K = \text{const.}$$

$$(x_A)_n = \left(\frac{b}{b + sK}\right)^n (x_A)_0 \tag{8-3}$$

where b = mass of raffinate solvent
s = mass of extract solvent
n = number of stages

In Eq. (8-3) the masses b and s are assumed the same for each stage.

Example 8-1. One hundred pounds of a solution of acetic acid (solute) and water (raffinate solvent) containing 40 wt % acetic acid is to be extracted two times using methylisobutylketone as the solvent. The extraction is isothermal at 25°C. Seventy-five pounds of ketone is to be used in each extraction.

(**a**) Determine the quantities and compositions of the various streams. (**b**) Find how much solvent would be required if the same final raffinate concentration were to be obtained with a single extraction.

The tie line data[1] are plotted in Fig. 8-4*a* as the binodal curve.

Solution.

a. Prepare the equilibrium curve by plotting y_A versus x_A (Fig. 8-4*b*).

Stage 1.

Overall balance: $$R_0 + S_1 = E_1 + R_1 = M_1$$

or $$100 + 75 = E_1 + R_1 = 175 = M_1$$

Solute balance: $$R_0(x_A)_0 + S_1(y_A)_{S_1} = M_1(x_A)_{M_1}$$

Hence $$100(0.40) + 75(0) = 175(x_A)_{M_1}$$

Thus $$(x_A)_{M_1} = \tfrac{40}{175} = 0.229$$

Solvent balance: $$R_0(x_S)_0 + S_1(y_S)_1 = M_1(x_S)_{M_1}$$

Hence $$100(0) + 75(1) = 175(x_S)_{M_1}$$

or $$(x_S)_{M_1} = \tfrac{75}{175} = 0.429$$

With the help of these coordinates, locate M_1 on $\overline{R_0S}$ as shown in

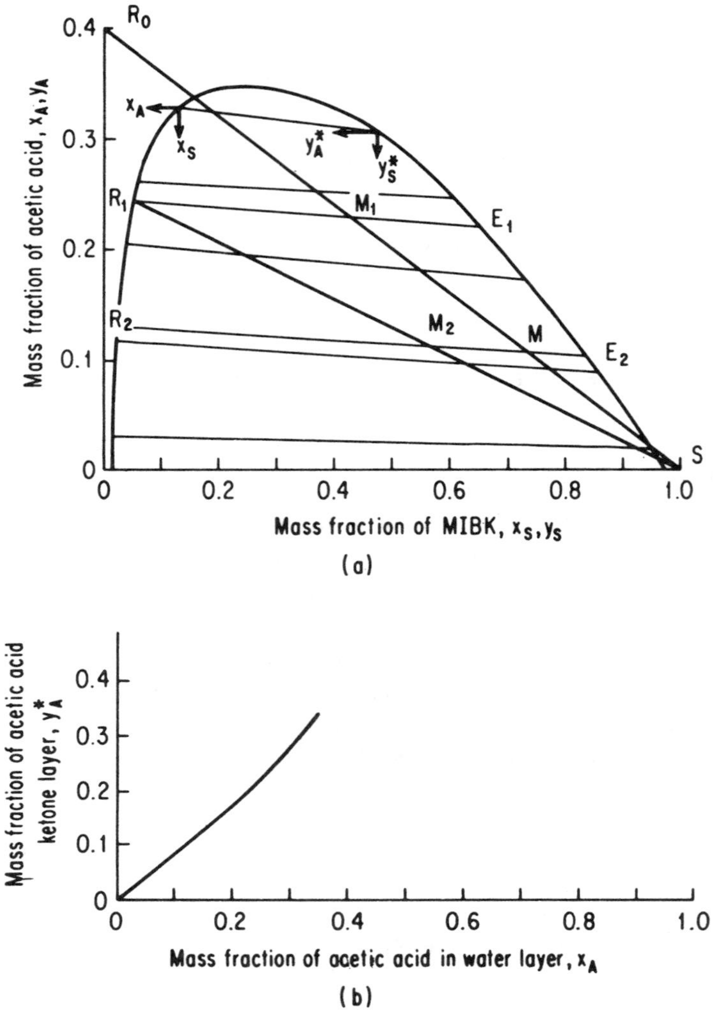

Figure 8-4. Multistage extraction on right-triangular diagram: (*a*) solution of Example 8-1; (*b*) auxiliary diagram for locating tie lines.

Fig. 8-4*a*. [*Note*: The location of M_1 needs $\overline{R_0S}$ and one of the coordinates $(x_A)_{M_1}$ and $(x_S)_{M_1}$.]

Now with the help of the distribution and equilibrium curves, locate the line $\overline{R_1E_1}$ passing through M_1 cutting the binodal curve in R_1 and E_1.

First calculate E_1 and R_1 by the mixture rule:

$$\frac{R_1}{E_1} = \frac{\overline{M_1E_1}}{\overline{M_1R_1}} = \frac{0.65 - 0.429}{0.429 - 0.05} = 0.58$$

$$R_1 = 0.58E_1 \qquad R_1 + E_1 = 175$$

$$1.58E_1 = 175 \qquad E_1 = 111 \text{ lb}$$

$$R_1 = 175 - 111 = 64 \text{ lb}$$

Stage 2. Overall material balance gives

$$R_1 + S_2 = R_2 + E_2 = M_2 = 63 + 75 = 138 \text{ lb}$$

By the solute balance,

$$R_1(x_A)_1 + S_2(y_A)_{S_2} = M_2(x_A)_{M_2}$$

or

$$63(0.245) + 0 = 138(x_A)_{M_2}$$

Therefore

$$(x_A)_{M_2} = \frac{63(0.245)}{138} = 0.112$$

Join R_1 and S and locate M_2 on it (Fig. 8-4*a*). Locate R_2 and E_2 on the binodal curve by drawing the tie line through M_2. (*Note:* This involves locating the tie line by interpolation.)

$$E_2 = \frac{M_2[(x_A)_{M_2} - (x_A)_2]}{(y_A)_2 - (x_A)_2}$$

$$= \frac{138(0.112 - 0.131)}{0.104 - 0.131} = 97 \text{ lb}$$

Therefore

$$R_2 = 138 - 97 = 41 \text{ lb}$$

b. The amount of the solvent for the same raffinate concentration with a single extraction is calculated in the following manner. The final raffinate concentration is $(x_A)_2 = 0.131$. The mixture point will be the intersection point of the tie line $\overline{R_2E_2}$ and $\overline{R_0S}$. By material balance,

$$R_0 + S = R + E = M$$

and by solute balance,

$$R_0(0.4) + S(0) = M(x_A)_M$$

$$(x_A)_M = 0.107 \qquad \text{from Fig. 8-6}a$$

$$M = R_0 + S = 100 + S$$

Therefore $\quad 100(0.4) + 8(0) = 100(0.107) + S(0.107)$

or

$$S = \frac{100(0.4 - 0.107)}{0.107 - 0} = 273.8 \text{ lb}$$

Thus, for the same raffinate concentration, a single-stage extraction requires 273.8 lb compared with 150 lb of solvent for the two-stage extraction.

Application to Multistage Countercurrent System

The calculations are made in the same manner as in leaching. Either stepwise calculations or graphical methods are used. Figure 8-5 illustrates a continuous countercurrent multistage extraction system. Overall material balance gives

$$E_{n+1} + R_0 = E_1 + R_n \tag{8-4}$$

Solute balance gives

$$E_{n+1}(y_A)_{n+1} + R_0(x_A)_0 = E_1(y_A)_1 + R_n(x_A)_n \tag{8-5}$$

Extract solvent balance gives

$$E_{n+1}(y_S)_{n+1} + R_0(x_S)_0 = E_1(y_S)_1 + R_n(x_S)_n \tag{8-6}$$

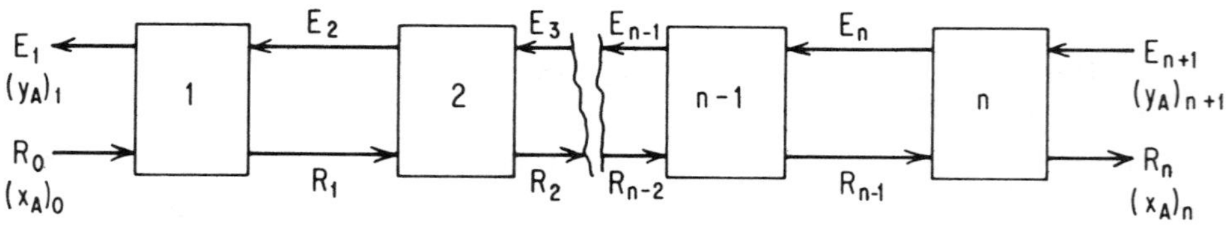

Figure 8-5. Multistage countercurrent liquid-liquid extraction system.

The procedure used to obtain the number of the equilibrium stages is illustrated by Example 8-2.

Example 8-2. Two thousand pounds per hour of acetic acid is countercurrently extracted with isopropyl ether from an acetic acid–water solution containing 30% acid so as to reduce the acid concentration to 2% in the solvent-free raffinate. Determine (**a**) minimum amount of the solvent per lb of feed solution and (**b**) the number of theoretical stages if the solvent-to-feed-solution ratio used is 1.6 times the minimum. The tie-line data[2] are plotted in Fig. 8-6 on a triangular diagram.

Solution.

a. Composition of R_0 is $(x_A)_0 = 0.3$. Composition of R'_n (free of solvent) equals 0.02. Composition $(y_S)_{n+1}$ equals 1.0.

These points are plotted as R_0, R'_n, and S. Join R'_n to S. The point of intersection of $\overline{R'_nS}$ with the solubility curve at R_n gives the composition of the raffinate stream R_n. The tie line passing through R_0 which cuts

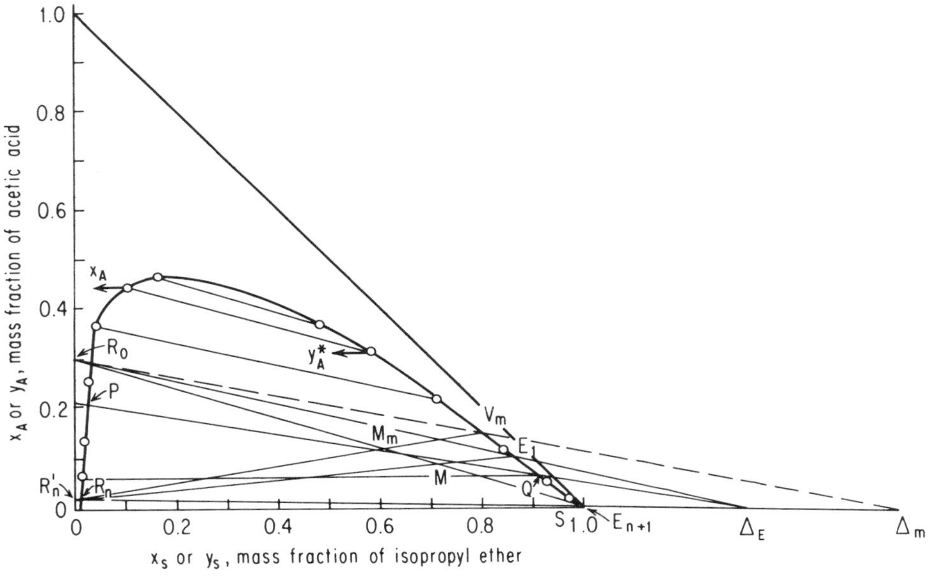

Figure 8-6. Determination of minimum reflux ratio (Example 8-2).

$\overline{R_nS}$ to the right nearest* to S gives the Δ for the minimum solvent-to-feed-solution ratio. Let V_m be the intersection point of the binodal curve and $\overline{R_0\,\Delta_m}$. Join V_m and R_n. From M_m, the intersection point of $\overline{V_mR_n}$ and $\overline{R_0S}$, the minimum solvent-to-feed-solution ratio is

$$\frac{\overline{R_0M_m}}{\overline{M_mS}} = \frac{8.2}{5.0} = 1.64$$

Thus, 1.64 lb solvent per lb feed solution is the minimum solvent.

b. If 1.6 times the minimum solvent is used,

$$\text{Actual solvent} = 1.64(1.6)$$
$$= 2.624 \text{ lb per lb of feed solution}$$

The composition of the actual addition point is calculated as

$$(y_A)_M = \text{mass fraction solute A} = \frac{0.3}{1 + 2.624} = 0.083$$

$$(x_S)_M = \text{mass fraction solvent B} = \frac{2.624}{3.624} = 0.724$$

This point is located on $\overline{R_0S}$. $\overline{R_nM}$ is extended to intersect the solubility curve in E_1. This intersection point gives the composition of the strong solution leaving the system. Join R_0E_1 and extend to cut $\overline{R_nS}$ extended in Δ_E. Using the Δ_E point and the solubility curve, it is possible to graphically obtain the equilibrium stages by following the difference point and the tie lines, but when the number of the stages is somewhat large, the lines get crowded. It is therefore more convenient to prepare an auxiliary diagram, as in Fig. 8-7. To get the operating line on this diagram, few lines are drawn from Δ_E to cut the solubility curve. For example, a line from Δ_E cuts the solubility curve in P and Q. y_A and x_A are read. Various corresponding values of y_A and x_A are obtained in this manner and tabulated below.

x_A	0.02	0.055	0.09	0.150	0.205	0.250	$0.3 = (x_A)_0$
y_A	0	0.010	0.02	0.04	0.06	0.08	$0.1 = (y_A)_1$

*It would have been farthest from S if the point of intersection were on the left side of the diagram.

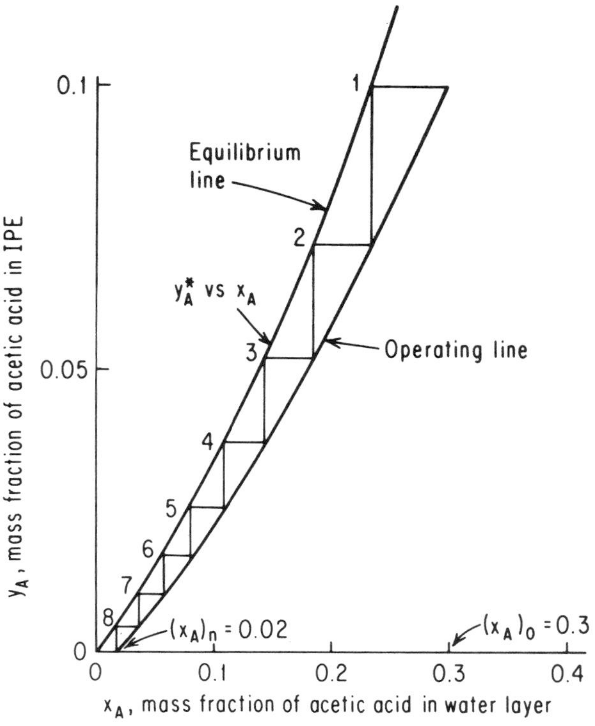

Figure 8-7. Determination of equilibrium stages on the distribution diagram (Example 8-2).

These are plotted as y_A versus x_A to give the operating line (Fig. 8-7). The equilibrium curve is plotted using the tie-line data, and the stages are then stepped as in the McCabe-Thiele diagram. When this is done, the number of theoretical stages from the graph is 7.7.

References

1. T. K. Sherwood, J. E. Evans, and J. V. K. Longeor, *Ind. Eng. Chem.*, vol. 31, 1939, p. 1144.
2. R. E. Treybal, *Mass Transfer Operations*, 3d ed., McGraw-Hill, New York, 1980, p. 494.

Chapter

9

Psychrometry and Humidification

In dealing with processes involving exchange of mass and energy between a liquid and an essentially insoluble gas, the principles of psychrometry are useful. Important operations to consider are the cooling and humidification and dehumidification of a gas. A familiarity with the use of the psychrometric chart for the air–water-vapor system at atmospheric pressure is also essential.

Definitions in Psychrometry

The following is a review of the important definitions used in psychrometry.

Humidity Y. Sometimes this is also called *specific humidity* or humidity ratio. When P is the total pressure which is low enough (near atmospheric) for the gas mixture to be considered ideal, the humidity is given by

$$Y = \frac{M_A p_A}{M_B(P - p_A)} \tag{9-1}$$

where M_A = molecular weight of vapor
M_B = molecular weight of dry gas
p_A = partial pressure of vapor in gas, atm
P = total pressure, atm
Y = lb of vapor per lb of vapor-free (dry) gas

P_A = saturation pressure

Saturation Humidity Y_s. When $p_A = P_A$, the vapor pressure of the liquid at the prevailing temperature, the gas is saturated with the vapor and the saturation humidity is given by

$$Y_s = \frac{M_A P_A}{M_B(P - P_A)} \tag{9-2}$$

Molal Absolute Humidity Y'. This is defined as the number of moles of vapor per mole of the dry gas or

$$Y' = \frac{p_A}{P - p_A} \tag{9-3}$$

Thus from Eqs. (9-1) and (9-3),

$$Y = Y' \frac{M_A}{M_B} \tag{9-4}$$

Percent Relative Humidity or Relative Humidity. This is a term which expresses the actual humidity of a gas at a given temperature as a percentage of its saturation humidity at the same dry-bulb temperature or

$$\text{Percent relative humidity} = 100\frac{p_A}{P_A} \tag{9-5}$$

Percentage Saturation or Percent Absolute Humidity. This is equal to the ratio of the actual humidity to the saturation humidity. Thus

$$\text{Percent absolute humidity} = \begin{cases} 100\dfrac{Y}{Y_s} & \text{mass basis} \quad (9\text{-}6) \\ 100\dfrac{Y'}{Y'_s} & \text{molal basis} \quad (9\text{-}7) \end{cases}$$

Dew-Point Temperature. This is the temperature at which the vapor-gas mixture becomes saturated when the gas mixture is cooled at constant total pressure.

Humid Volume V_H. This is the total volume of a unit mass of dry (vapor-free) gas plus the vapor it contains. By the gas laws, V_H is related to the humidity and temperature by the equation

$$V_H = 359\left(\frac{1}{M_{\mathrm{B}}} + \frac{Y}{M_{\mathrm{A}}}\right)\left(\frac{t_{\mathrm{G}} + 460}{492}\right)\frac{1}{P} \qquad \mathrm{ft^3/lb\ dry\ gas} \qquad (9\text{-}8)$$

where t_{G} is gas temperature, °F, and P is the total pressure, atm. For a saturated gas, $Y = Y_s$ and the volume is the saturated volume V_s.

Humid Heat C_S. This is the heat required to raise the temperature of 1 lb dry gas plus the vapor it contains by 1°F, at constant pressure. Thus

$$C_S = C_{P\mathrm{B}} + C_{P\mathrm{A}}Y \qquad (9\text{-}9)$$

where $C_{P\mathrm{B}}$ and $C_{P\mathrm{A}}$ are specific heats of the gas and vapor, respectively, at constant pressure, Btu/lb · °F.

Relative Enthalpy or Humid Enthalpy. This is the enthalpy of 1 lb of the dry carrier gas together with the amount of vapor it contains. By choosing t_0 as the base or reference temperature for both the components and basing the enthalpy of the component A or the liquid A at t_0, the enthalpy of the gas-vapor mixture at t_{G} (the temperature of the gas) is given by

$$H = C_{P\mathrm{B}}(t_{\mathrm{G}} - t_0) + Y\lambda_0 + C_{P\mathrm{A}}Y(t_{\mathrm{G}} - t_0) \qquad \mathrm{Btu/lb\ dry\ gas} \qquad (9\text{-}10)$$

where λ_0 is the latent heat of evaporation, Btu/lb, of the liquid A at t_0. From the definition of humid heat, the above expression can be written as

$$H = C_S(t_{\mathrm{G}} - t_0) + Y\lambda_0 \qquad (9\text{-}11)$$

Psychrometric Relations for Air–Water-Vapor Mixtures

The special reduced forms of Eqs. (9-1) to (9-11) for the air–water-vapor system are given below.

Humidity:

$$Y = 0.62\frac{p_A}{P - p_A} \quad \text{lb water vapor/lb dry air} \tag{9-12}$$

Saturation humidity:

$$Y_s = 0.62\frac{P_A}{P - P_A} \quad \text{lb water vapor/lb dry air} \tag{9-13}$$

Molal absolute humidity:

$$Y' = \frac{p_A}{P - p_A} \quad \text{mol water vapor/mol dry air} \tag{9-14}$$

$$Y = (0.62)Y' \tag{9-15}$$

Humid volume:

$$V_H = (0.0252 + 0.0405Y)(t_G + 460)\frac{1}{P} \quad \text{ft}^3\text{/lb dry air} \tag{9-16}$$

Humid heat:

$$C_S = 0.24 + 0.45Y \quad \text{Btu/lb dry air} \cdot {}^\circ\text{F} \tag{9-17}$$

Relative enthalpy:

$$H = (0.24 + 0.45Y)(t_G - 32) + 1075.2Y \text{ Btu/lb dry air} \tag{9-18}$$

Reference temperature = 32°F for both air and water

M_A(water vapor) = 18.02 M_B (dry air) = 28.97 ≐ 29

λ_0 = 1075.2 Btu/lb latent heat of vaporization of water at 32°F

C_P of water vapor = 0.45 Btu/lb °F

C_p of air = 0.24 Btu/lb · °F dry air

Wet-Bulb Temperature t_w

This is the steady-state, dynamic equilibrium temperature reached by a small mass of a liquid evaporating under adiabatic conditions into a continuous large stream of gas. At the wet-bulb temperature, the vapor is transferred to the surrounding gas by a process of diffusion through the gas film surrounding the small mass of the liquid while the sensible heat is transferred from the air to the liquid. This sensible heat is used up in the evaporation of the liquid which escapes to the gas-vapor mixture through the film. When the rate of heat transfer from the gas to the liquid equals the rate of heat required for the evaporation, the liquid attains the wet-bulb temperature. For this condition, the following equation can be established:

$$t_G - t_w = \frac{\lambda_w k_G M_B P}{h_G}(Y_w - Y)$$

But $M_B P k_G = k_Y$ where k_Y is the mass transferred, $\text{lb/h} \cdot \text{ft}^2 \cdot \Delta Y$, and k_G is the mass transferred, $\text{lb} \cdot \text{mol/h} \cdot \text{ft}^2 \cdot \text{atm}$. Therefore,

$$Y_w - Y = \frac{(h_G/k_Y)(t_G - t_w)}{\lambda_w} \tag{9-20}$$

In Eq. (9-20), h_G/k_Y is called the psychrometric ratio, $t_G - t_w$ is the wet-bulb depression, and λ_w is the latent heat of vaporization at t_w. h_G is the gas film coefficient of the heat transfer, $\text{Btu/h} \cdot \text{ft}^2 \cdot {}^\circ\text{F}$. The ratio[1] h_G/k_Y is given by

$$\frac{h_G}{k_Y} = \begin{cases} 0.294\,\text{Sc}^{0.56} & \text{for air–water-vapor mixtures} \\ C_S\left(\dfrac{\text{Sc}}{\text{Pr}}\right)^{0.56} & \text{for other gases and liquids} \end{cases} \tag{9-21}$$

where

$$\text{Sc} = \text{Schmidt number} = \frac{\mu}{\rho D} \tag{9-22}$$

and

$$\text{Pr} = \text{Prandtl number} = \frac{c_p \mu}{k} \tag{9-23}$$

The ratio Sc/Pr is called the *Lewis number*. For the air–water-vapor mixtures,

$$\mathrm{Sc} \doteq \mathrm{Pr} \qquad \text{and} \qquad \frac{h_G}{k_Y} = C_S = 0.24 \tag{9-24}$$

Adiabatic Saturation Temperature t_s

When a limited amount of a gas is in contact with a large amount of a liquid adiabatically, the temperature of the liquid will remain constant at t_s, the adiabatic saturation temperature and the gas will be humidified and cooled. With sufficient intimate contact, the gas will leave at t_s in a condition of saturation. Since t_G and p_A are not constant, the steady-state equations cannot be written. However, an enthalpy balance may be written with t_s as datum and using the inlet values t_G and Y as follows:

$$(C_{PB} + YC_{PA})(t_G - t_s) + Y\lambda_{as} = Y_{t_s}\lambda_{as} \tag{9-25}$$

which can be rewritten as

$$Y_{t_s} - Y = C_S\left(\frac{t_G - t_s}{\lambda_{as}}\right) \tag{9-26}$$

Equations (9-20) and (9-26) are identical if $h_G/k_Y = C_S$. This relation is approximately true in the case of water-vapor–air systems, and hence the wet-bulb and adiabatic saturation temperatures are effectively equal for the air–water-vapor mixtures below a temperature of 150°F.

The Psychrometric Chart

The properties for any given gas-vapor mixture, as defined before, can be conveniently represented on a chart called the *psychrometric chart*. One has to deal with the air-water system more frequently. Because of the importance of air–water-vapor systems, complete charts are available[2] in both the British engineering and SI units. One version of the chart in the fps units is given in Fig. 9-1 for ready reference.

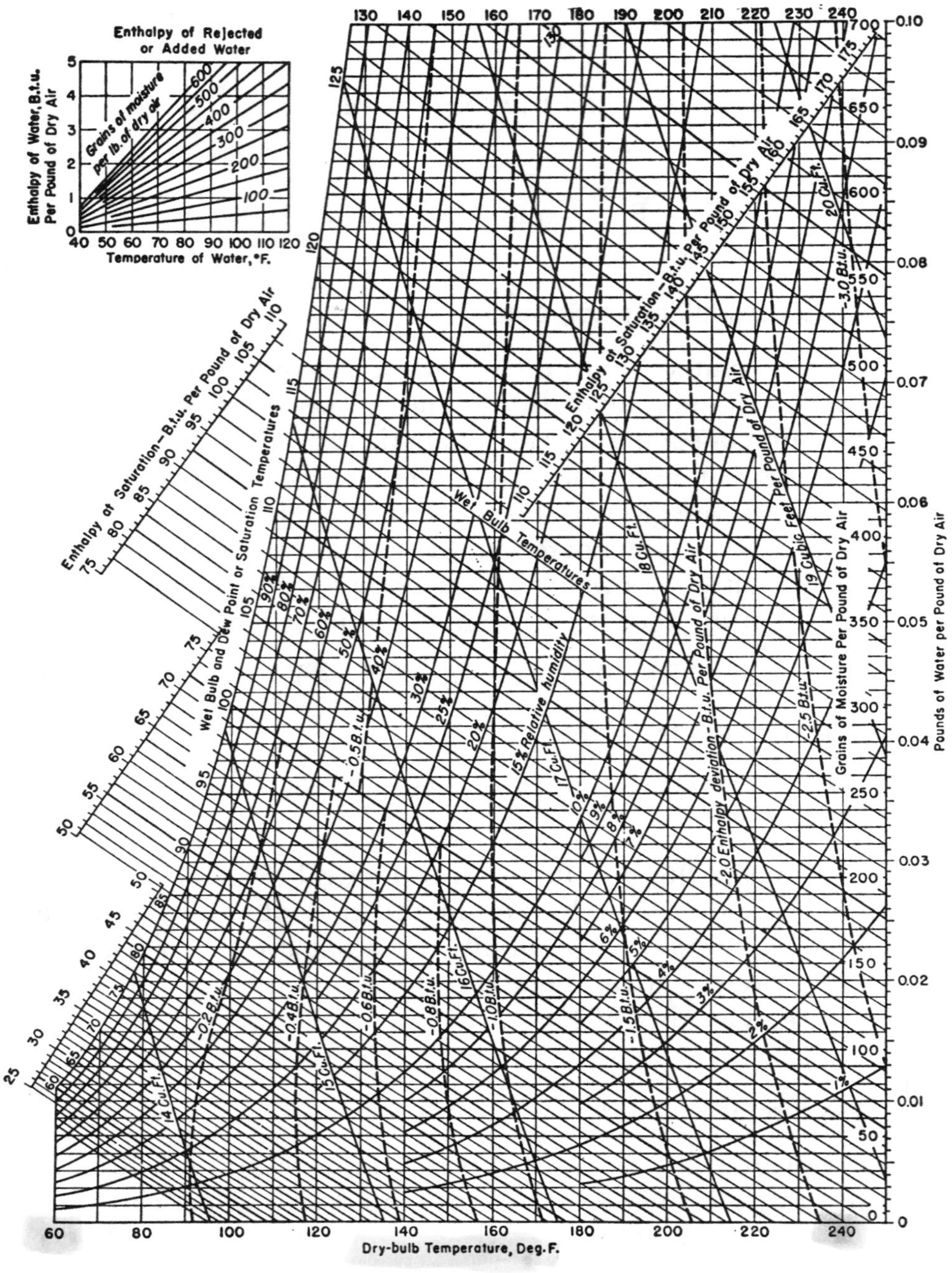

Figure 9-1. Psychrometric chart at high temperatures for air–water-vapor system at 1 atm. Barometric pressure: 29.92 inHg. (*Reproduced by permission of Carrier Corporation, Copyright 1980, Carrier Corporation.*)

Example 9-1. Air at dry-bulb temperature of 75°F and wet-bulb temperature of 65°F is heated to 250°F and fed to a dryer. If the air leaves the dryer 90 percent saturated and if the dryer operation is adiabatic, obtain with the use of the relative-humidity chart the following: (**a**) dew point, humidity, percent relative humidity, and humid volume of fresh air; (**b**) heat needed to heat 5000 ft^3 of fresh air from 75 to 250°F, (**c**) temperature at which air leaves the dryer and its humidity, and (**d**) water evaporated in the dryer per 100 lb dry air.

Solution.

a. For the air-water mixtures, the adiabatic and wet-bulb temperatures may be considered the same. To find the humidity of the fresh air, start at the intersection point of ordinate at 65°F, move to the right along the adiabatic line, and locate the intersection point of the adiabatic cooling line and the ordinate at 75°F. This point gives the initial condition of the air. At this point, from the humidity chart,

$$\text{Absolute humidity of air} = 0.0109 \text{ lb/lb dry air}$$

From the above point, move leftward along the constant-humidity line and locate the intersection point of the saturation curve and the constant-humidity line. Read the saturation temperature or dew point as 59.6°F. The point for the initial condition lies between the 50 and 60 percent relative humidities. By a rough interpolation, relative humidity is 58 percent. The humid volume of the fresh air lies between 13.5 and 14 ft^3/lb. Again by a rough interpolation, humid volume is 13.7 ft^3/lb.

b.

$$\text{Mass of 5000 ft}^3 \text{ of air} = \frac{5000}{13.7} = 365 \text{ lb}$$

From Eq. (9-17),

$$\begin{aligned}\text{Humid heat of initial air} &= 0.24 + 0.45(0.0109)\\ &= 0.245 \text{ Btu/lb dry air}\cdot{}^\circ\text{F}\end{aligned}$$

The heat needed to raise the temperature of air from 75 to 250°F is 365(0.245)(250 − 75) = 15,649 Btu.

c. Starting at 250°F and absolute humidity equal to 0.0109 lb/lb dry air, follow an adiabatic line toward the saturation curve. The intersection of the adiabatic line passing through the startup point and the saturation curve gives the following values:

$$\text{Saturation temperature} = 102°\text{F}$$

$$\text{Saturation humidity} = 0.046 \text{ lb/lb dry air}$$

Air leaving the dryer is 90 percent saturated. Therefore, the actual humidity of the air exiting from the dryer is

$$Y = 0.9(0.046) = 0.0414 \text{ lb/lb dry air}$$

Again from the humidity chart, the intersection point of constant humidity (0.0414 lb / lb dry air) and the adiabatic cooling line gives

$$\text{Dry bulb temperature of exit air} = 122°\text{F}$$

d. Water evaporated in the dryer/100 lb dry air is

$$100(0.0414 - 0.0109) = 3.05 \text{ lb}$$

Cooling Towers

Mass balance over a differential volume $S\,dZ$ of the cooling tower (Fig. 9-2*a*) gives

$$dL' = G'\,dY \tag{9-27}$$

and enthalpy balance gives

$$d(L'h) = L'C_L dt_L = G'\,dH \tag{9-28}$$

Integration over the tower from $Z = 0$ to $Z = Z$ gives

$$H_2 - H_1 = \frac{L'C_L}{G'}(t_{L2} - t_{L1}) \tag{9-29}$$

This equation defines the operating line for the tower in the H-versus-t_L coordinate system. By a consideration of the simultaneous mass and heat transfer from the interface to the bulk of a gas-vapor mixture and assuming the Lewis relationship, it can be shown that the number of transfer units is given by

$$(\text{NTU})_G = \int_{H_1}^{H_2} \frac{dH}{H_i - H} = \frac{k_G a M_B PSZ}{G'} \tag{9-30}$$

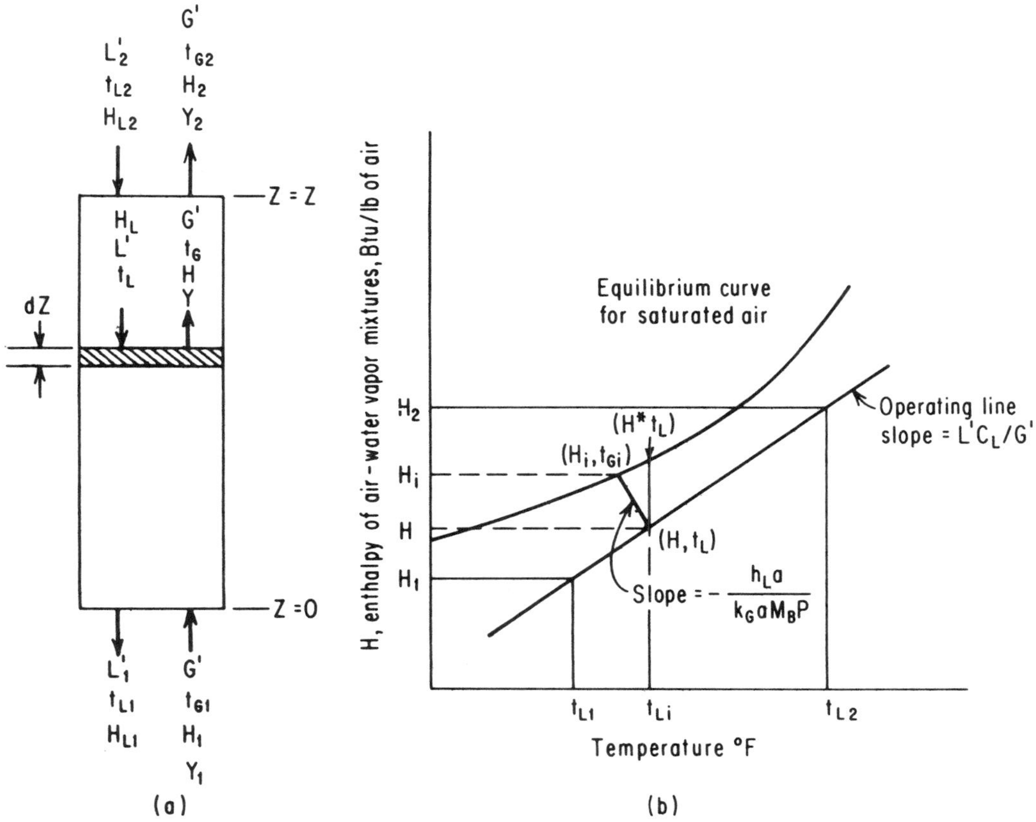

Figure 9-2. Cooling tower: (*a*) stream terminology; (*b*) operating diagram.

Then the height of transfer unit is given by

$$(\mathrm{HTU})_G = \frac{Z}{(\mathrm{NTU})_G} = \frac{G'}{k_G a M_B P S} \qquad (9\text{-}31)$$

where L' = superficial mass velocity of liquid, $\mathrm{lb/h \cdot ft^2}$
G' = superficial mass velocity of dry gas, $\mathrm{lb/h \cdot ft^2}$
C_L = specific heat of liquid at constant pressure, Btu/lb · °F
t_{L1} = temperature of water leaving the tower, °F

t_{L2} = temperature of water entering the tower, °F
Z = height of packing, ft
S = cross section of tower, ft^2
$k_G a$ = mass transfer coefficient, $lb \cdot mol/h \cdot ft^3 \cdot atm$
a = interfacial surface per unit volume of packing, ft^2/ft^3

Determination of the $(NTU)_G$ by Eq. (9-30) requires the value of the interface gas enthalpy H_i corresponding to the enthalpy H of the bulk of the gas at any given tower cross section. The expression relating these quantities can be derived by a consideration of the rate of heat transfer to the interface from the liquid and is given by

$$\frac{H_i - H}{t_{Gi} - t_L} = -\frac{h_L a}{k_G a M_B P} = -\frac{h_L a}{k_Y a} \qquad (9\text{-}32)$$

If $h_L a$ and $k_G a$ values are available, it is possible to determine H_i values. For this, the saturation curve and the operating line are plotted first. At a series of points between the top and bottom, lines are drawn with a slope of $-h_L a / k_G a M_B P$ to cut the saturation curve and the operating line (Fig. 9-2*b*). The corresponding values of H_i and H are read and tabulated. The $(NTU)_G$ can then be found out by graphical integration from a plot of $1/(H_i - H)$ versus H. Knowing the $(NTU)_G$, Z can be calculated.

$$\int_{H_1}^{H_2} \frac{dH}{H_i - H} = \frac{H_2 - H_1}{(H_i - H)_{av}} = N_{tG} \qquad (9\text{-}33)$$

The term $(H_2 - H_1)/(H_i - H)_{av}$ is termed the number of gas enthalpy transfer units N_{tG}. The height of the column is then given by

$$Z = H_{tG} N_{tG}$$

where the height of gas enthalpy transfer unit is

$$H_{tG} = \frac{G'}{k_Y a} \qquad (9\text{-}34)$$

The overall number of transfer units and overall height of transfer units are given by

$$N_{tOG} = \int_{H_1}^{H_2} \frac{dH}{H^* - H} = \frac{k_G a Z}{G'} = \frac{Z}{H_{tOG}} \tag{9-35}$$

In this case, the resistance in liquid phase is assumed negligible; therefore, H^* can be used in place of H_i, the enthalpy at the interface.

Equation (9-35) can also be used in the form

$$\frac{K_G a Z}{L'} = \int_{t_{L1}}^{t_{L2}} \frac{dt_L}{\mathrm{H}^* - \mathrm{H}} \tag{9-36}$$

which is the frequently used form in the cooling-tower industry.

Example 9-2. Water is cooled from 110 to 85°F by countercurrent contact with air in an induced-draft cooling tower at 1 atm. The air enters at dry-bulb temperature of 85°F, and the design wet-bulb temperature for the location is 75°F. The water rate is 1250 lb/h · ft² of the tower cross section and the dry air rate is 1.5 times the minimum. Calculate the height of transfer unit $(\mathrm{NTU})_G$ and the packed height Z. The tower is packed with wooden slats.*

Solution.

Step 1. Plot saturation enthalpy vs. temperature (Fig. 9-3).

Step 2. Enthalpy of air at dry-bulb temperature of 85°F wet-bulb temperature of 75°F = 38.6 Btu/lb (from the humidity chart). This point is located on the graph as point A whose coordinates are

$$t_{L1} = 85°\mathrm{F} \qquad H_1 = 38.6\ \mathrm{Btu/lb}$$

Step 3. From point A, draw a tangent AB' to the enthalpy curve.

$$\text{Slope of tangent} = \frac{89 - 38.6}{110 - 85} = 2.016$$

$$\frac{L'C_L}{G'_{min}} = 2.016 \qquad \text{or} \qquad G'_{min} = \frac{1250}{2.016} = 620\ \mathrm{lb/h \cdot ft^2}$$

*The transfer coefficients are given by $h_L a = 0.03(L')^{0.51}(G')$, $k_G a = 0.04(L')^{0.26}(G')^{0.72}$.

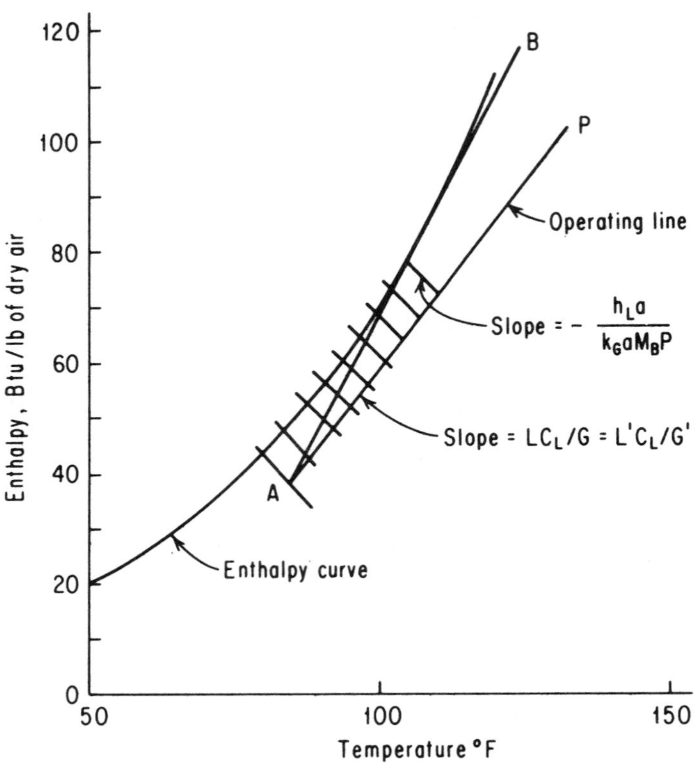

Figure 9-3. Determination of minimum L/G ratio (Example 9-2).

$C_L = 1$ for water. Actual $G' = 620(1.5) = 930$ lb/h · ft^2. Then the slope of the operating line = 1250(1)/930 = 1.344. The operating line AP with slope 1.344 is drawn through A. Then

$$H_2 - H_1 = \frac{L'C_L}{G'}(t_{L2} - t_{L1})$$

$$H_2 = H_1 + \frac{LC_L}{G}(t_{L2} - t_{L1})$$

$$= 38.6 + \frac{1250(1)}{930}(110 - 85)$$

$$= 38.6 + 33.6 = 72.2 \text{ Btu/lb}$$

TABLE 9-1 Calculation of $1/(H_i - H)$

H	H_i	$H_i - H$	$1/(H_i - H)$
72.2	78.4	6.2	0.161
68.0	73.5	5.5	0.181
64.0	69.1	5.1	0.196
56.0	60.5	4.5	0.222
52.0	56.4	4.4	0.227
48.0	52.5	4.5	0.222
42.6	47.6	5.0	0.200
38.6	43.8	5.2	0.192

Step 4. Calculation of transfer coefficients:

$$-\frac{h_L a}{k_G a M_B P} = -\frac{0.03(1250)^{0.51}(930)}{0.04(1250)^{0.26}(930)^{0.72}(29)(1)} = -1.042$$

Draw a series of lines, each of slope −1.042 to cut the operating line and the equilibrium curve, read from the graph the corresponding values of H_i and H, and prepare Table 9-1.

Make a plot of $1/(H_i - H)$ versus H and determine the area under the curve between H_2 and H_1 or by Simpson's rule,

$$(\text{NTU})_G = \int_{H_1}^{H_2} \frac{dH}{H_i - H} = \frac{16.8}{3}[0.192 + 4(0.236) + 0.161]$$

$$= 7.3 \text{ transfer units}$$

Step 5. Calculation of $(\text{HTU})_G$ and Z:

$$k_G a = 0.04(1250)^{0.26}(930)^{0.72} = 35 \text{ mol/h}\cdot\text{ft}^3\cdot\text{atm}$$

$$(\text{HTU})_G = \frac{930}{35(29)(1)} = 0.92 \text{ ft}$$

$$Z = (\text{NTU})_G(\text{HTU})_G = 7.3(0.92) = 6.72 \text{ ft}$$

Example 9-3. In Example 9-2, evaluate the following: (**a**) number of gas enthalpy transfer units; (**b**) height of gas enthalpy transfer unit; and (**c**) overall packing height on the basis of the gas-phase transfer units.

Solution. The number of gas enthalpy transfer units is

$$\frac{H_2 - H_1}{(H_i - H)_{av}} = N_{tG}$$

$$H_2 - H_1 = 72.2 - 38.6 = 33.6 \text{ Btu}$$

$$(H_i - H)_1 = 5.2 \text{ Btu} \qquad (H_i - H)_2 = 6.2 \text{ Btu}$$

$$(H_i - H)_{av} = \frac{6.2 - 5.2}{\ln (6.2/5.2)} = 5.69$$

$$N_{tG} = \frac{H_2 - H_1}{(H_i - H)_{av}} = \frac{33.6}{5.69} = 5.9 \text{ units}$$

$$H_{tG} = \frac{G'}{k_y a} = \frac{930}{k_G a M_B P} = \frac{930}{35(29)} = 0.92 \text{ ft}$$

$$Z = 5.90(0.92) = 5.43 \text{ ft}$$

References

1. R. E. Treybal, *Mass Transfer Operations,* McGraw-Hill, New York, 1980, p. 240.
2. *Chemical Engineers' Handbook*, 5th ed., R. H. Perry (ed.), McGraw-Hill, New York, 1973, p. 20-6.

Chapter

10

Drying

Drying generally refers to the removal of small amounts of volatile liquid, usually water, from solids by evaporation into a gas stream. Many types of drying equipment are available, e.g., tray, turbo, flash, fluid bed, spray, rotary, and thin film.

Equilibria

When a wet solid is brought in contact with air, the solid tends to lose moisture if the humidity of the air is lower than the humidity corresponding to the moisture content of the solid. If the air is more humid than the solid in contact with it, the solid will gain moisture until an equilibrium is attained.

Equilibrium Moisture. The portion of water in the wet solid that cannot be removed by the air in its contact is called the *equilibrium moisture*. Since an equilibrium is reached between the air and the solid, there is no humidity driving force between them.

Bound water in a solid exerts an equilibrium vapor pressure lower than that of pure water at the same temperature.

Unbound water exerts the equilibrium vapor pressure of pure water at the prevailing conditions.

Free Moisture. The free moisture is the amount of water in a wet solid in excess of the equilibrium moisture. Thus

$$X = X_T - X^*$$

where X = free moisture content of wet solid, lb/lb dry solid
X_T = total moisture content of wet solid, lb/lb dry solid
X^* = equilibrium moisture content of solid, lb/lb dry solid

Rate-of-Drying Curve

When the drying rates are determined and plotted against the

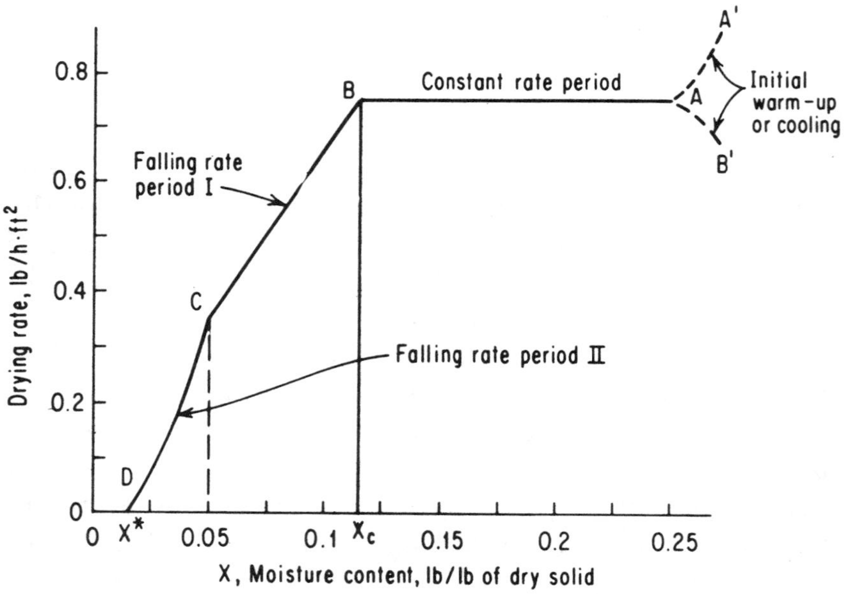

Figure 10-1. Rate of drying curve for solid (Example 10-1).

moisture content of the solid, a curve such as in Fig. 10-1 is obtained. There are usually two major parts to the rate curve, a portion over which the rate of drying is constant and another during which the drying rate is falling. Different solids and different drying conditions, however, may result in curves of different shapes in the falling-rate period.

In the constant-rate period, the evaporation rate is constant because of the presence of unbound moisture. The evaporation takes place from the surface. The constant drying rate can be expressed in terms of a constant mass-transfer coefficient as

$$N_c = k_Y(Y_s - Y) \tag{10-1}$$

where N_c = drying rate, lb/h · ft^2
k_Y = water evaporated, lb/h · ft^2 · ΔY
Y_s = saturation humidity at liquid-surface temperature, lb moisture/lb dry gas
Y = humidity of drying air or gas, lb moisture/lb dry air or gas

Under constant drying conditions k_Y is constant and, therefore, the rate of evaporation N_c is constant. If the solid is initially cold, its temperature slowly rises to t_s, the ultimate surface temperature (part *AB* of curve).

When the average value of the moisture content reaches a value of X_c equal to the critical moisture content, the drying rate begins to fall. The part *BC* represents the first falling-rate period in which unsaturated surface drying takes place. On further drying, diffusion from the interior of the solid controls the drying process, and the second falling-rate period is obtained (part *CD*). This part of the curve may not exist entirely in many cases, depending upon the solid and the drying conditions.

When the moisture content reaches the equilibrium moisture content value X^*, there is no more evaporation of water and the drying stops.

Rate of Drying

This can be expressed by the following equation

$$N = \frac{L_s\, dX}{A\, d\theta} \tag{10-2}$$

The time of drying θ is given by

$$\theta = \int d\theta = \frac{L_s}{A}\int_{X_2}^{X_1} \frac{dX}{N} \tag{10-3}$$

where L_s = dry solid, lb
M_D = dry solid lb/h · ft^2 for continuous drying
= L_s/A
A = drying surface, ft^2, or the cross section of bed for through circulation drying

Constant Rate Period. If X_1 and $X_2 > X_c$, the critical moisture content, then

$$\theta_c = \frac{L_s(X_1 - X_2)}{AN_c} \tag{10-4}$$

where N_c = constant rate of drying, lb/h · ft^2
A = drying surface or cross-sectional area perpendicular to gas flow, ft^2
θ_c = time of drying in constant rate period, h

Falling Rate Period. If X_1 and X_2 are both less than X_c, the integral in Eq. (10-3) has to be determined, graphically in the general case when the integral cannot be obtained mathematically, and by integration, in the special case when the rate of the drying curve is a straight-line function of X. Thus, if

$$N = mX + b \tag{10-5}$$

where m is the slope of the falling-rate curve and b is the intercept of the falling-rate curve. Then

$$\theta_f = \frac{L_s}{A}\int_{X_2}^{X_1} \frac{dX}{mX + b} = \frac{L_s}{mA} \ln \frac{mX_1 + b}{mX_2 + b} \tag{10-6}$$

but

$$N_1 = mX_1 + b \qquad N_2 = mX_2 + b \qquad m = \frac{N_1 - N_2}{X_1 - X_2} \tag{10-7}$$

Therefore

$$\theta_f = \frac{L_s(X_1 - X_2)}{A(N_1 - N_2)} \ln \frac{N_1}{N_2} = \frac{L_s(X_1 - X_2)}{AN_m} \tag{10-8}$$

where the log mean drying rate is given by

$$N_m = \frac{N_1 - N_2}{\ln (N_1/N_2)} \tag{10-9}$$

Very often the entire falling rate can be represented by a straight line or can be so approximated; then

$$\theta_f = \frac{L_s(X_c - X^*)}{N_c A} \ln \frac{X_1 - X^*}{X_2 - X^*} \tag{10-10}$$

Example 10-1. A batch of solid for which the drying rate curve of Fig. 10-1 applies is to be dried from a moisture content of 20 percent to a moisture content of 2 percent. Initial weight of the solids is 600 lb and the drying surface is 1 ft^2/10 lb dry weight. Estimate the total time of drying.

Solution. Initial moisture content $X_1 = 0.2/0.8 = 0.25$ lb/lb dry solid. Final moisture content $X_3 = 0.02/0.98 = 0.020$ lb/lb dry solid. From Fig. 10-1, critical moisture content, $X_c = 0.115$ lb/lb dry solid. Moisture content at the end of the first falling rate period is

$$X_2 = 0.05 \text{ lb/lb dry solid}$$

The rate of drying in constant-rate period from Fig. 10-1 is

$$N_c = 0.75 \text{ lb/h} \cdot \text{ft}^2$$

Constant-rate period:

$$\frac{L_s}{A} = \frac{\text{lb dry solid}}{\text{area}} = \frac{10}{1} = 10$$

$$\theta_c = \frac{L_s(X_1 - X_c)}{AN_c} = \frac{10(0.25 - 0.115)}{0.75} = 1.8 \text{ h}$$

Falling-Rate Period I.

$$N_c = 0.75 \text{ lb/h} \cdot \text{ft}^2 \qquad N_2 = 0.35 \text{ lb/h} \cdot \text{ft}^2$$

Applying Eq. (10-8) to the falling rate period gives

$$\theta_2 = \frac{L_s(X_1 - X_2)}{A(N_1 - N_2)} \ln \frac{N_1}{N_2}$$

$$= \frac{10(0.115 - 0.05)}{1(0.75 - 0.35)} \ln \frac{0.75}{0.35} = 1.24 \text{ h}$$

Falling-Rate Period II. Since the rate curve is not a straight line, graphical integration is required. Table 10-1 is prepared using the values of N read from Fig. 10-1.

TABLE 10-1 Calculation of $1/N$ at Various Values of X

X	N	$\frac{1}{N}$	X	N	$\frac{1}{N}$
0.05	0.35	2.94	0.045	0.28	3.571
0.04	0.22	4.55	0.035	0.16	6.25
0.03	0.116	8.62	0.025	0.078	12.821
0.02	0.035	28.57	0.0225	0.058	17.24

A plot of $1/N$ versus moisture content X is made (not shown here), and the area under the curve is determined to be 0.283. Therefore, the time of drying from $X_2 = 0.05$ to $X_3 = 0.02$ lb/lb dry solid is

$$\theta_f = \frac{10}{1} 0.283 = 2.83 \text{ h}$$

Then, total time of drying is 1.8 + 1.24 + 2.83 = 5.87 h.

References

1. R. E. Treybal, *Mass Transfer Operations*, McGraw-Hill, New York, 1980, pp. 655–710.

Chapter

11

Filtration

Given the filtration conditions and the specifications of a filter, the common problems to be solved are (1) the amount of the filtrate obtainable in a given time, (2) the amount of the solvent that can be passed through the cake in a given time, and (3) concentration of the recovered material in the wash solvent.

Filtration can be carried out in two ways: (1) constant-pressure filtration, i.e., the filtration rate varies, and (2) constant-rate filtration in which the ΔP varies. Two methods are available to treat filtration problems. They are (1) Ruth's equation[1] and (2) the Kozeny-Carman[2] relation.

Ruth's Equation

To establish Ruth's equation, the following variables are first defined:

$$V = \text{filtrate, ft}^3$$

$$\rho = \text{filtrate density, lb/ft}^3$$

$$W = \text{filterable solids, lb}$$

$$m = \frac{\text{wet cake}}{\text{dry cake}} \qquad \text{(washed)}$$

x = mass fraction of the solids in a slurry

Using these definitions, the following relations are developed:

$$\text{Liquid in slurry} = \frac{W}{x}(1 - x) \qquad \text{lb} \tag{11-1a}$$

$$\text{Liquid retained by cake} = (m - 1)W \qquad \text{lb} \tag{11-1b}$$

from which

$$\text{Filtrate} = V\rho = \frac{W}{x}(1 - x) - W(m - 1) \quad \text{lb} \tag{11-1c}$$

$$W = \frac{V\rho x}{1 - mx} \qquad \text{lb} \tag{11-1d}$$

The Hagen-Poiseuille equation for the laminar flow of fluids in tubes of circular cross section is given by

$$\Delta P = \frac{32Lu\mu}{g_c D^2} \tag{11-2}$$

Assuming Eq. (11-2) applies to the filtrate flow, it can be shown that the rate of filtration is given by

$$u = \frac{1}{A}\frac{dV}{d\theta} = \frac{\Delta P_C}{32\mu L/g_c D_e^2} \qquad \text{or} \qquad \frac{dV}{d\theta} = \frac{A}{\mu}\frac{\Delta P_C}{R_C} \tag{11-3}$$

where A = total filtering area, ft^2
θ = time of filtration, h
μ = viscosity of filtrate, lb/ft · h
L = thickness of cake, ft
ΔP_C = pressure drop across cake, lbf/ft^2
R_C = resistance of cake = $32L/g_c D_e^2$, $h^2 \cdot lbf/ft^2 \cdot lb$
D_e = equivalent diameter, ft; for the cake of flow area A, ft^2

g_c = Newton's law proportionality factor = 4.18×10^8 ft · lb/lbf · h^2

In actual practice, the ΔP_C across the cake is not easily measurable and, therefore, the resistance of the whole filter (not only of the cake) with a pressure drop ΔP is considered.

Then
$$R_T = R_1 + R_C + R_2 \tag{11-3a}$$

where R_T = total resistance
R_1 = resistance of the filter medium
R_2 = resistance of the slurry leads and channels

R_2 is generally neglected. This reduces Eq. (11-3a) to

$$R_T = R_1 + R_C \tag{11-3b}$$

Assume R_1 to be the resistance of a fictitious cake of mass W_1 lb and to have been deposited by the filtrate of volume V_1 ft^3. Then

$$R_T = R_1 + R_C = \frac{(W + W_1)\alpha}{A}$$

also
$$R_C = \frac{W\alpha}{A} \quad \text{and} \quad R_1 = \frac{W_1\alpha}{A} \tag{11-4}$$

where α is the average specific resistance of cake, h^2 · lbf/lb^2.

Then
$$W + W_1 = \frac{x}{1 - mx}(V + V_1) \quad \text{and} \quad \frac{dV}{d\theta} = \frac{A}{\mu}\frac{\Delta P}{R_T} \tag{11-5}$$

After eliminating R_T and $W + W_1$ from Eqs. (11-4) and (11-5), one obtains

$$(V + V_1)\, dV = (V + V_1)\, d(V + V_1) = \frac{A^2\, \Delta P(1 - mx)}{\mu \rho x \alpha}\, d\theta \tag{11-6}$$

Constant-Pressure Filtration. Integration of Eq. (11-6) yields (assuming ΔP constant):

$$\int_0^{V+V_1} (V + V_1)\, d(V + V_1) = \frac{A^2\, \Delta P(1 - mx)}{\mu \rho x \alpha} \int_{-\theta_1}^{\theta} d\theta \tag{11-6a}$$

where θ_1 is the imaginary time to lay down the fictitious cake. From Eq. (11-6*a*) the completion of integration gives

$$(V + V_1)^2 = \frac{2A^2\,\Delta P(1 - mx)}{\mu\rho x\alpha}(\theta + \theta_1) \qquad \text{at const. } \Delta P \tag{11-7}$$

Equation (11-7) can be written as

$$(V + V_1)^2 = C(\theta + \theta_1) \tag{11-8}$$

where

$$C = \frac{2A^2\,\Delta P(1 - mx)}{\mu\rho x\alpha} \qquad \text{ft}^6/\text{h} \tag{11-8a}$$

α will be constant for the noncompressible cakes. The constants C, θ_1, and V_1 are obtained by the use of the equation obtained by differentiating Eq. (11-8); thus

$$\frac{d\theta}{dV} = \frac{2V}{C} + \frac{2V_1}{C} \tag{11-8b}$$

A plot of $\Delta\theta/\Delta V$ versus V gives the slope $= 2/C$ and the intercept $2V_1/C$, from which the constants V_1 and C can be calculated.

Constant-Rate Filtration. When the filtration is conducted at a constant filtration rate, the ΔP changes during the filtration and is given by

$$\Delta P = \frac{\rho x\mu\alpha}{A^2(1 - mx)}\left(\frac{dV}{d\theta}\right)_C^2(\theta + \theta_1) \tag{11-9}$$

A plot of ΔP versus θ is a straight line which gives

$$\text{Slope} = \frac{\rho x\mu\alpha}{A^2(1 - mx)}\left(\frac{dV}{d\theta}\right)_C^2$$

and

$$\text{Intercept} = \frac{\rho\mu x\alpha\theta_1\left(\dfrac{dV}{d\theta}\right)_C^2}{A^2(1 - mx)} \tag{11-10}$$

from which the constants α and θ_1 can be calculated.

Rate of Washing

If displacement washing is assumed, the rate of washing is constant and equal to the rate of filtration at the end of the filtration cycle. Thus, the rate of washing is given by

$$\frac{dV}{d\theta} = \frac{C}{2(V + V_1)} \tag{11-10a}$$

where $dV/d\theta$ is the rate of the filtration at $\theta = \theta_f$ or at the end of the filtration period.

Kozeny-Carman Equation[2]

By mass balance,

$$(1 - X)LA\rho_s = V\rho m = \frac{(V + XLA)\rho x}{1 - x} \tag{11-11}$$

where L = thickness of the cake, ft
A = area of filter medium, ft^2
X = porosity of the cake = volume of void space/total volume of the cake
ρ_s = density of solids in the cake, lb/ft^3
ρ = density of the filtrate, lb/ft^3
m = mass ratio of dry cake/filtrate
V = volume of filtrate, ft^3

The velocity through a bed (assuming laminar flow) is given by

$$u = \frac{1}{A}\frac{dV}{d\theta} = \frac{K\overline{\ell w_f}\rho}{L\mu} = \frac{K\,\Delta P_C}{L\mu} \tag{11-12}$$

where $\overline{\ell w_f}$ = frictional losses/unit mass of fluid, ft · lbf/lb
u = velocity, ft/h
K = permeability of cake, $lb \cdot ft^3/lbf \cdot h^2$
ΔP_C = pressure drop through the cake, lbf/ft^2

Solving Eq. (11-11) for L, substituting in Eq. (11-12), and defining $\alpha = 1/K\rho_s(1 - X)$,

$$\frac{dV}{d\theta} = \frac{A^2\,\Delta P_C}{Vm\alpha\rho\mu} = \frac{\Delta P_C}{V\alpha m\mu\rho/A^2} \tag{11-13}$$

$$= \frac{\Delta\,P_C A^2}{2VC_V} \tag{11-13a}$$

where $$C_V = \frac{\alpha m\rho\mu}{2} \tag{11-14}$$

However, in actual practice, the total ΔP needs to be used since ΔP_C is not known. If it is assumed that the fictitious equivalent resistance of an equivalent cake thickness L_e which forms an equivalent filtrate volume V_e is R_e, then equivalent resistance is given by

$$R_e = \frac{V_e\alpha m\rho\mu}{A^2} \quad \text{or} \quad \frac{2C_V V_e}{A^2} \tag{11-15}$$

Therefore $$\frac{dV}{d\theta} = \frac{A^2\,\Delta P}{\alpha m\rho\mu(V + V_e)} = \frac{A^2\,\Delta P}{2C_V(V + V_e)} \tag{11-16}$$

where ΔP is the total resistance. For a constant-pressure filtration, the integration of Eq. (11-16) gives

$$\theta = \frac{C_V(V^2 + 2VV_e)}{A^2\,\Delta P} \tag{11-17}$$

Solving Eq. (11-11) for V, substituting for V in (11-12), defining $C_L = \mu[\rho_s(1 - x)(1 - X) - \rho x X]/2K\rho x$ and an equivalent cake thickness L_e,

$$\frac{dL}{d\theta} = \frac{\Delta P}{2C_L(L + L_e)} \tag{11-18}$$

At constant porosity,

$$\theta = \frac{C_L(L^2 + 2LL_e)}{\Delta P} \tag{11-19}$$

Constant-Pressure Filtration. If the rate data are at constant ΔP, the plot of $\Delta\theta/\Delta V$ versus V gives

$$\text{Slope} = \frac{2C_V}{A^2\,\Delta P} \qquad \text{Intercept} = \frac{2C_V V_e}{A^2\,\Delta P} \tag{11-20}$$

from which the constants C_V and V_e are calculated.

Constant-Rate Filtration. If the data are taken at constant rate of filtration, V_e is constant, and then

$$\Delta P = \frac{2C_V}{A^2}\left(\frac{dV}{d\theta}\right)_C V + \frac{2C_V}{A^2}\left(\frac{dV}{d\theta}\right)_C V_e \tag{11-21}$$

A plot of ΔP versus V enables one to find the constants in the filtration equation. Thus

$$\text{Slope} = 2\frac{C_V}{A^2}\left(\frac{dV}{d\theta}\right)_C \qquad \text{intercept} = 2\frac{C_V}{A^2}\left(\frac{dV}{d\theta}\right)V_e \tag{11-21a}$$

If the area of the filter and the rate of filtration are known, C_V and V_e can be calculated using Eq. (11-21*a*).

The permeability K of the Kozeny-Carman equation (11-12) and the specific resistance α of Ruth's equation (11-4) are related by

$$K = \frac{AL}{\alpha W} \tag{11-22}$$

Equation (11-10) can be written in a simplified form as

$$\frac{d\theta}{dV} = \frac{V + V_1}{C'} \qquad \text{where } C' = \frac{C}{2} \qquad \text{hence } \frac{dV}{d\theta} = \frac{C'}{V + V_1} \tag{11-23}$$

Example 11-1. Volume of the filtrate collected is 1 gal when the filtration rate is 1.5 gpm, and it is 5 gal when the filtration rate is 0.6 gpm. Calculate the volume collected when the filtration rate is 0.75 gpm. What is the total volume collected in 10 min? Assume constant-pressure filtration.

Solution. From Eq. (11-23),

$$d\theta = \frac{1}{C'}(V + V_1)dV$$

By integration,

$$\theta = \frac{1}{C'}\int (V + V_1)\, dV = \frac{1}{C'}\left(\frac{V^2}{2} + V_1 V\right) \qquad (a)$$

Calculate constants C' and V_1. When $dV/d\theta = 1.5$ gpm, $V = 1$ gal. Therefore,

$$\frac{3}{2} = \frac{C'}{1 + V_1} \qquad \text{or} \qquad 3 + 3V_1 = 2C' \qquad (b)$$

When $dV/d\theta = 0.6$ gpm, $V = 5$ gal, and

$$\frac{3}{5} = \frac{C'}{5 + V_1} \qquad \text{or} \qquad 15 + 3V_1 = 5C' \qquad (c)$$

Solving (*b*) and (*c*) for C' and V_1, $C' = 4$ and $V_1 = \frac{5}{3}$.

When the filtration rate is $\frac{3}{4}$ gpm, the filtrate volume collected is obtained by the substitution of $dV/d\theta = \frac{3}{4}$, $V_1 = \frac{5}{3}$, and $C' = 4$ in Eq. (11-23) and is given by

$$\frac{3}{4} = \frac{4}{V + \frac{5}{3}} \qquad \text{or} \qquad V = \frac{11}{3} = 3.67 \text{ gal}$$

The time required to collect 3.67 gal is

$$\frac{1}{4}\left[\frac{11^2}{2(9)} + \frac{5}{3}\left(\frac{11}{3}\right)\right] = 3.21 \text{ min}$$

Total Volume Collected at the End of 10 min. Substituting $\theta = 10$ min in the integrated equation (*a*) gives

$$10 = \tfrac{1}{4}\left(\tfrac{1}{2}V^2 + \tfrac{5}{3}V\right)$$

By simplification, $240 = 3V^2 + 10V$ or $3V^2 + 10V - 240 = 0$; i.e., $V^2 + \frac{10}{3}V - 80 = 0$. From which,

$$V = \tfrac{1}{2}\left[-\tfrac{10}{3} + \sqrt{(\tfrac{10}{3})^2 + 320}\right] = 7.43 \text{ gal}$$

Example 11-2. E. L. McMillen and H. A. Webber[3] obtained the following data in the constant-pressure filtration of a slurry of $CaCO_3$ in water. The mass fraction of solids in the slurry to the press was 0.139. The filter area was 1 ft^2. Mass ratio of wet cake to dry cake was 1.47 and dry-cake density was 73.5 lb/ft^3. Rate data obtained at 50 psi were as follows.

Time, s	0	19	68	142	241	368	702
Filtrate, lb	0	5	10	15	20	25	35

$\rho_{solids} = 164$ lb/ft^3 $\qquad \mu_{filtrate} = 0.86$ cP

Calculate the specific resistance of the cake α, the filter medium resistance R_1, and the porosity of the cake.

Solution. Equation (11-8), which is in terms of volume, can be converted to units of pounds with the use of the density ρ as follows:

$$\frac{\rho\, d\theta}{3600\, \rho\, dV} = \frac{2\rho V}{C\rho} + \frac{2V_1\rho}{\rho C}$$

in which θ is in seconds. Since $\rho\, dV = d(\rho V) = dW'$ and $\rho V = W'$, the above reduces to

$$\frac{d\theta}{dW'} = \frac{7200W'}{\rho^2 C} + \frac{7200W_1'}{\rho^2 C}$$

In the above equation, W' is the filtrate, lb, and W_1' is the filtrate, lb, corresponding to the fictitious cake. Thus at constant ΔP, a plot of $d\theta/dW'$ versus W' will be a straight line. To make this plot for the example, the calculations of Table 11-1 are first made.

TABLE 11-1 Calculations of the Filtration Rates and Average Filtrate Flows

θ	W'	$\Delta\theta$	$\Delta W'$	$\dfrac{\Delta\theta}{\Delta W'}$	$\overline{W'} = \dfrac{(W')_1 + (W')_2}{2}$
(Given)	(Given)	←		Calculated	→
s	lb	s	lb	lb/s	lb
0	0	....	....	....	....
19	5	19	5	3.8	2.5
68	10	49	5	9.8	7.5
142	15	74	5	14.8	12.5
241	20	99	5	19.8	17.5
368	25	127	5	25.4	22.5
524	30	156	5	31.2	27.5
702	35	178	5	35.6	32.5

A plot of $\Delta\theta/\Delta W'$ versus $\overline{W'} = \frac{1}{2}(W_1' + W_2')$ is made in Fig. 11-1. From this figure, intercept = 1.2 and

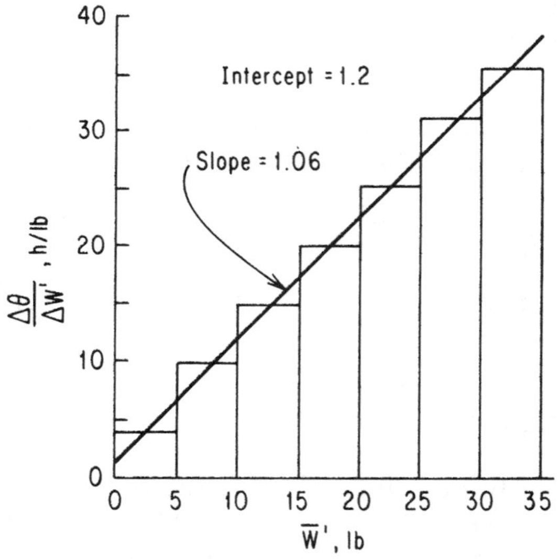

Figure 11-1. A plot of $\Delta\theta/\Delta W'$ versus W' (Example 11-2).

$$\text{Slope} = \frac{35.6 - 3.8}{32.5 - 2.5} = \frac{31.8}{30} = 1.06 = \frac{7200}{\rho^2 C}$$

Therefore $$C = \frac{7200}{\rho^2(1.06)} = \frac{7200}{62.4^2(1.06)} = 1.744 \text{ ft}^6/\text{h}$$

$$\text{Intercept} = 1.2 = \frac{7200 W'_1}{C\rho^2}$$

Therefore $$W'_1 = \frac{1.2C\rho^2}{7200} = \frac{1.2(1.744)(62.4)^2}{7200} = 1.13 \text{ lb}$$

$$V_1 = \frac{1.13}{\rho} = \frac{1.13}{62.4} = 0.018 \text{ ft}^3$$

Calculation of θ_1.

$$\int_{-\theta_1}^{\theta} d\theta = \frac{7200}{\rho^2 C}\int_0^{(W' + W'_1)} (W' + W'_1)\, dW'$$

or $$\theta + \theta_1 = \frac{3600}{\rho^2 C}(W' + W'_1)^2$$

Substitution of the values at $\theta = 19$ s, and $W' = 5$ lb in the above equation gives

$$19 + \theta_1 = \frac{3600}{(62.4)^2(1.744)}(5 + 1.13)^2$$

from which

$$\theta_1 = 0.92 \text{ s} = 2.56 \times 10^{-4} \text{ h}$$

Specific Resistance. From Eq. (11-8*a*),

$$\frac{2A^2\, \Delta P\, (1 - mx)}{\mu\rho x\alpha} = C$$

Therefore $$\alpha = \frac{2A^2\, \Delta P\, (1 - mx)}{\mu\rho x C}$$

$$= \frac{2(1)^2(144)(50)[1 - 1.47(0.139)]}{2.07(62.4)(0.139)(1.744)}$$

$$= 365.9 \text{ h}^2 \cdot \text{lbf/lb}^2$$

Filter-Medium Resistance R_1. By definition [Eq. (11-4)], the filter-medium resistance is given by

$$R_1 = \frac{W'_1 \alpha}{A}$$

$$= \frac{(1.13 \text{ lb})(365.9 \text{ h}^2 \cdot \text{lbf/lb}^2)}{1 \text{ ft}^2}$$

$$= 413.5 \text{ h}^2 \cdot \text{lbf/ft}^2 \cdot \text{lb}$$

Porosity of Cake.

$$(1 - X)\,\rho_S = 73.5$$

Therefore

$$X = 0.552 \text{ or } 55.2 \text{ percent}$$

References

1. B. F. Ruth, *Ind. Eng. Chem.*, vol. 27, 1935, p. 708.
2. P. C. Carman, *Trans. Inst. Chem. Engrs. (London)*, vol. 16, 1938, pp. 168–188.
3. E. L. McMillen and H. A. Webber, *Trans. Am. Inst. Chem. Engrs.*, vol. 34, 1938, p. 213.

Chapter

12

Thermodynamics

In past P.E. examinations, there has been a special emphasis on the understanding and application of the basic principles of thermodynamics. Candidates are required to answer questions on the application of mass balance, first and second laws of thermodynamics, equations of state, thermodynamic property evaluation for real substances, and chemical equilibria.

Thermodynamic System

A system is a portion of the universe selected or set apart for study. It may be a specified volume in space or a given quantity of matter. Anything outside this system is termed its *surroundings*. In practice, the surroundings refer to the immediate surroundings of the system. A system with constant mass is called a *closed system*; without a constant mass, a system is called an *open system*. Boundaries of a system can be real or imaginary, stationary or movable.

Thermodynamic Properties

Any system characteristic that is observable and/or measurable is called the *system property*, e.g., temperature, pressure, mass, area, volume, surface tension. A thermodynamic state of the system is a particular condition of the system which is characterized and fixed by certain properties of the system. Properties are distinguished as either average or point and fundamental or derived. Temperature and pressure are fundamental point properties; mass and volume are not point properties. An extensive property of a system is one which can be obtained by summing up the properties (same) of its parts. Intensive properties of a system do not depend upon the extent (mass) of the system, e.g., temperature, pressure, specific volume.

Mass-Balance Equation

Application of the principle of accounting or balance to mass of a system over a chosen accounting period gives the mass-balance equation

$$M_E - M_B = \Sigma M_I - \Sigma M_O + \Sigma M_P - \Sigma M_C \qquad (12\text{-}1)$$

where M_E = mass of system at end of accounting period
M_B = mass of system at beginning of accounting period
ΣM_I = sum of all masses that entered system during accounting period
ΣM_O = sum of all masses that left system during accounting period
ΣM_P = mass changes resulting from physical or chemical transformations
ΣM_C = mass changes connected with atomic transmutations and relativistic effects (This term is zero in ordinary physical and chemical transformations.)

In the application of the mass-balance equation to a system under consideration, the following points are worth noting: (1) If the system undergoes physical changes only, $\Sigma M_P = 0$ and $\Sigma M_C = 0$. (2) If the system undergoes a chemical transformation,

$\Sigma M_C = 0$ and ΣM_P may be zero or nonzero depending upon the system chosen. For the system as a whole, $\Sigma M_P = 0$. (Law of conservation of mass applies if the mass balance is considered.) If one of the component species is chosen as a system, $\Sigma M_P \neq 0$. If a mole balance is considered, moles may not be conserved. (3) For a steady-state system, $M_E = M_B$, i.e., mass within the confines of the system remains constant.

Definitions of Energy

Certain energy terms are required to postulate the first law of thermodynamics.

Internal Energy *U*. Internal or intrinsic energy is the energy associated with a system because of component molecules having kinetic energies from translational, rotational, and vibrational motions and having potential energy from intermolecular forces. The molecular level (microscopic) kinetic and potential energies are grouped together as the internal energy of the system and is denoted by U. For a closed, i.e., constant-mass system, it is a unique state property of the system.

Kinetic Energy KE. All energy associated with the macroscopic motion of the mass of a system is called *kinetic energy*. It can be expressed as $\text{KE} = \phi(M, u) = Mu^2/2g_c$, where M and u are the mass and velocity of the system, respectively.

Potential Energy PE. Potential energy is the energy associated with a system macroscopically due to the system interactions with the gravitational, electric, and magnetic fields that exist in the surroundings of the system.

Heat Energy Q. Heat Q is the energy transferred across the boundaries of a system under the influence of a temperature difference during the accounting period; it is not associated with the mass transferred across the boundaries of the system.

Work *W*. This is defined as the force moving through a distance or

$$W = (\text{force})(\text{distance})$$

The definition is extended to include all forms of work, such as mechanical, electrical, etc. The mechanical equivalence of work and heat was established by Joule. Work is a form of energy which is transferred across the boundaries of a system during an accounting period and is not associated with either the mass transfer across the boundary or the temperature difference between the surroundings and the system.

If fluid pressure is the only force acting on a system, the work is given by

$$\delta W = PA\, d\left(\frac{V}{A}\right) \tag{12-2}$$

where P = pressure
V = volume
A = area

Since A is constant,

$$\delta W = P\, dV \tag{12-2a}$$

and

$$W = \int_{V_1}^{V_2} P\, dV \tag{12-2b}$$

Both the heat Q and work W are energies in transition and are not system state properties. Q is taken as positive when it is transferred to the system and negative when it leaves the system. Work is positive when it is transferred from the system to the surroundings and negative when it is done on the system.

Enthalpy Function *H*. The sum $U + PV$ is given the symbol H and is called the *enthalpy* of the system. Since U, P, and V are system state properties, enthalpy H is also a state property.

First Law of Thermodynamics

The first law of thermodynamics or the law of conservation of energy in the most general form can be written for a given accounting period as

$$(U + \text{PE} + \text{KE})_E - (U + \text{PE} + \text{KE})_B = \Sigma(H + \text{PE} + \text{KE})_t + Q - W - \Sigma E_c \qquad (12\text{-}3)$$

where $(U + \text{PE} + \text{KE})_E$ = sum of internal, potential, and kinetic energies associated with system at end of accounting period

$(U + \text{PE} + \text{KE})_B$ = sum of internal, potential, and kinetic energies associated with system at beginning of accounting period

Q = heat transferred across boundaries of system during accounting period

W = all work that crosses boundaries of system during accounting period

$\Sigma(H + \text{PE} + \text{KE})_t$ = energy transfer associated with mass transfer

ΣE_c = energy converted to mass within system boundaries due to atomic transmutations ($\Sigma E_c = 0$ during ordinary physical and chemical changes.)

Entropy and Second Law of Thermodynamics

Mass and energy balances alone are not always sufficient to solve all the problems involving energy flows. In energy balance equations both work and heat are considered to be forms of energy. However, they are energies of different quality. Whereas work can be transformed into another form of work or heat quantitatively, it is known from practical experience that heat cannot be converted to work quantitatively and that heat cannot be transferred from a lower temperature to a higher temperature without the aid of an external agent. Heat is therefore an energy of a lower or degraded quality when compared to work, and it depends upon temperature. The higher the temperature at which it is available in reference to a given lower temperature (usually that of surroundings), the higher is its quality. An additional state variable to account for the unidirectional nature of the heat flow and the less efficient conversion of heat into work is provided by the postulation of the second law of thermodynamics.

Second Law of Thermodynamics

No practical engine (not even a reversible one) can convert heat into work quantitatively, or it is impossible for a self-acting machine unaided by an external agency to transfer heat from a lower temperature to a higher one.

The second law of thermodynamics limits the efficiency of a process involving the conversion of heat into work. The Carnot principle states that the efficiency of a reversible engine depends only upon the temperature levels at which the engine absorbs and rejects heat and is independent of the medium used. Thus, if a reversible engine abstracts heat Q_1 at a higher tempeature T_1 and rejects heat Q_2 at the lower temperature T_2 while doing a net amount of work W_{rev} in the process, the efficiency of the engine is given by

$$\text{Efficiency } \eta_{\text{rev}} = \frac{W_{\text{rev}}}{Q_1} = \frac{Q_1 - Q_2}{Q_1} = \frac{T_1 - T_2}{T_1} \tag{12-4}$$

from which also follows the relation

$$\frac{Q_2}{Q_1} = \frac{T_2}{T_1} \tag{12-4a}$$

If $Q_2 = 0$, $T_2 = 0$ since $Q_1 \neq 0$ and $T_1 \neq 0$, which means the zero on the absolute scale of the temperature is the temperature of a heat reservoir to which a reversible heat engine rejects no heat because all the absorbed heat is converted to work. (This is true only of the Carnot reversible idealized heat engine.) The Carnot principle thus enables one to define zero on the temperature scale. It can also be deduced from the above that all reversible engines operating between the same temperature limits have the same efficiencies.

For an engine undergoing reversible Carnot cycle, it can be shown that

$$\frac{Q_1}{T_1} - \frac{Q_2}{T_2} = 0 \tag{12-5}$$

and for a differential reversible cycle, regardless of path,

$$\oint \frac{dQ_R}{T} = 0 \tag{12-6}$$

The integral $\int dQ_R/T$ is the same for any reversible path and is dependent upon only the initial and final states. This integral is termed the change in entropy which is a state property and was first introduced by Clausius. Thus

$$dS = \frac{dQ_R}{T} \tag{12-7}$$

where S is entropy $= \int dQ_R/T$.

In an engine which uses a finite temperature difference for the transfer of heat, the work done by the engine is given by

$$Q_1 - Q_2 - W = 0 \tag{12-8}$$

and

$$\frac{Q_1}{T_1} - \frac{Q_2}{T_2} + S_P = 0 \tag{12-9}$$

where S_P is the entropy production (increase in entropy). The work done by the engine is then given by

$$W = Q_1\left(\frac{T_1 - T_2}{T_1}\right) - T_2 S_p \tag{12-10}$$

and, therefore, to obtain maximum work from an engine operating between two temperature limits T_1 and T_2, all the processes must be carried out reversibly so that $S_p = 0$. In other words, the reversible process produces the maximum work and no actual process can be as efficient as the reversible one. By solving Eq. (12-9) for Q_2, one gets

$$Q_2 = Q_1\frac{T_2}{T_1} + T_2S_p \tag{12-11}$$

In the above equation, Q_2 can be zero only if $T_2 = 0$, and this leads to the principle of the second law of thermodynamics that it is impossible to convert heat quantitatively into work.

$$S_p \begin{cases} = 0 & \text{for a reversible process} \\ > 0 & \text{for an irreversible process} \end{cases} \tag{12-12}$$

These two relations are combined as one other statement of the second law of thermodynamics, viz.,

$$S_p = \Delta S \geqslant 0 \tag{12-13}$$

Lost work is given by

$$W_{\text{lost}} = T_s\,\Delta S \tag{12-14}$$

Since the entropy is an extensive state property and is characteristic of the state of the system, the principle of the balance or accounting can be applied to it over a selected accounting period. The entropy balance is given by

$$S_E - S_B = \sum S_t + \sum\frac{Q}{T_B} + S_p - S_c \tag{12-15}$$

$$S_p \geqslant 0 \tag{12-15a}$$

where S_E = entropy of system at end of accounting period
S_B = entropy of system at beginning of accounting period
ΣS_t = entropy associated with mass transfer across boundaries of system during accounting period

$\Sigma Q/T_B$ = entropy transfer not associated with mass transfer

S_p = entropy increase in a nonreversible process

S_p = 0 for a reversible process

S_c = entropy production associated with atomic transmutations

By application of the energy and entropy balances to a constant mass and composition system undergoing a reversible process with the pressure as the only external force, the following relation results:

$$dU = T\,dS - P\,dV \tag{12-16}$$

Using this relation and the definition of enthalpy H, the following relation is obtained:

$$dH = d(U + PV) = T\,dS + V\,dP \tag{12-17}$$

Two more thermodynamic properties are defined for convenience. The work function or Helmholtz function is

$$A = U - TS \tag{12-18}$$

Gibb's free energy is

$$G = H - TS \tag{12-19}$$

Differentiation of Eqs. (12-16) and (12-19), and the use of Eqs. (12-16) and (12-17) for substitution yields the relations

$$dA = P\,dV - S\,dT \tag{12-20}$$

$$dG = V\,dP - S\,dT \tag{12-21}$$

Thermodynamic Properties of Matter

The mass, energy, and entropy balances relate the mass, heat, and work to the changes in the thermodynamic state of a system. To make effective thermodynamic analysis of the practical systems and processes, values of the thermodynamic state properties such as internal energy, enthalpy, and entropy are required. However, these are not easily measurable properties. For some fluids, thermodynamic property data in the form of charts or tables are available. In many situations, only the *PVT* and heat-capacity data are available. The thermodynamic analysis thus involves computation of the required state properties using the available *PVT* and heat-capacity data, equations of state, and graphical as well as tabular data.

In manipulating the available data for a particular thermodynamic analysis, certain definitions and relations in the form of differentials are useful. These are available elsewhere.[1a]

PVT Relationships—Equations of State for Ideal Gases

An ideal gas is defined as one which meets the following conditions:

1. The equation of state is

$$PV = NRT$$

where P = pressure
V = volume
R = gas constant
N = number of mols
T = absolute temperature

2. The internal energy is independent of both the pressure and volume and is a function of the temperature alone.

The ideal-gas law applies reasonably well to all gases within certain limits of the pressure and temperature changes. Two other laws follow from the ideal-gas equation of state, viz.,

Boyle's law: $P_1V_1 = P_2V_2$ at const. T and N

Charles' law: $\dfrac{V_1}{T_1} = \dfrac{V_2}{T_2}$ at const. P and N

Dalton's law of partial pressures and Amagat's law of partial volumes should also be recalled in connection with the ideal-gas law. According to Avogadro's law, all perfect gases have the same number of molecules in any given volume at a given temperature and pressure.

Thermodynamic Relations for an Ideal Gas. By the application of the first law of thermodynamics to a close system of 1 mol of an ideal gas, the following relations result:*
Constant volume (isometric process):

$$d\bar{U} = dQ = C_V\, dT \qquad \text{or} \qquad \Delta\bar{U} = Q = \int C_V\, dT$$

Constant pressure (isobaric process):

$$d\bar{H} = dQ = C_p\, dT \qquad \Delta\bar{H} = Q = \int C_p\, dT$$

Constant temperature (isothermal process):

$$d\bar{U} = dQ - dW = 0 \qquad \text{since } d\bar{U} = C_V\, dT = 0$$

Thus

$$Q = W = \int P\, d\bar{V}$$

and integration results in

$$Q = W = RT \ln \frac{\bar{V}_2}{\bar{V}_1} = RT \ln \frac{P_1}{P_2} \qquad (12\text{-}22)$$

Adiabatic Process. For an adiabatic process,

$$dQ = 0 \qquad \text{and} \qquad d\bar{U} = -dW = -P\, d\bar{V}$$

In this case the following relation can be developed:

$$\frac{T_2}{T_1} = \left(\frac{\bar{V}_1}{\bar{V}_2}\right)^{k-1} = \left(\frac{P_2}{P_1}\right)^{(k-1)/k} \qquad \text{or} \qquad PV^k = \text{const.} \qquad (12\text{-}23)$$

where k is the ratio of the specific heats, C_p/C_v.

*A bar over the symbol indicates molar properties.

The work of the adiabatic process can be calculated by the relation

$$\overline{W} = \frac{RT_1}{k - 1}\left[1 - \left(\frac{P_2}{P_1}\right)^{(k-1)/k}\right] \tag{12-24}$$

Polytropic Compression. The polytropic compression is neither adiabatic nor isothermal. For this type of compression, the equation of state PV^n = constant applies, where n is given by

$$\frac{n-1}{n} = \frac{k-1}{k\eta_p}$$

where η_p is polytropic efficiency.

Polytropic work of an ideal-gas compression is given by

$$\overline{W}_p = \frac{RT_1}{n - 1}\left[1 - \left(\frac{P_2}{P_1}\right)^{(n-1)/n}\right] \tag{12-25}$$

Also

$$\frac{T_2}{T_1} = \left(\frac{P_2}{P_1}\right)^{(n-1)/n} = \left(\frac{\overline{V}_1}{\overline{V}_2}\right)^{n-1} \tag{12-26}$$

It can be easily shown that when an ideal gas at P_1, T_1, V_1, is compressed to P_2, V_2, T_2, the total entropy change is given by

$$\Delta\overline{S}_{\text{total}} = \int_{T_1}^{T_2} \frac{C_p\,dT}{T} - R\ln\frac{P_2}{P_1} \tag{12-26a}$$

Evaluation of Thermodynamic Properties for Real Fluids

For the use of the energy and entropy balances for real substances when thermodynamic tables are not available, changes in U, H, and S are calculated using equations for thermodynamic state properties. The information required for such calculations is: (1) equation of state or PVT data, (2) heat-capacity data, and (3) critical-constants data.

Equations of State. For real fluids, the ideal-gas law does not hold over the entire range of temperatures and pressures. Hence, other equations of state have been proposed. These are of the form: $\phi(P, V, T) = 0$. Some *PVT* relations follow.

Van der Waals' equation for 1 mol of gas is

$$\left(P + \frac{a}{\overline{V}^2}\right)(\overline{V} - b) = RT \tag{12-27}$$

where a and b are van der Waals' constants. In terms of critical constants, their values are

$$a = \frac{27R^2T_c^2}{64P_c} = \frac{9}{8}RT_c\overline{V}_c \qquad b = \frac{RT_c}{8P_c} = \frac{\overline{V}_c}{3} \tag{12-28}$$

where b is excluded volume per mole.

The Redlich-Kwong equation is

$$P = \frac{RT}{\overline{V} - b} - \frac{a}{T^{1/2}\overline{V}(\overline{V} + b)} \tag{12-29}$$

Values of Redlich-Kwong constants in terms of critical constants are

$$a = \frac{0.4278R^2T_c^{2.5}}{P_c} \qquad b = 0.0867\frac{RT_c}{P_c} \tag{12-30}$$

The ratios $P_r = P/P_c$, $T_r = T/T_c$, and $V_r = \overline{V}/\overline{V}_c$, $\rho_r = \rho/\rho_c$ are the reduced conditions of pressure, temperature, volume, and density, respectively.

Substituting values of the van der Waals constants into the van der Waals equation of state and using the definitions of the reduced variables, the following relation is obtained:

$$\left(P_r + \frac{3}{V_r^3}\right)\left(V_r - \frac{1}{3}\right) = \frac{8}{3}T_r \tag{12-31}$$

This shows that the fluids obeying van der Waals' equation will have the same reduced volume $V_r = \overline{V}/\overline{V}_c$ at given values of T_r and P_r. The fluids are said to be in the corresponding states when they are at the same fraction of the critical-state values.

Compressibility Factor Z. Volumetric behavior can be represented by defining a dimensionless, empirical parameter, called *compressibility factor* Z as

$$P\bar{V} = ZRT \qquad \text{where } Z = \phi\,(T, P) \tag{12-32}$$

Generalized compressibility charts giving values of Z as functions of P_r and T_r are available.[1b] The compressibility at the critical point Z_c is given by

$$Z_c = \frac{P_c\bar{V}_c}{RT_c} \qquad \text{or} \qquad Z = \phi\,(T_r, P_r) \tag{12-33}$$

which can be shown to be equal to 0.375 by the substitution of the van der Waals constants for P_c, T_c, and $\bar{V}_c$.

Although the value of Z_c is shown to be 0.375 as above, the actual values of Z_c lie between 0.22 and 0.33. Hence, a third parameter Z_c is introduced to represent Z values as

$$Z = \phi\,(T_R, P_r, Z_c) \tag{12-34}$$

Virial Equations of State. By a consideration of the intermolecular forces and using the methods of statistical mechanics, the $P\bar{V}T$ relationship of gases is presented by a virial equation of state as

$$Z = \frac{P\bar{V}}{RT} = 1 + \frac{B}{\bar{V}} + \frac{C}{\bar{V}^2} + \cdots \tag{12-35a}$$

where B is the second virial coefficient and C is the third virial coefficient and so on. Another form of virial equation is

$$Z = \frac{P\bar{V}}{RT} = 1 + B'P + C'P^2 + D'P^3 + \cdots \tag{12-35b}$$

For engineering applications, the virial equations are truncated, i.e., few terms are retained and others are ignored because in many practical cases a finite number of terms suffices to represent experimental data very well without sacrificing much accuracy. Two forms of truncated equations are

$$Z = \frac{P\bar{V}}{RT} = 1 + \frac{BP}{RT} \tag{12-36}$$

$$Z = \frac{P\overline{V}}{RT} = 1 + \frac{B}{\overline{V}} \tag{12-37}$$

These are applicable up to a pressure of 15 atm. Above this pressure and up to 50 atm, an equation truncated to 3 terms is used and is given by

$$Z = \frac{P\overline{V}}{RT} = 1 + \frac{B}{\overline{V}} + \frac{C}{\overline{V}^2} \tag{12-38}$$

Evaluation of $\Delta\overline{H}$, $\Delta\overline{U}$, and $\Delta\overline{S}$

Changes in the enthalpy, internal energy, and entropy can be evaluated by the use of thermodynamic relations.[2a]

Tabular Data. For many substances, tabulated property data are available.[1c]

Graphical Presentation of Property Data. In presenting the thermodynamic property data in graphical form, two variables are selected arbitrarily as the independent variables and the third variable is treated as a parameter. Most important charts are: (1) *PV* diagram with *T* as a parameter, (2) temperature-entropy (*TS*) diagram, (3) pressure-enthalpy (*PH*) diagram, and (4) enthalpy-entropy (*HS*) diagram. The enthalpy-entropy diagram is also called the *Mollier diagram* and is very useful in solving the turbine and compressor problems. Pressure-enthalpy or *PH* diagrams are useful in solving refrigeration problems; temperature-entropy (*TS*) diagrams are useful in the analysis of engines.

Example 12-1. A steam turbine is operating under the following conditions: steam to turbine at 900°F and 120 psia, velocity = 250 ft/s; steam exiting turbine at 700°F and 1 atm, velocity = 100 ft/s. Calculate the rate at which work can be obtained from this turbine if the steam flow is 25,000 lb/h and the turbine operation is steady state adiabatic.

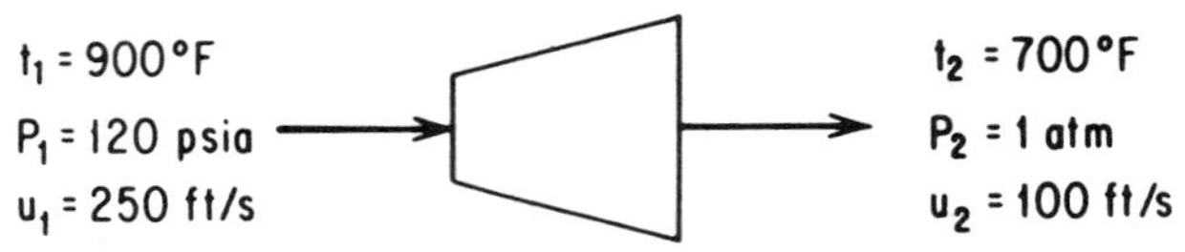

Figure 12-1. Sketch for Example 12-1.

Solution. The system is space bounded by turbine (i.e., turbine and its content). See Fig. 12-1. Mass balance is

$$\overset{3}{M_E} - \overset{3}{M_B} = M_I - M_O + \overset{2}{\Sigma M_P} - \overset{1}{\Sigma M_C}$$

1. No atomic transmutations
2. No chemical transformation
3. Steady state

Therefore, mass balance reduces to

$$M_I - M_O = 0 \qquad \text{or} \qquad M_I = M_O$$

Energy balance is

$$(U + \text{PE} + \text{KE})_E - (U + \text{PE} + \text{KE})_B$$
$$= (H + \text{PE} + \text{KE})_I - (H + \text{PE} + \text{KE})_O + Q - W - E_c$$

Therefore, energy balance is

$$0 = (H + \text{KE})_I - (H + \text{KE})_O - W$$

$$W = (H + \text{KE})_I - (H + \text{KE})_O$$

From the steam tables, $\underline{H}_I = 1478.8$ Btu/lb, $\underline{H}_O = 1383.2$ Btu/lb. Since the steam is still superheated (i.e., it is above its saturation temperature), there is no condensation. Now since $M_I = M_O = 25{,}000$ lb, the work W can be obtained as

$$W = (H_I - H_O) + (\text{KE}_I - \text{KE}_O)$$

$$= M(\underline{H}_I - \underline{H}_O) + \frac{Mu_1^2}{2g_c} - \frac{Mu_2^2}{2g_c}$$

$$= 25{,}000\left[(1478.8 - 1383.2)\frac{\text{Btu}}{\text{lb}} + \frac{250^2 - 100^2}{2(32.2)}\frac{\text{ft}^2/\text{s}^2(1\ \text{Btu})}{(\text{ft}\cdot\text{lb}/\text{lbf}\cdot\text{s}^2)778\ \text{ft}\cdot\text{lbf}}\right]$$

$$= 25{,}000(95.6 + 1.05)\ \text{Btu/h}$$

$$W = \frac{25{,}000(96.65)\ \text{Btu/h}}{2545\ (\text{Btu/h})/\text{hp}} = 949.4\ \text{hp}$$

Note: In the above equation and the following treatment, a bar under a symbol indicates properties per unit mass.

Example 12-2. Calculate the specific volume (ft^3/lb · mol) of SO_2 gas at 250 psia and 310°F using (**a**) ideal-gas law, (**b**) van der Waals' equation of state, (**c**) tabular data, and (**d**) generalized compressibility chart.

The following data are available for SO_2:

$$a = 1730\ \text{atm} \cdot \text{ft}^6/\text{lb} \cdot \text{mol}^2 \qquad b = 0.909\ \text{ft}^3/\text{lb mol}$$

$$T_c = 775.3°\text{R} \quad P_c = 1143.7\ \text{atm} \quad \overline{V}_c = 1.956\ \text{ft}^3/\text{lb} \cdot \text{mol} \quad Z_c = 0.29$$

Solution. $R = 10.73$ psia · ft^3/lb · mol · °R, $T = 460 + 310 = 770$°R, $P = 0$ psia.

a. *Volume by Ideal-Gas Law.*

$$\overline{V} = \frac{RT}{P}$$

$$= \frac{10.73(\text{psia} \cdot \text{ft}^3/\text{lb} \cdot \text{mol} \cdot °\text{R})(770°\text{R})}{250\ \text{psia}}$$

$$= 33.05\ \text{ft}^3/\text{lb} \cdot \text{mol}$$

b. *Volume by van der Waals' Equation of State.*

$$a = 1730\ \text{atm} \cdot \text{ft}^6/\text{lb} \cdot \text{mol}^2 \qquad b = 0.909\ \text{ft}^3/\text{lb} \cdot \text{mol}$$

$$P = \frac{250}{14.7} = 17\ \text{atm} \qquad R = 0.7302\frac{\text{atm} \cdot \text{ft}^3}{\text{lb} \cdot \text{mol} \cdot °\text{R}} \qquad T = 770°\text{R}$$

Then substituting in the van der Waals equation gives

$$\left(17 + \frac{1730}{\overline{V}^2}\right)(\overline{V} - 0.909) = 0.7302(770)$$

This is a cubic equation in $\overline{V}$, and it can be solved by iterative procedure or plotting a graph of values of the left side of the above equation at various values of $\overline{V}$. The iterative trials can be started with the value of $\overline{V}$ found by using the ideal-gas law. The calculations are shown in the table below.

$\overline{V}$	LHS	LHS – RHS
33.05	597.30	35.04
30.00	550.47	−11.50
31.00	565.72	3.717

By interpolation,

$$\overline{V} = 30 + \frac{0-(-11.5)}{3.717-(-11.5)}(31-30)$$

$$= 30 + \frac{11.5}{15.217}(1)$$

$$= 30.76 \text{ ft}^3/\text{lb}\cdot\text{mol}$$

When $\overline{V} = 30.76$, LHS – RHS = 0.2. This is very near to 0; therefore, $\overline{V} = 30.76$ ft^3/lb·mol is a satisfactory value.

c. *Volume by the Use of Tabular Data.* The following pertinent data are reproduced from the table[1e] of thermodynamic data for SO_2.

t, °F	200 psia $\underline{V}$ ft^3/lb	300 psia $\underline{V}$ ft^3/lb
300	0.590	0.378
320	0.610	0.392
310	0.600	0.385

Temperature interpolation gives volumes at 310°F (see table above). By pressure interpolation, volume at 250 psia and 310°F is

$$\underline{V} \text{ at 250 psia} = \tfrac{1}{2}(0.600 + 0.385) = 0.4925 \text{ ft}^3/\text{lb}$$

$$\overline{V} = 0.4925(64) = 31.52 \text{ ft}^3/\text{lb}\cdot\text{mol}$$

d. *Volume by Generalized Compressibility Factor Chart.*

$$T_r = \frac{T}{T_c} = \frac{770}{775.3}$$

$$P_r = \frac{P}{P_c} = \frac{250}{1143.7}$$

From the compressibility chart, $Z = 0.92$. Therefore

$$\overline{V} = \frac{ZRT}{P} = \frac{0.92(10.73)(770)}{250} = 30.4 \text{ ft}^3/\text{lb}\cdot\text{mol}$$

Example 12-3. What pressure will be developed when 1 lb · mol of carbon dioxide is stored in a cylinder of 3.0 ft³ volume at 150°F? Use the generalized chart. Critical constants for CO_2 are

$$T_c = 547.4°\text{R} \qquad P_c = 1071.6 \text{ psia} \qquad \overline{V}_c = 1.54 \text{ ft}^3/\text{lb}\cdot\text{mol}$$

Solution.

$$T = 460 + 150 = 610°\text{R}$$

$$V_r = \frac{\overline{V}}{V_c} = \frac{3}{1.54} = 1.95$$

$$T_r = \frac{610}{547.4} = 1.114$$

Since P_r is not known, it is not possible to enter the Z chart, and trial-and-error procedure is necessary. An equation relating P_r and Z is required. Obtain this as follows:

$$P = \frac{ZRT}{\overline{V}} = \frac{Z(10.73)(610)}{3} = 2181.8(Z)$$

Now $$P = P_c P_r = 1071.6 P_r$$

$$Z = \frac{P}{2181.8} = \frac{1071.6}{2181.8} P_r = 0.49115\, P_r$$

or $$P_r = Z(2.036)$$

A value of Z is assumed and T_r determined from Z chart using calculated P_r. Procedure is repeated till T_r calculated equals T_r = 1.114. The following table summarizes the trials:

Z	P_r	T_r (from Z Chart)
1	2.036	2.5
0.7	1.425	1.21
0.63	1.283	1.11

The last trial gives T_r = 1.11 which is very close to 1.1114. Therefore, $P = P_r P_c = 1.283(1071.6) = 1375$ psia.

Example 12-4. Superheated steam at 200 psia and 50°F superheat expands adiabatically and reversibly to 14.7 psia. Calculate the final enthalpy, specific volume, and entropy.

Solution. The initial conditions are 200 psia and 50°F superheat, i.e., 432°F. From the steam tables, at 200 psia and 432°F,

$$\underline{S}_1 = 1.58 \text{ Btu/lb}\cdot{}^\circ\text{R} \qquad \underline{V}_1 = 2.45 \text{ ft}^3\text{/lb} \qquad \underline{H} = 1226 \text{ Btu/lb}$$

Final condition:

$$\text{Pressure} = 14.7 \text{ psia}$$

$$\text{Initial entropy } \underline{S}_1 = \underline{S}_2 \text{ (final entropy)} = 1.58 \text{ Btu/lb}\cdot{}^\circ\text{R}$$

At 14.7 psia, this value lies between the entropies of the liquid and the vapor, and therefore the final condition of the steam is a mixture of the liquid and vapor, since $\underline{S}_1$ = 0.3121 Btu/lb·°R and $\underline{S}_g$ = 1.7568 Btu/lb·°R.

Let x be quality of the final steam. Then the entropy of the steam in the final condition

$$\underline{S}_f = 1.7568x + 0.3121(1 - x) = 1.58 \text{ Btu/lb}\cdot{}^\circ\text{R}$$

from which,

$$x = 0.878$$

$$\underline{V}_2 = x(\underline{V}_g) + (1 - x)\underline{V}_1$$

$$\underline{V}_2 = 0.878(26.78) + (1 - 0.878)(0.01672) = 23.515 \text{ ft}^3/\text{lb mixture}$$

$$\underline{H}_2 = 0.878(1150.4) + (1 - 0.878)(180.1) = 1032 \text{ Btu/lb mixture}$$

Note: In throttling expansion, the initial and final enthalpies are the same. This fact can be used to calculate the quality of a mixture after an isenthalpic expansion similar to one through a throttling valve.

Example 12-5. One pound of a mixture of steam and water at 160 psia is contained in a rigid vessel. Heat is added to the vessel until the contents reach the condition of 550 psia and 600°F. Calculate the amount of heat added.

Solution. Energy balance with the contents of the vessel as the system gives

$$(U + \text{PE} + \text{KE})_E - (U + \text{PE} + \text{KE})_B = \Sigma(H + \text{PE} + \text{KE})_t + Q - W - E_c$$

which, for the conditions of the example, reduces to

$$\underline{U}_E - \underline{U}_B = Q$$

Steam tables do not contain the values of the internal energy, and therefore $\underline{U}_E - \underline{U}_B$ is to be found from the enthalpies and $\Delta(P\underline{V})$ since $\Delta\underline{H} = \Delta\underline{U} + \Delta(P\underline{V})$. The values of $\underline{H}$, $\underline{V}_l$, and $\underline{V}_g$ at the two conditions (from steam tables) are:

Initial Condition	Final Condition
160 psia, $t = 363.5$°F	550 psia, 600°F
$\underline{V}_l = 0.01815$ ft^3/lb	$\underline{H}_E = 1294.3$ Btu/lb
$\underline{V}_g = 2.834$ ft^3/lb	$\underline{V}_g = 1.0431$ ft^3/lb
$\underline{h}_l = 335.93$ Btu/lb	Final condition superheated; therefore no liquid present
$\underline{H}_g = 1195.1$ Btu/lb	

Let x be the quality of the steam at beginning. Now since the vessel is rigid, initial volume equals final volume:

$$\underline{V}_l = 2.834x + 0.01815(1 - x) = 1.0431$$

From which

$$x = 0.364$$

$$\underline{H}_B = 1195.1(0.364) + 335.93(1 - 0.364) = 648.7 \text{ Btu/lb}$$

$$\text{Heat added} = \Delta\underline{U} = \Delta\underline{H} - \Delta(P\underline{V})$$

$$= (1294.3 - 648.7) - \frac{(550 - 160)(1.0431)(144)}{778}$$

$$= 570.3 \text{ Btu/lb}$$

Example 12-6. Ethane is compressed adiabatically from 200 psia and 80°F to 1050 psia. Calculate the reversible work for this process given that flow is 150 lb/min.

Solution. Choose as a system the space within the boundaries of the compressor and its inlet and outlet at the specified conditions. For the steady-state compression, the energy balance reduces to

$$W = H_I - H_O = M(\underline{H}_I - \underline{H}_O) = 150(\underline{H}_I - \underline{H}_O)$$

Similarly the entropy balance reduces to

$$S_I = S_O \qquad \text{or} \qquad M\underline{S}_I = M\underline{S}_O$$

i.e., $\underline{S}_I = \underline{S}_O$. $\underline{S}_I = 1.6395$ Btu/lb·°R. Use tables of ethane properties to obtain the required values: at inlet, $t = 80°\text{F}$, $P = 200$ psia, $\underline{H}_I = 446.5$ Btu/lb; at outlet $P = 1050$ psia, $\underline{S}_O = 1.6395$ Btu/lb·°R. Since there are no data for $P = 1050$ psia, a cross-interpolation is required to determine the outlet enthalpy and temperature. Reading entropy columns at 1000 psia, it is seen that temperature lies between 240 and 260°F.

Obtain by interpolation $\underline{H}$ and $\underline{S}$ at 1050 psia:

$$\underline{S}_{1050} = (-1.6281 + 1.5735)\tfrac{50}{500} + 1.6281 = 1.62264 \qquad \text{at } 240°\text{F}$$

$$\underline{S}_{1050} = (-1.6472 + 1.5964)\tfrac{50}{500} + 1.6472 = 1.64212 \qquad \text{at } 260°\text{F}$$

$$\underline{H}_{1050} = (-493.5 + 470.3)\tfrac{50}{500} + 493.5 = 491.2 \qquad \text{at } 240°\text{F}$$

$$\underline{H}_{1050} = (-507.2 + 486.4)\tfrac{50}{500} + 507.2 = 505.12 \qquad \text{at } 260°\text{F}$$

Now obtain values of $\underline{H}$ and t at $\underline{S} = 1.6395 = \underline{S}_O$.

t, °F	$\underline{H}$, Btu/lb	$\underline{S}$, Btu/lb·°R
240	491.2	1.62284
260	505.12	1.64212

$$\underline{H}_O = \frac{1.6395 - 1.6228}{1.6421 - 1.6228}(505.12 - 491.2) + 491.2 = 503.23 \text{ Btu/lb}$$

$$t_0 = \frac{1.6395 - 1.6228}{1.6421 - 1.6228}(260 - 240) + 240 = 257.3°\text{F}$$

$$-W = M(\underline{H}_O - \underline{H}_I) = 150(503.23 - 446.5) = 8509.5 \text{ Btu/min}$$

$$W = -8509.5 \text{ Btu/min reversible work}$$

Note: The sign is negative because work is done on the fluid.

It is sometimes convenient to estimate the difference between the value of a property for an actual fluid and that for an ideal gas at the same temperature and pressure. The deviations of the enthalpy, entropy, and specific heat are important and are given by Hougen and Watson.[2b]

It is also possible to express the differences in terms of the reduced variables by the use of the principle of the corresponding states. Hougen, Watson, and Ragatz have given the charts and tables for the property departures. Using these, the changes in the enthalpy and entropy at two states could be evaluated using the data at low pressure.

Third Law of Thermodynamics

Nernst and Planck postulated that at absolute zero temperature, the absolute entropy of a pure perfectly crystalline substance free of any random arrangement is equal to zero. This was subsequently proved to be true by experimentation and is considered the third law of thermodynamics. The absolute entropy of a substance at constant pressure can be calculated by the equation

$$\bar{S} = \int_0^{T_f} \frac{C_{ps}}{T}\,dT + \frac{\Delta \bar{H}_f}{T_f} + \int_{T_f}^{T_b} \frac{C_{p\ell}}{T}\,dT + \frac{\Delta \bar{H}_v}{T_b} + \int_{T_b}^{T} \frac{C_{pg}}{T}\,dT \qquad (12\text{-}39)$$

where $\bar{S}$ = absolute entropy, Btu/lb · mol · °R
T_f = temperature of fusion, °R
$\Delta \bar{H}_f$ = heat of fusion, Btu/lb · mol
$C_{p\ell}$ = specific heat of liquid, Btu/lb · mol · °F
T_b = boiling point of liquid, °R
$\Delta \bar{H}_v$ = heat of vaporization, Btu/lb · mol · °R
C_{pg} = specific heat of vapor, Btu/lb · mol · °R

Example 12-7. Calculate the entropy change when 1 lb of SO_2 gas is heated from 70 to 2000°F at constant pressure. The molar heat capacity of SO_2 is given by

$$C_p = 6.157 + (1.38 \times 10^{-2})T + 0.9103 \times 10^{-5}T^2 + 2.057 \times 10^{-9}T^3$$

where T is in K.

Solution. Calculate first the entropy change for 1 g · mol using temperatures in Kelvins.

$$T_1 = \frac{460 + 70}{1.8} = 294.44 \text{ K}$$

$$T_2 = \frac{460 + 2000}{1.8} = 1366.7 \text{ K}$$

$$\Delta \bar{S} = \int_{T_1}^{T_2} \frac{C_p\,dT}{T}$$

$$= \int_{294.44}^{1366.7} \left(\frac{6.157}{T} + 1.384 \times 10^{-2} - 0.9103 \times 10^{-5}T + 2.057 \times 10^{-9}T^2 \right) dT$$

$$= 23.32 \text{ g} \cdot \text{cal/g} \cdot \text{mol} \cdot \text{K}$$

$$= 23.32 \text{ Btu/lb} \cdot \text{mol} \cdot {}^\circ\text{R} \quad \text{or} \quad \frac{23.32}{64} = 0.3644 \text{ Btu/lb} \cdot {}^\circ\text{R}$$

Example 12-8. Calculate Q, W, ΔU, ΔH, and ΔS for (**a**) the system, (**b**) the surroundings, and (**c**) the universe for freezing of 0.025 kg of supercooled liquid SO_2 at 185.7 K and 0.1 MPa. Data for SO_2 at 0.1 MPa are as follows: melting point = 197.7 K, latent heat of fusion = 7.4×10^3 kJ/kmol at melting point; c_p of liquid SO_2 = 1.3 kJ/kg · K; c_p of solid SO_2 = 0.96 kJ/kg · K.

Solution. Assume 1 kmol of SO_2 as a basis. Visualize the freezing process as follows:

$$\underset{\text{at } 185.7\text{ K}(T_1)}{\text{Liquid } SO_2} \longrightarrow \underset{\text{at } 197.7\text{ K}(T_2)}{\text{liquid } SO_2} \longrightarrow \underset{\text{at } 197.7\text{ K}(T_2)}{\text{solid } SO_2} \longrightarrow \underset{\text{at } 185.7\text{ K}(T_3)}{\text{solid } SO_2}$$

For a constant-pressure process, the energy balance reduces to

$$\Delta U = Q - W$$

$$= Q - P\,\Delta V$$

or

$$Q = \Delta(U + PV) = \Delta H$$

Since ΔV is negligible for this cooling process,

$$Q = \Delta U = \Delta H \qquad \text{and} \qquad W = P\,\Delta V \doteq 0$$

Now

$$Q = \Delta\overline{H} = \int_{T_1}^{T_2} C_{p\ell}\,dT + \Delta\overline{H}_f + \int_{T_2}^{T_3} C_{ps}\,dT = \Delta\overline{U}$$

$$= C_{p\ell}(T_2 - T_1) + \Delta\overline{H}_f + C_{ps}(T_3 - T_2)$$

$$= 64(1.3)(197.7 - 185.7) - 7.4 \times 10^3 + 64(0.96)(185.7 - 197.7)$$

$$= -7138.88 \text{ kJ/kmol} = -7.138 \times 10^6 \text{ J/kmol}$$

$$\Delta\overline{S} = \int \frac{dQ}{T} = \int_{T_1}^{T_2} C_{p\ell}\frac{dT}{T} + \frac{\Delta H_f}{T_f} + \int_{T_2}^{T_3} \frac{C_{ps}\,dT}{T}$$

$$= C_{p\ell} \ln \frac{T_2}{T_1} + \frac{\Delta H_f}{T_f} + C_{ps} \ln \frac{T_3}{T_2}$$

$$= 64(1.3) \ln \frac{197.7}{185.7} - \frac{7.4 \times 10^3}{197.7} + 64(0.96) \ln \frac{185.7}{197.7}$$

$$= -36.068 \text{ kJ/kmol} \cdot \text{K}$$

For 0.025 kg of SO_2,

$$\Delta S = -36.068\left(\frac{0.025}{64}\right) = -0.014 \text{ kJ/K}$$

$$Q = \frac{-7.138 \times 10^6(0.025)}{64} = -2788.2 \text{ J}$$

For the surroundings

$$\Delta S = +\frac{2788.2}{185.7} = 0.015 \text{ kJ/K}$$

For the universe,

$$\begin{aligned} \Delta S &= \Delta S(\text{system}) + \Delta S(\text{surroundings}) \\ &= -0.014 + 0.015 = 0.001 \text{ kJ/K} \end{aligned}$$

Fugacity and Fugacity Coefficients

At constant T,

$$dG_T = \overline{V}\, dP = RT\,\frac{dP}{P} \qquad \text{for ideal gas} \qquad (12\text{-}40)$$

$$= RT\, d \ln P \qquad (12\text{-}40a)$$

For nonideal substances, a fugacity for a pure component i is defined by

$$d\overline{G}_i = RT\, d \ln f_i \qquad \text{at const. } T \qquad (12\text{-}41)$$

A further restriction on the definition of f_i is

$$\lim_{P\to 0} \frac{f}{P} = 1 \qquad (12\text{-}42)$$

The ratio f/P is called the fugacity coefficient, for which a generalized chart is available.[1d] It is denoted by ν. Now

$$RT\, d \ln f = d\overline{G} = \overline{V}\, dP = \frac{ZRT}{P}\, dP \qquad (12\text{-}43)$$

Therefore

$$d \ln f = Z\, d \ln P \qquad \text{at const. } T \tag{12-44}$$

Subtracting $d \ln P$ from both sides,

$$d \ln \frac{f}{P} = (Z - 1) d \ln P = (Z - 1)\frac{dP}{P} \qquad \text{at const. } T \tag{12-45}$$

Integration between zero pressure P^* and P gives

$$\ln \frac{f}{P} - \ln \frac{f^*}{P^*} = \int_{P^*}^{P} (Z - 1)\frac{dP}{P} \tag{12-46}$$

Since $f^* / P^* = 1$,

$$\ln \frac{f}{P} = \int_{P^*}^{P} (Z-1)\frac{dP}{P} = -\frac{1}{RT}\int\left(\frac{RT}{P} - \overline{V}\right)dP \tag{12-47}$$

Isothermal Effect of Pressure on $\overline{G}$, $\overline{S}$, $\overline{U}$

The isothermal effect of pressure on $\overline{G}$, $\overline{S}$, and $\overline{U}$ is given by the following equations:

$$(\overline{G}_2 - \overline{G}_1)_T = RT \ln \frac{f_2}{f_1} = RT \ln \frac{v_2 P_2}{v_1 P_1} \tag{12-48}$$

$$(\overline{S}_2 - \overline{S}_1)_T = \frac{(\overline{H}_2 - \overline{H}_1)_T - (\overline{G}_2 - \overline{G}_1)_T}{T} \tag{12-49}$$

$$\left(\overline{U}_2 - \overline{U}_1\right)_T = \left(\overline{H}_2 - \overline{H}_1\right)_T - \left(P_2\overline{V}_2 - P_1\overline{V}_1\right)_T \tag{12-50}$$

$$= (\overline{H}_2 - \overline{H}_1)_T - RT(Z_2 - Z_1)_T \tag{12-50a}$$

Example 12-9*a*. Estimate the fugacity of SO_2 at 1000 psia and 460°F. Use the fugacity coefficient chart for SO_2. $P_c = 1143.7$ psia, $T_c = 775.3$°R.

Solution.

$$P_r = \frac{1000}{1143.7} = 0.8744 \qquad T_r = \frac{460 + 460}{775.3} = 1.187$$

From the fugacity coefficient chart,

$$\nu = 0.86 = \frac{f}{P}$$

Hence $f = 0.86(1000) = 860$ psi.

Example 12-9*b*. Determine the fugacity and fugacity coefficient for SO_2 at 400°F and 800 psia from the enthalpy and entropy values. The enthalpy and entropy data for SO_2 are given below.

Properties of SO_2 at 400°F

P **(pressure), psi**	$\underline{V}$ **(volume), ft³/lb**	$\underline{H}$ **(enthalpy), Btu/lb**	$\underline{S}$ **(entropy), Btu/lb·°R**
10	14.35	266.7	0.5506
800	.1446	249.5	0.4004

Solution.

$$d\overline{G}_T = RT\, d \ln f_T \qquad \text{or} \qquad d \ln f_T = \frac{d\overline{G}_T}{RT}$$

Integration between high P and a low pressure (reference state) gives

$$RT \ln \frac{f}{f^*_T} = (\overline{G} - \overline{G}^*)$$

and since $\overline{G} = \overline{H} - T\overline{S}$,

$$\ln \frac{f}{f^*} = \frac{1}{R}\left[\frac{\overline{H} - \overline{H}^*}{T} - (\overline{S} - \overline{S}^*)\right]$$

If the reference state is a low pressure, then $f^* = P^*$ and

$$\ln \frac{f}{P^*} = \frac{1}{R}\left[\frac{\overline{H} - \overline{H}^*}{T} - (\overline{S} - \overline{S}^*)\right]$$

Assuming SO_2 behaves as an ideal gas at 10 psia,

$$P^* = 10 \text{ psia} \qquad \underline{H}^* = 266.7 \text{ Btu/lb} \qquad \underline{S}^* = 0.5506 \text{ Btu/lb}\cdot{}^\circ\text{R}$$

For superheated SO_2 at 800 psi and 400°F,

$$\underline{H} = 249.5 \text{ Btu/lb} \qquad \underline{S} = 0.4004 \text{ Btu/lb}\cdot{}^\circ\text{R}$$

Hence

$$\ln \frac{f}{P^*} = \frac{64}{1.987}\left(\frac{249.5 - 266.7}{400 + 460} - 0.4004 + 0.5506\right) = 4.194$$

Thus

$$\frac{f}{P^*} = e^{4.194} = 66.3$$

Therefore

$$f = 66.3(10) = 663 \text{ psia}$$

and fugacity coefficient

$$\frac{f}{P} = \frac{663}{800} = 0.829$$

Heats of Mixing and Enthalpy of a Solution

Enthalpy change that takes place on dissolving a component in another substance is called the *heat of mixing*. The enthalpy change of mixing depends upon the nature of the solute and solvent, temperature, and initial and final concentrations of the solution.

Standard integral heat of solution is the change in the enthalpy of the system when 1 mol of a solute is mixed with n_1 moles of a solvent at constant temperature of 25°C and constant pressure of 1 atm. The enthalpy of the solution H_s is given by

$$H_s = n_1\overline{H}_1 + n_2\overline{H}_2 + n_2\,\Delta\overline{H}_{s2} \qquad (12\text{-}51)$$

where H_s = enthalpy of $n_1 + n_2$ moles of solution of components 1 and 2 at temperature T relative to temperature T_0 (reference temperature)

$\overline{H}_1, \overline{H}_2$ = molal enthalpies of components 1 and 2 at temperature T relative to the temperature T_0

$\Delta\overline{H}_{s2}$ = integral heat of solution of component 2 at temperature T

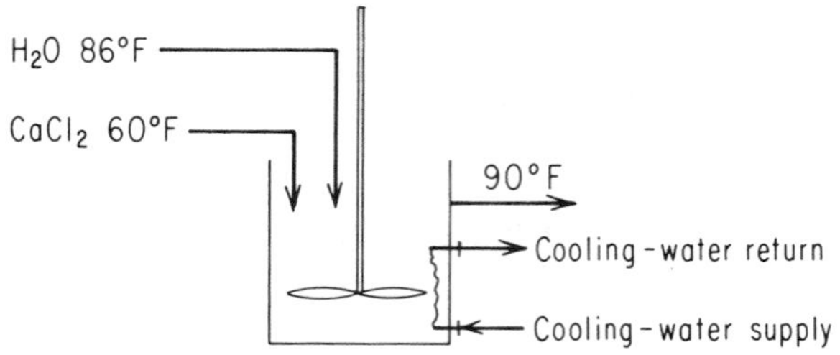

Figure 12-2. Sketch for Example 12-10.

Integral heats of solution are available elsewhere.[3a] Heats of solution involved in the mixing of liquids are called heats of mixing.

Example 12-10. How much heating or cooling must be done if 50 gpm of 20 wt % $CaCl_2$ brine is to be produced at 90°F by feeding to the dissolver solid anhydrous $CaCl_2$ at 60°F and water at 86°F? The dissolver is equipped with a 25-hp agitator which has an efficiency of 80 percent. See Fig. 12-2.

Solution. Energy balance on the contents of the tank results in

$$0 = n_1\overline{H}_1 + n_2\overline{H}_2 - (n_1 + n_2)\,\Delta H_s + Q_s - Q_t$$

where Q_s is the heat generated because of the agitator and Q_t is the heat caused by heating or cooling. It is required first to determine $\overline{H}_1$, $\overline{H}_2$, and $\overline{H}_s$. For calculation of enthalpies, assume 25°C as reference temperature. Obtain specific heats of solution[3b] of $CaCl_2$ and enthalpy[1f] of water. For each 100 lb of 20% $CaCl_2$ solution,

$$\text{Moles of } CaCl_2 = \tfrac{20}{111} = 0.18 \qquad \text{Moles of water} = \tfrac{80}{18} = 4.44$$

$$\text{Molecular weight of solution} = \frac{100}{4.44 + 0.18} = 21.645$$

$$\frac{\text{Moles of water}}{\text{Moles of } CaCl_2} = \frac{4.44}{0.18} = 24.7 \text{ mol}$$

From *Chemical Process Principles,*[3a]

$$\Delta\overline{H}_s = -17{,}500 \text{ cal/g}\cdot\text{mol at } 25^\circ\text{C} = -31{,}500 \text{ Btu/lb}\cdot\text{mol}$$

From steam tables, enthalpy of water at 77°F(25°C) is

$$45.03(18) = 810.5 \text{ Btu/lb} \cdot \text{mol}$$

(Datum temperature for enthalpies of water is 32°F.) The specific heat of $CaCl_2$ solid is

$$0.164 \text{ Btu/lb} \cdot {}^\circ\text{F} = 18.20 \text{ Btu/lb} \cdot \text{mol} \cdot {}^\circ\text{F}$$

The enthalpy of $CaCl_2$ at 25°C(77°F) is

$$C_p \, \Delta t = 18.2(77 - 32) = 819 \text{ Btu/lb} \cdot \text{mol}$$

The enthalpy of $CaCl_2$ solution at 60°F is

$$819 - 17(18.2) = 509.6 \text{ Btu/lb} \cdot \text{mol}$$

(Datum temperature for $CaCl_2$ is also taken as 32°F.) Enthalpy of $CaCl_2$ solution at 25°C or 77°F equals

$$(n_1 + n_2)\overline{H}_s = n_1\overline{H}_1 + n_2\overline{H}_2 + n_2 \, \Delta\overline{H}_s$$

$$= 4.44(810.5) + 0.18(819) + 0.18(-31{,}500) = -1924 \text{ Btu}$$

or
$$\overline{H}_s = \frac{-1924}{0.18 + 4.44} = -416.4 \text{ Btu/lb} \cdot \text{mol solution}$$

Specific heat of $CaCl_2$ solution[3b] equals

$$0.77 \text{ Btu/lb} \cdot {}^\circ\text{F} = 0.77(21.645)$$
$$= 16.7 \text{ Btu/lb} \cdot \text{mol of solution} \cdot {}^\circ\text{F}$$

Therefore
$$(\overline{H}_s)_{90} = -416.4 + \int_{77}^{90} C_p \, dT$$

$$= -416.4 + 16.7(90 - 77) = -199.3 \text{ Btu/lb} \cdot \text{mol}$$

50 gal of 20% solution of specific gravity 1.1721 at 90°F is to be prepared per minute.

$$\text{Solution to be prepared} = 50 \text{ gpm} \times (8.33)(1.1721 \text{ lb/min})$$
$$= 488.18 \text{ lb/min}$$
$$= \frac{488.18}{21.645} = 22.6 \text{ lb} \cdot \text{mol/min}$$

$$\text{Moles of water} = \frac{4.44}{4.62}(22.6) = 21.71 \text{ lb} \cdot \text{mol}$$

$$\text{Moles of } CaCl_2 = 0.88 \text{ lb} \cdot \text{mol}$$

The enthalpy of water at 86°F is

$$54.03(18) = 972.5 \text{ Btu/lb} \cdot \text{mol}$$

By heat balance,

$$0 = 21.71(972.5) + 0.88(509.6) - 22.6(-199.3) + 42.4(25)(0.8) + Q_t$$

$$Q_t = -26{,}914 \text{ Btu/min}$$

The minus sign indicates that the heat has to be removed and cooling is required.

Thermodynamic Properties at Phase Transition

At phase transition, $d\overline{G}_{TP} = 0$. Hence

$$d\overline{G}_{TP} = -\overline{S}dT + \overline{V}dP = 0 \tag{12-52}$$

For two phases in equilibrium, there results

$$(-\overline{S}\,dT + \overline{V}\,dP)_1 = (-\overline{S}\,dT + \overline{V}\,dP)_2 \tag{12-53}$$

For a liquid at its boiling point and in equilibrium with its vapor, one can establish (since $\overline{V}_g >> \overline{V}_\ell$)

$$\frac{dP}{dT} = \frac{\Delta\overline{H}_v}{T\,\Delta\overline{V}} \tag{12-54}$$

which is called the Clapeyron equation. $\Delta\overline{H}_v$ is the latent heat of vaporization. If the vapor phase is ideal,

$$\Delta\overline{V}_g = \overline{V}_g = \frac{RT}{P}$$

$$\text{Therefore} \quad \frac{dP}{dT} = \frac{\Delta\overline{H}_v}{T(RT/P)} \quad \text{or} \quad \frac{d\ln P}{dT} = \frac{\Delta\overline{H}_v}{RT^2} \tag{12-55}$$

If $\Delta\overline{H}_v$ is independent of temperature, Eq. (12-55) can be integrated to

$$\ln\frac{P_2}{P_1} = \frac{\Delta\overline{H}_v}{R}\left(\frac{1}{T_1} - \frac{1}{T_2}\right) \tag{12-56}$$

Power Cycles

A cyclic process or a cycle consists of a series of operations repeated in the same order. Cycles of heat engines used to convert heat into work consist of the following elements: (1) working fluid which receives heat at a higher temperature and rejects it at a lower temperature, (2) a high-temperature reservoir from which heat is added to the working fluid, (3) a low-temperature receiver to which heat is rejected, and (4) a heat engine or heat pump to convert heat into work or work into heat. During each cycle, the working substance goes through a series of operations and returns to its initial condition. In many cases, the same mass of the fluid may not be involved although the states of the cyclic operations are repeated. A reversible cyclic process is composed of individual operations which by themselves are reversible.

The Carnot reversible cycle is a hypothetical device which uses a perfect gas as the working fluid. It is taken as a standard in evaluating the efficiency of the actual power cycles used to convert heat into work. The Carnot cycle consists of the following four reversible steps (see Fig. 12-3):

1. Reversible isothermal expansion from V_1 to V_2 of the working fluid by absorbing heat Q_1 at constant temperature T_1, from the heat source. Work obtained in this step is

$$\int_{V_1}^{V_2} P\, dV$$

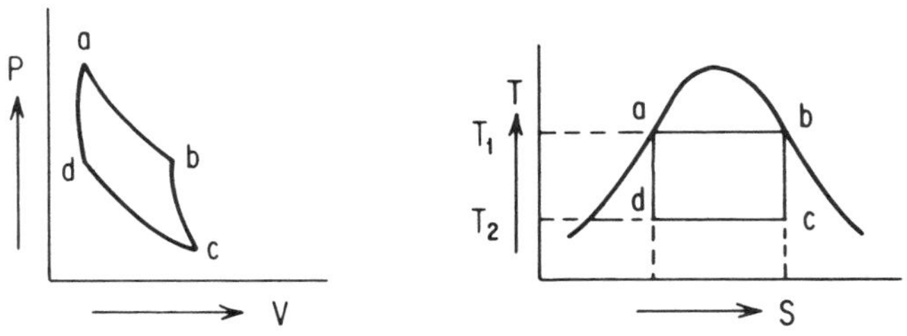

(a) (b)

Figure 12-3. Carnot cycle (*a*) on *PV* diagram, (*b*) *TS* diagram.

2. This step comprises a reversible adiabatic expansion of the fluid from V_2, P_2, T_1 to the state P_3, V_3, T_2. Work obtained in this expansion is

$$\int_{V_2}^{V_3} P\,dV$$

This is obtained by direct conversion of the internal energy into work.

3. This step consists of a reversible isothermal compression from V_3 to V_4 at T_2. Work of compression is

$$-\int_{V_3}^{V_4} P\,dV$$

and amount $-Q_2$ flows into the low-temperature (T_2) receiver.

4. Isentropic (adiabatic) compression to P_1, T_1, and V_1. The reversible work done on the gas in this step is

$$-\int_{V_4}^{V_1} PdV$$

Since the fluid returns to its original state after completing the reversible cycle, $\Delta U = 0$. Since no heat was gained or lost during the adiabatic steps, the net work $(W_{rev})_{net}$ is given by

$$W_{rev} = \sum \oint P\,dV = Q_1 - Q_2$$

Since total entropy change during the reversible cycle is zero, one obtains

$$\Delta S_T = \Delta S_1 + \Delta S_2 = 0$$

from which

$$\frac{Q_1 - Q_2}{Q_1} = \frac{T_1 - T_2}{T_1} = \frac{W_{net}}{Q_1} \tag{12-57}$$

W_{net}/Q_1 is called the *thermodynamic ideal efficiency* of the cycle. The 100 percent efficiency for the Carnot cycle is possible only if the temperature of the receiver is absolute zero or heat is absorbed by the engine at an infinite temperature.

Example 12-11. A working fluid goes through a Carnot cycle between the temperature limits of 533.2 and 333.2 K. At higher temperature 527 kJ are supplied to the cycle. Find the thermal efficiency, work done, amount of heat rejected, change in entropy during the isothermal process, and total entropy change.

Solution. $T_1 = 533.2$ K, $T_2 = 333.2$ K

1. Since the cycle is Carnot reversible cycle, thermal efficiency e is

$$e = \frac{T_1 - T_2}{T_1} = \frac{533.2 - 333.2}{533.2}(100) = 37.5 \text{ percent}$$

2. Work done and amount of heat rejected:

$$\frac{Q_1 - Q_2}{Q_1} = \frac{T_1 - T_2}{T_1}$$

$$W_{net} = Q_1 - Q_2 = \frac{T_1 - T_2}{T_1}Q_1$$

$$= 0.375(527) = 197.6 \text{ kJ}$$

$$Q_2 = \text{heat rejected} = Q_1 - W_{net} = 527 - 197.6 = 329.4 \text{ kJ}$$

3. Entropy change in isothermal step:

$$\Delta S = \frac{Q_1}{T_1} = \frac{527}{533.2} = 0.988 \text{ kJ/K}$$

for total mass at higher temperature

$$\Delta S = \frac{Q_2}{T_2} = -\frac{329.3}{333.2} = -0.988 \text{ kJ/K} \qquad \text{at lower temperature}$$

4. Since cycle is reversible, $\Delta S_T = 0$ or

$$\frac{Q_1}{T_1} - \frac{Q_2}{T_2} = 0.988 - 0.988 = 0$$

Refrigeration

The purpose of refrigeration is to maintain a given system at a temperature below that of its surroundings. Heat is abstracted from the low-temperature region and rejected at a higher temperature. The refrigeration methods of more importance are (1) vapor-compression refrigeration and (2) absorption refrigeration.

Definitions. A ton of refrigeration is the removal of heat at a rate of 200 Btu/min, 12,000 Btu/h, or 288,000 Btu/d. Coefficient of performance is the ratio of heat abstracted at low temperature divided by the work done in compression:

$$\beta = \frac{\text{heat removed from low-temperature region}}{\text{work of compression}} \tag{12-58}$$

Carnot Refrigerator (Reversed Carnot Cycle). An ideal refrigeration cycle is the reversed ideal Carnot cycle, and its performance depends only upon the temperature at which heat is abstracted and rejected.

Carnot refrigeration cycle steps are shown on *PV* and *TS* diagrams in Fig. 12-4. For a Carnot cycle, the coefficient of performance β is given by

$$\beta = \frac{\text{heat abstracted in evaporator}}{\text{work done on the fluid}} = \frac{Q_1}{-W} = \frac{Q_1}{Q_2 - Q_1}$$

$$= \frac{T_1\,\Delta S}{(T_2 - T_1)\,\Delta S} = \frac{T_1}{T_2 - T_1} \tag{12-59}$$

The actual refrigeration cycles are not as efficient as Carnot's reversed cycle because of the irreversibilities in the actual processes.

Actual vapor compression refrigeration cycles are carried out in two ways. One uses free expansion through a throttle valve

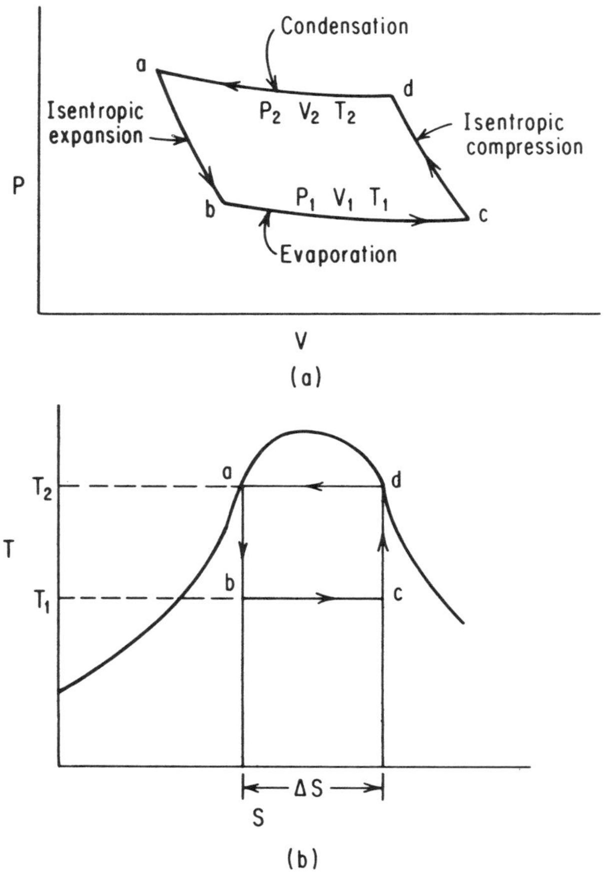

Figure 12-4. Carnot reversed cycle on (*a*) *PV* diagram, (*b*) *TS* diagram.

and the other adiabatic expansion through a turbine. Free expansion is isenthalpic while the turbine expansion is more or less isentropic.

Vapor-Compression Cycle with Turbine Expansion. The vapor-compression cycle using turbine expansion is shown in Fig. 12-5. Both isentropic compression and expansion are less than theoretically efficient in practice, and hence *AB* and *CD* (Fig. 12-5)

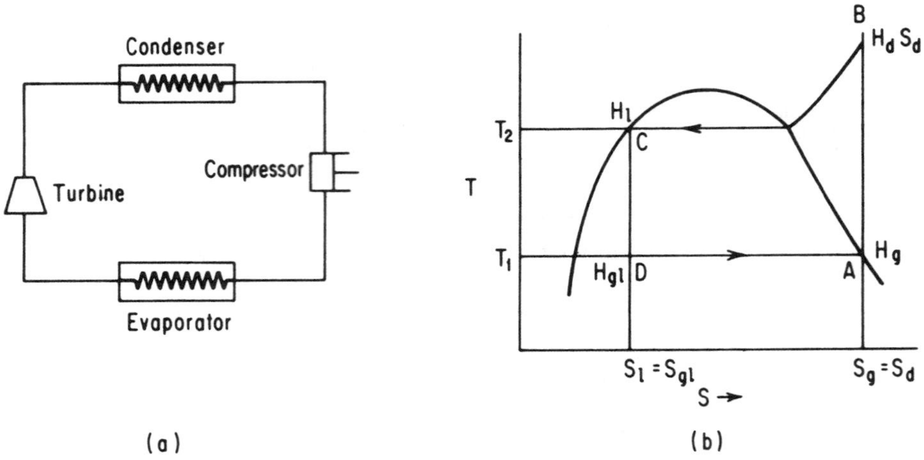

Figure 12-5. Vapor compression refrigeration cycle with turbine expansion: (*a*) system representation; (*b*) *TS* diagram.

will not be vertical on the *TS* diagram. In the ideal case,

$$S_\ell = S_{g\ell} \qquad \text{and} \qquad S_d = S_g \tag{12-60}$$

where g = vapor at compressor inlet
d = vapor at compressor discharge
$g\ell$ = liquid-vapor mixture entering evaporator
ℓ = cooled liquid leaving condenser

Heat absorbed at low temperature,

$$Q_1 = \Delta H = H_g - H_{g\ell} \tag{12-61}$$

Heat rejected at condenser,

$$Q_2 = H_d - H_\ell \tag{12-62}$$

$$\text{Work done} = (H_d - H_\ell) - (H_g - H_{g\ell}) \tag{12-63}$$

and therefore β, the coefficient of performance, is given by

$$\beta = \frac{H_g - H_{g\ell}}{(H_d - H_\ell) - (H_g - H_{g\ell})} \tag{12-64}$$

After turbine expansion, the refrigerant is a liquid-vapor mixture. Its entropy is given by

$$S = S_{g\ell} = S_g - (\Delta S_{\text{vap}})x \tag{12-65}$$

where x is the quality of refrigerant after turbine expansion. Similarly,

$$H_{g\ell} = H_g - (\Delta H_{\text{vap}})x \tag{12-66}$$

where ΔS_{vap} and ΔH_{vap} represent the changes of the entropy and enthalpy per unit mass at the low pressure and temperature. If x is eliminated from the above equations,

$$\begin{aligned} H_{g\ell} &= H_g - \frac{\Delta H}{\Delta S_{\text{vap}}}(S_g - S_{g\ell}) \\ &= H_g - \left(\frac{\Delta H}{\Delta S}\right)_{\text{vap}}(S_g - S_\ell) \end{aligned} \tag{12-67}$$

from which

$$H_{g\ell} = H_g - T_1(S_g - S_{g\ell}) \tag{12-68}$$

This equation allows calculation of $H_{g\ell}$ provided S_g is saturation vapor entropy. $H_{g\ell}$ could also be obtained by a trial-and-error procedure from

$$(xS'_\ell) + (1 - x)S'_g = S_{g\ell} = S_\ell \tag{12-69}$$

where S'_ℓ and S'_g are the entropies of the liquid and vapor in the liquid-vapor mixture after turbine expansion.

Vapor Compression with Free Expansion. The basic single-stage cycle consists of steps, as shown in Fig. 12-6. The calculation of the steps are as follows:

1. Net refrigeration effect is given by

$$Q_1 = \underline{H}_g - \underline{H}_{g\ell} = \underline{H}_g - \underline{H}_\ell \qquad \text{since } \underline{H}_\ell - \underline{H}_{g\ell} \tag{12-70}$$

where $\underline{H}_g$ = enthalpy of vapor leaving evaporator, Btu/lb

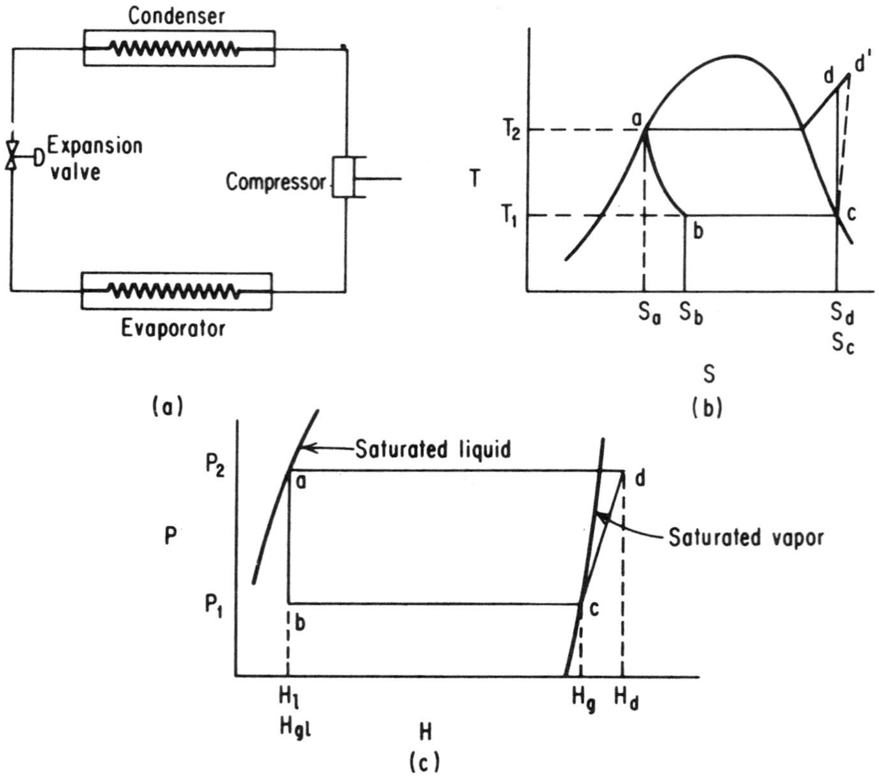

Figure 12-6. Vapor compression refrigeration cycle with free expansion: (*a*) system sketch; (*b*) *TS* diagram; (*c*) *PH* diagram.

$\underline{H}_\ell$ = enthalpy of liquid leaving condenser, Btu/lb
$\underline{Q}_1$ = refrigeration effect, Btu/lb

2. Weight rate of flow in lb/min · ton is

$$W = \frac{200 \text{ Btu/min} \cdot \text{ton}}{\underline{Q}_1 \text{ Btu/lb}} \tag{12-71}$$

$$\text{Volume of vapor to be handled} = W\underline{V}_g \tag{12-72}$$

where $\underline{V}_g$ is the specific volume of vapor entering compressor.

$$\text{Heat of compression} = \underline{H}_d - \underline{H}_g$$

where $\underline{H}_d$ is the enthalpy of vapor discharged by compressor, Btu/lb, and $\underline{H}_g$ is the enthalpy of vapor at compressor suction, Btu/lb.

$$\text{Work of compression} = -W_s$$
$$= (\underline{H}_d - \underline{H}_g)W \qquad \text{Btu/min}\cdot\text{ton} \quad (12\text{-}73)$$

$$1 \text{ hp} = 42.4 \text{ Btu/min}$$

$$\frac{\text{hp}}{\text{ton}} = \frac{\text{work of compression (Btu/min}\cdot\text{ton)}}{42.4 \text{ Btu/min}}$$

$$= \frac{4.713}{\beta} \qquad (12\text{-}74)$$

$$\text{Condenser heat load} = Q_2 = \underline{H}_d - \underline{H}_\ell \text{ Btu/lb}$$

This is also equal to the refrigeration effect plus the heat of compression:

$$\underline{H}_g - \underline{H}_\ell + \underline{H}_d - \underline{H}_g = \underline{H}_d - \underline{H}_\ell \text{ Btu/lb} \qquad (12\text{-}75)$$

$$\beta = \frac{\text{net refrigeration effect, Btu/lb}}{\text{heat of compression, Btu/lb}} \qquad (12\text{-}76)$$

$$= \frac{\underline{H}_g - \underline{H}_\ell}{\underline{H}_d - \underline{H}_g} \qquad (12\text{-}77)$$

Examples are given to illustrate the methods of calculation.

Example 12-12. A refrigerator is to be maintained at 10°F. Cooling water is available at 86°F. The evaporator and condenser are of sufficient capacity so that a 10°F approach is possible in each unit. Freon 12 is to be used as the refrigerant. Capacity of the unit is to be 5 tons. Assume the compression to be isentropic. Compute: (**a**) the pounds of refrigerant to be circulated, (**b**) the coefficient of performance β, and (**c**) volumetric displacement of compressor, for (*i*) vapor compression cycle with turbine expansion and (*ii*) vapor compression cycle with free expansion. (*iii*) What is β for Carnot reversed cycle?

Solution. Properties of Freon 12 required for the example are given in Table 12-1.

TABLE 12-1 Properties of Freon 12

Temp. °F	Pressure psi	Saturated Liquid $\underline{H}$ Btu/lb	Saturated Liquid $\underline{S}$ Btu/lb·°R	Saturated Vapor $\underline{H}$ Btu/lb	Saturated Vapor $\underline{S}$ Btu/lb·°R	Saturated Vapor $\underline{V}_g$ ft³/lb
0	23.95	8.52	0.019323	77.27	0.16888	1.6089
96	124.7	30.14	0.061536	86.69	0.16333	
110	....	....	....	89.23	0.16776	
120	....	....	....	91.01	0.17087	

Case i. With a condensed liquid,

$$\underline{H}_\ell = 30.14 \text{ Btu/lb} \qquad \underline{S}_\ell = 0.061536 \text{ Btu/lb}\cdot{}^\circ\text{R}$$

Liquid vapor mixture after turbine expansion is isentropic expansion; therefore $\underline{S}_\ell = \underline{S}_{g\ell} = 0.061536$ Btu/lb·°R. Hence

$$\underline{H}_{g\ell} = \underline{H}_g - T_1(\underline{S}_g - \underline{S}_{g\ell})$$
$$= 77.27 - 460(0.16888 - 0.061536) = 27.89 \text{ Btu/lb}$$

Since the evaporator temperature is known, $\underline{H}_{g\ell}$ can also be obtained by calculating x, the quality of refrigerant after expansion and using x to calculate the enthalpy of the liquid-vapor mixture.

Compressor Discharge. By following the *PH* diagram[1g] at constant entropy $\underline{S}_g = 0.16888$ (saturated vapor at 0°F), the compressor discharge temperature is closer to 110°F but less than 120°F. H can be read from the *PH* diagram, but a more accurate value can be obtained by interpolation from the tables. Thus

$$\underline{H}_d = 89.23 + \frac{0.16888 - 0.16776}{0.17087 - 0.16776}(91.01 - 89.23)$$
$$= 89.87 \text{ Btu/lb}$$

$$\beta = \frac{\underline{H}_g - \underline{H}_{g\ell}}{(\underline{H}_d - \underline{H}_\ell) - (\underline{H}_g - \underline{H}_{g\ell})}$$
$$= \frac{77.27 - 27.89}{(89.87 - 30.14) - (77.27 - 27.89)} = 4.8$$

$$\text{Refrigerant circulation} = \frac{12{,}000 \text{ Btu/h}\cdot\text{ton (5 ton)}}{\underline{H}_g - \underline{H}_{g\ell}}$$

$$= \frac{60{,}000}{77.27 - 27.89} = 1215 \text{ lb/h}$$

$$\text{Volumetric displacement of compressor} = \tfrac{1215}{60}(1.6089) = 32.5 \text{ ft}^3\text{/min}$$

Case ii. Vapor compression cycle with free expansion: In this case $\underline{H}_g = 77.27$ Btu/lb.

$$\underline{H}_{g\ell} = \underline{H}_\ell = 30.14 \text{ Btu/lb}$$

$$\beta = \frac{\underline{H}_g - \underline{H}_\ell}{\underline{H}_d - \underline{H}_g} = \frac{77.27 - 30.14}{89.86 - 77.27} = 3.74$$

$$\text{Refrigerant circulation} = \frac{60{,}000}{77.27 - 30.14} = 1273 \text{ lb/h}$$

$$\text{Volmetric displacement} = \frac{1273(\underline{V}_g)}{60} = \frac{1273(1.6089)}{60} = 34.14 \text{ ft}^3\text{/min}$$

Case iii. Carnot refrigerator. In this case the coefficient of performance is

$$\beta = \frac{460}{(460 + 96) - 460} = 4.792$$

References

1. *Chemical Engineers' Handbook*, 5th ed., R. H. Perry (ed.), McGraw-Hill, 1973: (*a*) p. 4-45; (*b*) pp. 3-232, 3-233; (*c*) pp. 3-150 to 3-210; (*d*) p. 4-52; (*e*) p. 3-202; (*f*) pp. 3-205 to 3-207; (*g*) p. 3-192.
2. O. A. Hougen, K. M. Watson, and R. A. Ragatz, *Chemical Process Principles Part II*, 2d ed., Wiley, New York, 1954: (*a*) pp. 533–545; (*b*) pp. 595–619.
3. O. A. Hougen, K. M. Watson, and R. A. Ragatz, *Chemical Process Principles Part I*, 2d ed., Wiley, New York, 1954: (*a*) pp. 320–321, (*b*) p. 269.

Chapter

13

Chemical Kinetics

Thermodynamics of Chemical Reactions

Heats of Reaction. Heat of reaction is the energy absorbed or given out by the system when the products of the reaction are restored to the same temperature and pressure as the reactants. If the pressure and temperature are the same for both the reactants and the products, the heat of reaction is the enthalpy change ΔH_R. *Exothermic* reactions are those in which heat is evolved. In *endothermic* reactions heat is absorbed.

The standard heat of reaction is defined as the change in enthalpy resulting from a reaction which takes place at 1 atm pressure and 25°C. In this case both the reactants and the products are at 1 atm and 25°C.

The heats of reactions can be calculated from the heats of formation and heats of combustion. These are given in standard handbooks. The heat of reaction is dependent on the temperature as follows:

$$\left(\frac{\delta\,\Delta H}{\delta\,T}\right)_p = \Delta C_p \tag{13-1}$$

If the heat of reaction is known at a base or reference temperature (usually 25°C = 298 K), the heat of reaction at any other temperature may be calculated by

$$\Delta H_T = \Delta H_{T_0} + \int_{T_0}^{T} \Delta C_p \, dT \qquad (13\text{-}1a)$$

where ΔH_{T_0} is the heat of reaction at T_0 and ΔC_p is the difference in the sum of the molal heat capacities of the products and the sum of the molal heat capacities of the reactants, each multiplied by the stoichiometric coefficient n_i or

$$\Delta C_p = \Sigma(n_i C_{pi})_{\text{products}} - \Sigma(n_i C_{pi})_{\text{reactants}} \qquad (13\text{-}1b)$$

where C_{pi}'s, the molal heat capacities, are known as functions of temperature.

When the mean molal heat capacities can be used for the reactants and the products, the heat of reaction at another temperature is given by

$$\Delta H_T = \Delta H_{T_0} + [\Sigma(n_i C_{pi})_{\text{products}} - \Sigma(n_i C_{pi})_{\text{reactants}}](T - T_0) \qquad (13\text{-}2)$$

Hess's Law. The total change in the enthalpy of a system is dependent on the temperature, pressure, state of aggregation, and state of combination at the beginning and at the end; it is independent of the number of intermediate reactions. This principle is useful in calculating the heat of a reaction from the known heats of reactions, if that reaction can be expressed as the end reaction.

Example 13-1. Calculate the heat of reaction for the following at 25°C and 1 atm.

$$CaC_2(s) + 2H_2O(\ell) \longrightarrow Ca(OH)_2(s) + C_2H_2(g)$$

From the tables in Ref. 1*a*,

$CaC_2(s)$:	$\Delta H_f = -15{,}000$ cal/g · mol
$H_2O(\ell)$:	$\Delta H_f = -68{,}317$ cal/g · mol
$Ca(OH)_2(s)$:	$\Delta H_f = -235{,}800$ cal/g · mol

Solution. The heat of formation of acetylene is to be calculated from its heat of combustion.

$$C_2H_2(g) + 2\tfrac{1}{2}O_2(g) \longrightarrow 2CO_2(g) + H_2O(\ell) \quad \Delta H_c = -310{,}615 \text{ cal/g}\cdot\text{mol}$$

$CO_2(g)$: $\Delta H_f = -94{,}052 \text{ cal/g}\cdot\text{mol}$

$H_2O(\ell)$: $\Delta H_f = -68{,}317 \text{ cal/g}\cdot\text{mol}$

$O_2(g)$: $\Delta H_f = 0 \text{ cal/g}\cdot\text{mol}$

Therefore $\quad \Sigma(\Delta H_f)_P - \Sigma(\Delta H_f)_R = \Delta H_c$

Then $[2(-94{,}052) + (-68{,}317)] - [(\Delta H_f)_{C_2H_2(g)} + (0)_{O_2}] = -310{,}615$

$$\text{or} \quad -(\Delta H_f)_{C_2H_2(g)} = -310{,}615 + 2(94{,}052) + 68{,}317$$

$$= -54{,}194 \text{ cal/g}\cdot\text{mol } C_2H_2(g)$$

$$(\Delta H_f)_{C_2H_2(g)} = 54{,}194 \text{ cal/g}\cdot\text{mol}$$

$$(\Delta H_R)_{25} = -235{,}800 + 54{,}194 - [2(-68{,}317) + (-15{,}000)]$$

$$= -29{,}972 \text{ cal/g}\cdot\text{mol } CaC_2$$

Example 13-2*a*. Calculate the heat of formation of liquid carbon tetrachloride.

Solution. The following equations are written first:

$$CCl_4(\ell) + 2H_2O(aq) \longrightarrow CO_2(g) + 4HCl(aq) \qquad \Delta H_R = -84{,}170 \text{ cal} \quad (I)$$

$$H_2(g) + \tfrac{1}{2}O_2(g) \longrightarrow H_2O(\ell) \qquad \Delta H_1 = -68{,}317.4 \text{ cal} \quad (II)$$

$$C(\beta) + O_2(g) \longrightarrow CO_2(g) \qquad \Delta H_2 = -94{,}052 \text{ cal} \quad (III)$$

$$\tfrac{1}{2}H_2(g) + \tfrac{1}{2}Cl_2(g) \longrightarrow HCl(aq) \qquad \Delta H_3 = -40{,}023 \text{ cal} \quad (IV)$$

To apply Hess's law, multiply Eq. (IV) by 4 and Eq. (II) by 2 to obtain

$$CCl_4(\ell) + 2H_2O(aq) \longrightarrow CO_2(g) + 4HCl(aq) \qquad \Delta H_R = -84{,}170 \text{ cal} \quad (a)$$

$$2H_2(g) + O_2(g) \longrightarrow 2H_2O(\ell) \qquad 2(\Delta H_1) = -136{,}635 \text{ cal} \quad (b)$$

$$C(\beta) + O_2(g) \longrightarrow CO_2(g) \qquad \Delta H_2 = -94{,}052 \text{ cal} \qquad (c)$$

$$2H_2(g) + 2Cl_2(g) \longrightarrow 4HCl(aq) \qquad 4(\Delta H_3) = -160{,}092 \text{ cal} \qquad (d)$$

From (a), (b), (c), and (d), by algebraic addition $(c) + (d) - (a) - (b)$ gives

$$-CCl_4(\ell) - 2H_2O(\ell) \longrightarrow -CO_2(g) - 4HCl(aq) \qquad -\Delta H_R \qquad (-a)$$

$$-2H_2(g) - O_2(g) \longrightarrow -2H_2O(\ell) \qquad -2\ \Delta H_1 \qquad (-b)$$

$$C(\beta) + O_2(g) \longrightarrow CO_2(g) \qquad \Delta H_2 \qquad (+c)$$

$$2H_2(g) + 2Cl_2(g) \longrightarrow 4HCl(aq) \qquad 4\ \Delta H_3 \qquad (+d)$$

or $$-CCl_4 + C(\beta) + 2Cl_2(g) \longrightarrow 0 \qquad \Delta H_2 + 4\ \Delta H_3 - \Delta H_R - 2\ \Delta H_1$$

or $$C(\beta) + 2Cl_2(g) \longrightarrow CCl_4$$

$$\Delta H_f = \Delta H_2 + 4\ \Delta H_3 - \Delta H_R - 2\Delta H_1$$
$$= -94{,}052 - 160{,}092 + 84{,}170 + 136{,}635$$
$$= -33{,}339 \text{ cal/g}\cdot\text{mol}$$

Thus, the heat of formation of CCl_4 is -33.34 kcal/g · mol.

Heat of Reaction from Heats of Combustion. The heats of reaction can also be calculated from the heats of combustion of the reactants and products with the following relation:

$$(\Delta H)_{\text{reaction}} = \Sigma(\Delta H_c)_{\text{reactants}} - \Sigma(\Delta H_c)_{\text{products}} \qquad \text{at } 25°\text{C, 1 atm} \qquad (13\text{-}3)$$

Example 13-2b. Calculate the heat of reaction using the heats of combustion and the heats of formation for the following:

$$C_2H_5OH(\ell) + O_2(g) \longrightarrow CH_3COOH(\ell) + H_2O(\ell) \qquad \text{at } 25°\text{C}$$

Solution.

C_2H_5OH: $\Delta H_c = -326{,}700$ cal/g · mol at 25°C

CH_3COOH: $\Delta H_c = -208{,}340$ cal/g·mol at 25°C

$$(\Delta H_R)_{25°C} = \Sigma(\Delta H_c)_{\text{reactants}} - \Sigma(\Delta H_c)_{\text{products}} \quad \text{at } 25°C$$

$$= -326{,}700 + 208{,}340$$

$$= -118{,}360 \text{ cal/g·mol } C_2H_5OH$$

From heats of formation,

$$\Delta H_R = \Sigma(\Delta H_f)_P - \Sigma(\Delta H_f)_R$$

$C_2H_5OH(\ell)$: $\Delta H_f = -66{,}200$ cal

$O_2(g)$: $\Delta H_f = 0$

$CH_3COOH(\ell)$: $\Delta H_f = -116{,}400$ cal

$H_2O(\ell)$: $\Delta H_f = -68{,}317$ cal

$$\Delta H_R = -116{,}400 - 68{,}317 + 66{,}200$$

$$= -118{,}517 \text{ cal/g·mol } C_2H_5OH$$

Effect of Pressure on Heat of Reaction. The effect of pressure on the heat of reaction of a gaseous system depends upon the degree of deviation of the components from ideal-gas behavior. If both the reactants and products behave as ideal gases, there is no effect. Even for nonideal behavior, the effect of pressure is generally very small.

Chemical Equilibrium

The standard free-energy change between the free energies of the products and reactants at standard state $\Delta G°$ is given by

$$\Delta G° = -RT \ln K \tag{13-4}$$

where K = equilibrium constant
T = absolute temperature
R = gas law constant

For a reaction of the type $aA + bB \longrightarrow cC + dD$, the equilibrium constant K is defined by the relation

$$K = \frac{a_C^c a_D^d}{a_A^a a_B^b} \tag{13-5}$$

where a's are the equilibrium activities. Activity a_i in terms of fugacity is given by

$$a_i = \frac{f_i}{f_i^\circ} \tag{13-6}$$

If the standard state chosen is unit fugacity or $f_i^\circ = 1$,

$$K = \frac{f_C^c f_D^d}{f_A^a f_B^b} \tag{13-7}$$

For ideal-gas behavior,

$$K_P = \frac{p_C^c p_D^d}{p_A^a p_B^b} \tag{13-8}$$

where the p's are the partial pressures. Since $p_i = y_i P_T$ where P_T is the total pressure and y_i is the mole fraction of component i, K_P in the case of the gaseous reactions is given by

$$K_P = \frac{(y_c P_t)^c (y_d P_t)^d}{(y_a P_t)^a (y_b P_t)^b} \tag{13-9}$$
$$= K_y P_t^{[(c+d)-(a+b)]}$$

If the reaction takes place at 1 atm the pressure term in the above equation is unity and then

$$K_p = K_y \tag{13-9a}$$

The Van't Hoff equation relates K to the heat of reaction by

$$\frac{d \ln K}{dT} = \frac{\Delta H_T^\circ}{RT^2} \tag{13-10}$$

where ΔH_T° is the standard-state enthalpy change for the reaction. If ΔH_T° is approximately independent of temperature, then

Eq. (13-10) can be integrated to give

$$\ln \frac{K_2}{K_1} = \frac{-\Delta H_T^\circ}{R}\left(\frac{1}{T_2} - \frac{1}{T_1}\right) \tag{13-11}$$

where K_2 is the equilibrium constant at temperature T_2 and K_1 is the equilibrium constant at temperature T_1. If ΔH° varies with temperature and if the heat capacities of the reactants and products are expressed by equations of the type

$$C_p^\circ = \alpha + \beta T + \gamma T^2 + \cdots$$

$$\Delta C_p^\circ = \Delta\alpha + \Delta\beta T + \Delta\gamma T^2 + \cdots$$

and the reaction is of the type

$$n_a\text{A} + n_b\text{B} + \cdots \longrightarrow n_r\text{R} + n_s\text{S} + \cdots$$

where
$$\Delta\alpha = (\Sigma n\alpha)_{\text{products}} - (\Sigma n\alpha)_{\text{reactants}}$$
$$\Delta\beta = (\Sigma n\beta)_{\text{products}} - (\Sigma n\beta)_{\text{reactants}}$$
$$\Delta\gamma = (\Sigma n\gamma)_{\text{products}} - (\Sigma n\gamma)_{\text{reactants}}$$

then substituting ΔC_P° in Eq. (13-1) and integrating yields

$$\Delta H_T^\circ = I_H + \Delta\alpha T + \Delta\beta \tfrac{1}{2}T^2 + \Delta\gamma\tfrac{1}{3}T^3 + \cdots \tag{13-12}$$

where I_H is the constant of integration.

When use is made of Eqs. (13-10) and (13-12), K as a function of the temperature is given by integrating Eq. (13-10) as

$$\ln K = -\frac{I_H}{RT} + \frac{\Delta\alpha}{R}\ln T + \frac{\Delta\beta}{R}\tfrac{1}{2}T + \frac{\Delta\gamma}{R}\tfrac{1}{6}T^2 + \cdots + I \tag{13-13}$$

Knowing the K at one temperature, it is possible to evaluate I in the above equation, and therefore the evaluation of the K is possible at any other temperature. Also since $\Delta G_T^\circ = -RT \ln K$, one can get

$$\Delta G_T^\circ = I_H - \Delta\alpha T \ln T - \tfrac{1}{2}\Delta\beta T^2 - \tfrac{1}{6}\Delta\gamma T^3 - IRT \tag{13-14}$$

or
$$\Delta G_T^\circ = I_H + I_G T - \Delta\alpha T \ln T - \tfrac{1}{2}\Delta\beta T^2 - \tfrac{1}{6}\Delta\gamma T^3 \tag{13-14a}$$

where
$$I_G = -(IR) \tag{13-14b}$$

Knowing ΔH_{298}, the heat of the reaction at 25°C, and the specific heats of the components, it is possible to evaluate the constants in the above equations. Note the following conditions:

If $\Delta G^\circ < 0$, i.e., negative, spontaneous reaction takes place in the standard states.

If $\Delta G^\circ = 0$, there is equilibrium in standard states.

If $\Delta G^\circ > 0$, i.e., positive, then $K < 1$, and therefore the reactants in the standard state will not react to produce the products in the standard states.

Example 13-3. Calculate the equilibrium constant K and equilibrium conversion for the following reaction:

$$C_2H_6(g) \longrightarrow C_2H_4(g) + H_2(g)$$

at a temperature of 1000 K, if the reaction takes place at 1 atm pressure. Data on ΔG°, ΔH°, and C_p are given in Table 13-1.

TABLE 13-1 Data on ΔG°_{298}, ΔH°_{298}, and C_p Values

Component	ΔG°_{298} kJ/mol	ΔH°_{298} kJ/mol	C_p kJ/mol · K
C_2H_6	−32.886	−84.667	$0.0096 + 8.37 \times 10^{-5}T$
C_2H_4	68.124	52.3	$0.0117 + 12.55 \times 10^{-5}T$
H_2	0	0	$0.0289 + 1.67 \times 10^{-5}T$

Solution. Calculate ΔG°_{298} and ΔH°_{298} for the reaction

$$\Delta G^\circ_{298} = \Sigma(\Delta G^\circ_{298})_{\text{products}} - \Sigma(\Delta G^\circ_{298})_{\text{reactants}}$$

where P and R denote the products and reactants, respectively.

$$\Delta G^\circ_{298} = 68.124 + 0 - (-32.886)$$
$$= 101.01 \text{ kJ/mol}$$
$$\Delta H^\circ_{298} = 52.3 + 0 - (-84.667)$$
$$= 136.967 \text{ kJ/mol}$$

$$C_p = \begin{cases} 0.0289 + 1.67 \times 10^{-5}T & \text{for } H_2 \\ 0.0117 + 12.55 \times 10^{-5}T & \text{for } C_2H_4 \\ -0.0096 - 8.37 \times 10^{-5}T & \text{for } -C_2H_6 \end{cases}$$

By addition, $\Delta C_p = 0.031 + 5.85 \times 10^{-5}T$

Calculation of I_H *and* $\Delta G°$. From the above ΔC_p equation, $\Delta\alpha = 0.031$, $\Delta\beta = 5.85 \times 10^{-5}$, $\Delta\gamma = 0$. From Eq. (13-12),

$$\Delta H_T^\circ = I_H + (\Delta\alpha T + \Delta\beta \tfrac{1}{2}T^2 + \cdots)$$

or

$$I_H = \Delta H_T^\circ - (\Delta\alpha T + \Delta\beta \tfrac{1}{2}T^2 + \cdots)$$

At 298 K,

$$\Delta H_T^\circ = 136.967 \text{ kJ/mol}$$

$$I_H = 136.967 - [0.031(298) + 5.85 \times 10^{-5}(\tfrac{1}{2})(298)^2]$$

$$= 125.131 \text{ kJ/mol}$$

At $T = 298$ K, using Eq. (13-14*a*),

$$101.01 = 125.131 - 0.031(298) \ln 298 - 2.925 \times 10^{-5}(298)^2 + I_G(298)$$

which gives

$$I_G = 0.1044$$

$$\Delta G_T^\circ = 125.131 - 0.031T \ln T - 2.925 \times 10^{-5}T^2 + 0.1044T$$

At $T = 1000$ K,

$$\Delta G_{1000}^\circ = 125.131 - 0.031(1000) \ln 1000 - 2.925 \times 10^{-5}(1000)^2 + 0.1044(1000)$$

$$= -13.859 \text{ kJ/g}\cdot\text{mol}$$

Then

$$\ln K = \frac{-\Delta G^\circ}{RT} = -\frac{-13.859 \times 10^3}{8.314(1000)} = 1.667$$

and solving for K gives $K = 5.296$.

Calculation of Conversion.

Assuming components behave as ideal gases ($P_T = 1$ atm),

$$K = \frac{\bar{f}_{H_2}\bar{f}_{C_2H_4}}{\bar{f}_{C_2H_6}} = \frac{y_{H_2}\, y_{C_2H_4}}{y_{C_2H_6}} \frac{P_T P_T}{P_T}$$

$$= \frac{y_{H_2}\, y_{C_2H_4}}{y_{C_2H_6}} \qquad \text{since } P_T = 1 \text{ atm}$$

At equilibrium, let X be the number of moles of C_2H_6 converted. The moles of each component in the reaction mixture are:

	Initial	**Equilibrium**
C_2H_6	1	$1 - X$
C_2H_4	0	X
H_2	0	X

Therefore, total moles = $1 + X$ at equilibrium. The mole fractions of the components are then calculated as

$$y_{C_2H_6} = \frac{1 - X}{1 + X} \qquad y_{H_2} = \frac{X}{1 + X} \qquad y_{C_2H_4} = \frac{X}{1 + X}$$

Therefore

$$K = \frac{[X/(1 + X)][X/(1 + X)]}{(1 - X)/(1 + X)}$$

$$= \frac{X^2}{1 - X^2} = 5.296$$

$$X^2 = \frac{5.296}{6.296}$$

and $X = 0.917$ (positive acceptable root of the quadratic)

Therefore, the conversion of C_2H_6 to C_2H_4 is 91.7 percent.

Rate of Reaction

The rate of a reaction is the number of units of mass of some participating reactant which is transformed into a product per unit time and per unit volume of the system.

$$-r = -\frac{1}{V}\frac{dn}{dt} \tag{13-15}$$

where r is the reaction rate in moles per unit time times the unit volume and n is the number of moles of the reactant present at time t. At constant volume, Eq. (13-15) becomes

$$-r = -\frac{d(n/V)}{dt} = -\frac{dC}{dt} \tag{13-16}$$

where C is the concentration of the reactant at time t. If x moles are transformed in time t,

$$x = n_{A0} - n_A \tag{13-17}$$

and then

$$-r = +\frac{1}{V}\frac{dx}{dt} \tag{13-17a}$$

If X_A is the fractional conversion of the component A in time t,

$$n_A = n_{A0} - n_{A0}X_A$$

and

$$-r = +\frac{1}{V}\frac{dX_A}{dt} \tag{13-18}$$

For many reactions, it is possible to express the effect of the moles of the components independently, so that

$$-r = -\frac{1}{V}\frac{dn_A}{dt} = kf(n_A, n_B, \ldots) \tag{13-19}$$

in which k is the specific reaction rate or the rate constant.

Law of Mass Action

This states that the rate of chemical reaction is proportional to the active masses of the participants (Guldberg and Waage law).

The Rate Equation

Rate equations for some reactions at constant volume and temperature are given in Table 13-2.

TABLE 13-2 Reactions at Constant Volume and Temperature*

Reaction: $aA + bB + cC \rightarrow \text{products}$

Rate equation: $$\frac{dx}{dt} = k(n_{A0} - x)^p \left(n_{B0} - \frac{bx}{a}\right)^q \left(n_{C0} - \frac{cx}{a}\right)^r$$

Order	Reaction	Rate Equation	Integral
0	A → products	$\frac{dx}{dt} = k$	$x - x_0 = k(t - t_0)$
$\frac{1}{2}$	A → products	$\frac{dx}{dt} = k(n_{A0} - x)^{1/2}$	$(n_{A0} - x_0)^{1/2} - (n_{A0} - x)^{1/2} = \frac{k(t - t_0)}{2}$
1	A → products	$\frac{dx}{dt} = k(n_{A0} - x)$	$\ln \frac{n_{A0} - x_0}{n_{A0} - x} = k(t - t_0)$
2	2A → products	$\frac{dx}{dt} = k(n_{A0} - x)^2$	$\frac{1}{n_{A0} - x} - \frac{1}{n_{A0} - x_0} = k(t - t_0)$
2	A + B → products	$\frac{dx}{dt} = k(n_{A0} - x)(n_{B0} - x)$ $(n_{A0} \neq n_{B0})$	$\frac{1}{n_{B0} - n_{A0}} \ln \frac{(n_{A0} - x_0)(n_{B0} - x)}{(n_{B0} - x_0)(n_{A0} - x)} = k(t - t_0)$
3	3A → products	$\frac{dx}{dt} = k(n_{A0} - x)^3$	$\left(\frac{1}{n_{A0} - x}\right)^2 - \left(\frac{1}{n_{A0} - x_0}\right)^2 = 2k(t - t_0)$

*n_{A0}, n_{B0}, and n_{C0} are initial moles of A, B, and C, respectively.

Molecularity and Order of Reaction

Molecularity of an elementary reaction is the number of the molecules involved in the reaction. This applies only to the elementary reactions, in which case the molecularity can be one, two, or three.

In the case of many complex reactions, the rate equation is found empirically. Hence the exponents of the concentration terms in the rate equation are not related to the stoichiometric coefficients and are different. The sum of these empirically determined exponents is called the *overall order of the reaction*. The order of the reaction is given by

$$n = p + q + r \tag{13-20}$$

where n is the overall order of reaction. (It may be a fractional number.) p is the order of the reaction with respect to the reactant A, q with respect to B, and r with respect to C, respectively. Orders with respect to other components are defined the same way.

Effect of Temperature on Rate of Reaction

The effect of temperature on the reaction rate is expressed by the equation

$$\frac{d \ln k}{dT} = \frac{E}{RT^2} \tag{13-21}$$

By integration of the above, one obtains the Arrhenius relation

$$k = \alpha e^{-E/RT} \tag{13-22}$$

where E is the activation energy and α is the frequency factor.

The collision theory yields the following relation:

$$k = \alpha T^{1/2} e^{-E/RT} \tag{13-23}$$

whereas the transition theory predicts

$$k = \alpha T e^{-E/RT} \tag{13-24}$$

or, in general,

$$k = \alpha\, T^m e^{-E/RT} \tag{13-25}$$

Differentiating the logarithms of both sides of Eq. (13-25) gives

$$\frac{d \ln k}{dT} = \frac{m}{T} + \frac{E}{RT^2} = \frac{mRT + E}{RT^2} \tag{13-26}$$

Since $mRT << E$ for most of the reactions, the above gives

$$\frac{d \ln k}{dT} = \frac{E}{RT^2} \tag{13-26a}$$

which shows Arrhenius theory is a good approximation to both the collision and transition-state theories.

Example 13-4. Determine the activation energy and frequency factor given the following data for the bimolecular formation of methyl ether in an ethyl alcohol solution.

T,°C	0	6	12	18	24	30
$k \times 10^5$ L/g·mol·s	5.6	11.8	24.5	48.8	100	208

Solution.

$$k = \alpha e^{-E/RT}$$

$$\ln k = \ln \alpha - \frac{E}{RT}$$

or

$$\log k = \log \alpha - \frac{E}{2.3RT}$$

Thus if log k is plotted against $1/T$ (T is in Kelvins), a straight line is obtained. The slope of this line is $-E/(2.3R)$, and the intercept is log α. Thus E and α can be determined from a plot of log k versus $1/T$. For the given data, the following table is prepared.

T,°K	273	279	285	291	297	303
$\frac{1}{T} \times 10^3$	3.663	3.584	3.509	3.436	3.367	3.300
$k \times 10^5$	5.6	11.8	24.5	48.8	100	208

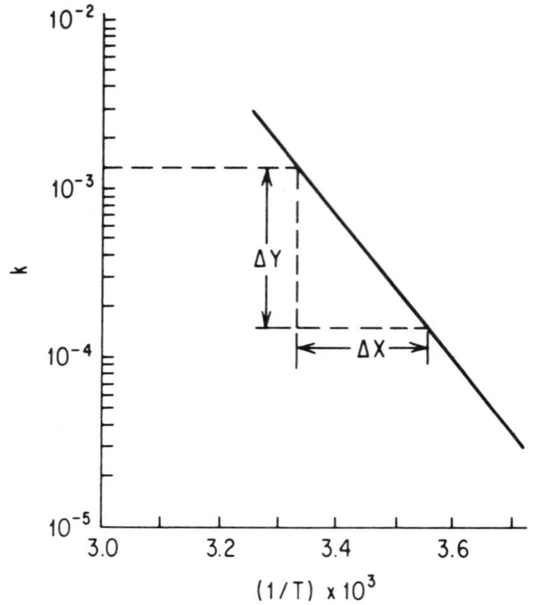

Figure 13-1. Plot of k versus $1/T$ (Example 13-4).

A plot of k versus $1/T$ is made on a semilogarithmic paper as shown in Fig. 13-1.

From the graph, the slope is

$$-\frac{E}{2.3R} = \frac{\log(1.3 \times 10^{-3}) - \log(1.5 \times 10^{-4})}{(3.34 - 3.56) \times 10^{-3}}$$

$$= \frac{-\log 8.67}{0.22 \times 10^{-3}} = -4263$$

Therefore

$$E = -2.3(R)(-4263) = 2.3(1.987)(4263) = 19{,}482 \text{ cal/g}\cdot\text{mol}$$

Now

$$\log k = -\frac{E}{2.3R}\frac{1}{T} + \log \alpha$$

At $1/T = 3.34 \times 10^{-3}$,

$$\log k = \log(1.3 \times 10^{-3}) = -2.88606$$

$$-2.88606 = -\frac{19{,}482(3.34 \times 10^{-3})}{2.3(1.987)} + \log \alpha$$

from which

$$\alpha = 2.25 \times 10^{11}$$

By calculating the values at other readings, an average α can be determined.

Example 13-5. The rate of a bimolecular reaction at 500 K is 10 times the rate at 400 K. Calculate the activation energy of this reaction by (**a**) the Arrhenius law, and (**b**) collision theory.

Solution.

a. Let k_1 be the reaction rate at 400 K and k_2 at 500 K

$$k_1 = \alpha \exp\left[-\frac{E}{R(400)}\right] \qquad \text{and} \qquad k_2 = \alpha \exp\left[-\frac{E}{R(500)}\right]$$

$$\frac{k_2}{k_1} = 10 = \exp\left[-\frac{E}{R}\left(\frac{1}{500} - \frac{1}{400}\right)\right] = \exp\left(\frac{E}{2000R}\right)$$

$$\ln 10 = \frac{E}{2000R}$$

$$E = 2000(1.987) \ln 10 = 9150 \text{ cal/g}\cdot\text{mol}$$

b. *Activation Energy by Collision Theory*

$$\frac{k_2}{k_1} = \left(\frac{500}{400}\right)^{1/2} \exp\left[-\frac{E}{R}\left(\frac{1}{500} - \frac{1}{400}\right)\right] = 10$$

from which

$$\exp\left(\frac{E}{2000R}\right) = 10\left(\frac{400}{500}\right)^{1/2} = 8.94427$$

and
$$\frac{E}{2000R} = \ln 8.94427$$

Hence
$$E = 2000(1.987)(\ln 8.94427) = 8707 \text{ cal/g}\cdot\text{mol}$$

Example 13-6. The third-order gas-phase reaction $2NO + O_2 \rightarrow 2NO_2$ has a specific reaction rate of

$$k_c = 2.65 \times 10^4 \text{ L}^2/(\text{g}\cdot\text{mol})^2\cdot\text{s}$$

at 30°C and 1 atm. Find k_p and k_n. Show clearly the conversion of units.

Solution.

$$k_c = (RT)^n k_p = \left(\frac{RT}{P_T}\right) k_n$$

Here $n = 3$, $P_T = 1$ atm, and $T = 273 + 30 = 303$ K.

$$k_p = \frac{k_c}{(RT)^n}$$

$$= \frac{2.65 \times 10^4 \text{ L}^2/(\text{g}\cdot\text{mol})^2\cdot\text{s}}{[(0.08206 \text{ L}\cdot\text{atm/g}\cdot\text{mol}\cdot\text{K})\,(303 \text{ K})]^3}$$

$$= \frac{2.65 \times 10^4}{0.08206^3}\,\frac{\text{L}^2}{(\text{g}\cdot\text{mol})^2\cdot\text{s}}\left[\frac{(\text{g}\cdot\text{mol})^3}{\text{L}^3\cdot\text{atm}^3\cdot 303^3}\right]$$

$$= 1.7239 \text{ g}\cdot\text{mol/L}\cdot\text{atm}^3\cdot\text{s}$$

and $$k_n = P_T^n k_p = 1^3 \text{ atm}^3(1.7239) \text{ g}\cdot\text{mol/L}\cdot\text{atm}^3\cdot\text{s}$$

$$= 1.7239 \text{ g}\cdot\text{mol/L}\cdot\text{s}$$

Interpretation of Kinetic Data and the Constants of the Rate Equation

Kinetic data are generally correlated by trial. However, in most cases, the stoichiometr y of the reaction suggests the form of the rate equation which should be tried first. When the mathematical equation is written, the next step is to find the constants in the rate equation. For this the following methods are used: (1) method of differentiation, (2) method of integrated equation, (3) method of halftimes, (4) method of reference curves, and (5) method of k calculation.

Simple Irreversible Reactions

Consider the general equation

$$a\text{A} + b\text{B} + \cdots \longrightarrow c\text{C} + d\text{D} + \cdots$$

At constant volume and temperature, the rate equation in terms of x (mols converted), i.e., $x = (n_{A0} - n_A)$, will be

$$\frac{dx}{dt} = k(n_{A0} - x)^a \left(n_{B0} - \frac{bx}{a} \right)^b \cdots$$
$$= k'(n_{A0} - x)^{a+b+\cdots} = k'(n_{A0} - x)^n$$

Method of Differentiation. In this method, the plot of log (dx/dt) versus log $(n_{A0} - x)$ is made. If the plot is a straight line, the slope of the curve is n and intercept is log k'. If the curve obtained is not a straight line, the reaction is probably complex.

Method of Integration. Integration of the rate equation yields ($n \neq 1$):

$$\left(\frac{1}{n_{A0} - x} \right)^{n-1} - \left(\frac{1}{n_{A0}} \right)^{n-1} = (n-1)kt \tag{13-27}$$

When $n = 1$, the solution of the rate equation is

$$\ln \left(\frac{n_{A0}}{n_{A0} - x} \right) = kt \tag{13-28}$$

Therefore, for a first-order irreversible reaction, a plot of ln $(n_{A0} - x)$ versus t will give a straight line from which the reaction rate constant k can be calculated.

In the case of the reactions of order 2 or greater, the method of integration involves a trial-and-error search for n.

Method of Halftimes. At 50 percent, the integrated equations are

$$\frac{dx}{dt} = k(n_{A0} - x) \qquad \text{First order: } t_{1/2} = \frac{\ln 2}{k} \tag{13-29}$$

$$\frac{dx}{dt} = k(n_{A0} - x)^2 \qquad \text{Second order: } t_{1/2} = \frac{1}{kn_{A0}} \tag{13-29a}$$

$$\frac{dx}{dt} = k(n_{A0} - x)^n \qquad n\text{th order: } t_{1/2} = \frac{2^{n-1}-1}{k(n-1)n_{A0}^{n-1}} \tag{13-29b}$$

Thus
$$\log t_{1/2} = \log \frac{2^{n-1} - 1}{k(n-1)} - (n-1) \log n_{A0} \qquad (13\text{-}29c)$$

Therefore, a plot of log $t_{1/2}$ versus log n_{A0} will give a straight line whose slope is $1 - n$, which determines n. k is then calculated from

$$k = \frac{2^{n-1} - 1}{t_{1/2}(n-1)n_{A0}^{n-1}} \qquad (13\text{-}29d)$$

However, data from several experiments as a function of the initial quantity n_{A0} must be available.

Method of Reference Curves. A reference plot[1] of the percent conversion x/n_{A0} versus $t/t_{0.9}$, the fraction of the time required for 90 percent conversion can be used to determine the order of a reaction. Each curve of this plot is determined by a unique value of n. The data for a reaction under investigation are plotted on the same scale as this plot, and the order of the reaction is found by superimposition.

Method of *k* Calculation. In this method a value for the order of the reaction is assumed and k values are calculated at various experimental data points. If the k values calculated are nearly constant, the assumed n value is correct.

Irreversible Reactions

Unimolecular First-Order Reactions. These are of the type

$$A \longrightarrow \text{products} \qquad (13\text{-}30)$$

Rate equation:
$$-r_A = -\frac{dC_A}{dt} = kC_A$$

Boundary condition: $C_A = C_{A0}$ at $t = 0$

Solution: $$\ln \frac{C_A}{C_{A0}} = -kt \qquad \text{or} \qquad \frac{C_A}{C_{A0}} = e^{-kt} \qquad (13\text{-}30a)$$

In terms of fractional conversion X_A, the rate equation is

$$\frac{dX_A}{dt} = k(1 - X_A) \qquad (13\text{-}30b)$$

With boundary condition $X_A = 0$ at $t = 0$, the solution of Eq. (13-30*b*) is

$$-\ln (1 - X_A) = kt \qquad (13\text{-}30c)$$

Only the simple first-order equations can be treated this way. Complex first-order reactions cannot be treated by the above method.

Bimolecular Second-Order Reactions. These are of the type

$$A + B \longrightarrow \text{products}$$

Rate equation: $$-r_A = -\frac{dC_A}{dt} = -\frac{dC_B}{dt} = kC_A C_B \qquad (13\text{-}31)$$

The amounts of A and B that have reacted are equal at any time t. Let these be

$$C_{A0}X_A = C_{B0}X_B$$

The rate equations can be written in terms of X_A as

$$-r_A = C_{A0}\frac{dX_A}{dt} = k(C_{A0} - C_{A0}X_A)(C_{B0} - C_{A0}X_A) \qquad (13\text{-}31a)$$

Let $M = C_{B0}/C_{A0}$; then

$$-r_A = C_{A0}\frac{dX_A}{dt} = kC_{A0}^2(1 - X_A)(M - X_A) \qquad (13\text{-}31b)$$

from which

$$\int_0^{X_A} \frac{dX_A}{(1 - X_A)(M - X_A)} = C_{A0}k\, dt \qquad (13\text{-}31c)$$

The solution if $M \neq 1$ is

$$\ln \frac{M - X_A}{M(1 - X_A)} = C_{A0}(M - 1)kt \qquad M \neq 1 \tag{13-31d}$$

or

$$\ln \frac{C_B}{MC_A} = (C_{B0} - C_{A0})kt \tag{13-32}$$

The plots of $\ln (C_B/C_A)$ versus t or $\ln (M - X_A)/M(1 - X_A)$ versus t are straight lines.

Special Case When *M* = 1. If $M = 1$, i.e., if at $t = 0$ the reactants are equal in molar concentration, the above equations are indeterminate. In this case, the rate equations become

$$-r = -\frac{dC_A}{dt} = -\frac{dC_B}{dt} = kC_A^2 = kC_B^2 \tag{13-33}$$

The boundary conditions are: at $t = 0$, $C_A = C_{A0} = C_{B0}$. The solution is

$$\frac{1}{C_A} - \frac{1}{C_{A0}} = kt \tag{13-33a}$$

In terms of X_A—fractional conversion,

$$-r_A = C_{A0}\frac{dX_A}{dt} = C_{B0}\frac{dX_B}{dt} = kC_{A0}^2(1 - X_A)^2$$

Therefore

$$\frac{dX_A}{dt} = kC_{A0}(1 - X_A)^2$$

With the boundary condition, $X_A = 0$ at $t = 0$, the solution is

$$\frac{1}{C_{A0}} \frac{X_A}{1 - X_A} = kt \tag{13-33b}$$

The above rate equations and solutions will also apply to the reactions of the type

$$2A \longrightarrow \text{products}$$

Empirical Rate Equations of the *n*th Order for Irreversible Reactions

Rate equation: $$-r_A = -\frac{dC_A}{dt} = kC_A^n \qquad (13\text{-}34)$$

Boundary conditions: $C_A = C_{A0}$ at $t = 0$

Solution: $$C_A^{1-n} - C_{A0}^{1-n} = (n-1)kt \qquad n \neq 1 \qquad (13\text{-}34a)$$

The value of n must be found by trial and error. In terms of X_A,

Rate equation: $$-r_A = C_{A0}\frac{dX_A}{dt} = kC_{A0}^n(1 - X_A)^n \qquad (13\text{-}34b)$$

Boundary condition: $X_A = 0$ $t = 0$

Solution: $$(1 - X_A)^{1-n} - 1 = (n-1)C_{A0}^{n-1}kt \qquad (13\text{-}34c)$$

Zero-Order Reaction

Rate equation: $$-r_A = -\frac{dC_A}{dt} = k \qquad (13\text{-}35)$$

Boundary condition: $C_A = C_{A0}$

Solution: $$C_{A0} - C_A = kt \qquad (13\text{-}35a)$$

or $$C_{A0}X_A = kt \qquad (13\text{-}35b)$$

In this case, the conversion is proportional to time.

Reactions in Parallel. Consider elementary reactions of the following type:

$$A \xrightarrow{k_1} B \qquad A \xrightarrow{k_2} C$$

The rate equations for these reactions are

$$-r_A = -\frac{dC_A}{dt} = k_1 C_A + k_2 C_A = (k_1 + k_2) C_A \qquad (13\text{-}36)$$

$$r_B = \frac{dC_B}{dt} = k_1 C_A \qquad (13\text{-}36a)$$

$$r_C = \frac{dC_C}{dt} = k_2 C_A \qquad (13\text{-}36b)$$

Boundary condition: $C_A = C_{A0}$ at $t = 0$

$C_B = C_{B0}$ $C_C = C_{C0}$ at $t = 0$

Solution: $$-\ln \frac{C_A}{C_{A0}} = (k_1 + k_2)t \qquad (13\text{-}36c)$$

and in terms of X_A,

$$-\ln (1 - X_A) = (k_1 + k_2)t \qquad (13\text{-}36d)$$

If $-\ln (C_A/C_{A0})$ or $-\ln (1 - X_A)$ is plotted versus t, the slope of the resulting straight line is $k_1 + k_2$. Also

$$\frac{r_B}{r_C} = \frac{dC_B}{dC_C} = \frac{k_1}{k_2}$$

from which $$\frac{C_B - C_{B0}}{C_C - C_{C0}} = \frac{k_1}{k_2} \qquad (13\text{-}36e)$$

or $$C_B - C_{B0} = \frac{k_1}{k_2} C_C - \frac{k_1}{k_2} C_{C0} \qquad (13\text{-}36f)$$

Thus the slope of the straight-line plot of C_B versus C_C gives the ratio k_1/k_2. Knowing $k_1 + k_2$ and k_1/k_2 gives the individual values of k_1 and k_2.

Homogeneous Catalyzed Reactions. A catalyzed reaction can be represented as follows:

$$\mathrm{A} \xrightarrow{k_1} \mathrm{P}$$
$$\mathrm{A} + \mathrm{C} \xrightarrow{k_2} \mathrm{P} + \mathrm{C}$$

where C and P represent the catalyst and product, respectively. The reaction rates are

$$-\left(\frac{dC_{\mathrm{A}}}{dt}\right)_1 = k_1 C_{\mathrm{A}} \qquad (13\text{-}37)$$

$$-\left(\frac{dC_{\mathrm{A}}}{dt}\right)_2 = k_2 C_{\mathrm{A}} C_{\mathrm{C}} \qquad (13\text{-}37a)$$

The overall rate of disappearance of A then is

$$-\frac{dC_{\mathrm{A}}}{dt} = k_1 C_{\mathrm{A}} + k_2 C_{\mathrm{A}} C_{\mathrm{C}} \qquad (13\text{-}37b)$$

The catalyst concentration remains unchanged.

Boundary condition: $C_{\mathrm{A}} = C_{\mathrm{A0}}$ at $t = 0$ (13-37*c*)

Solution: $$-\ln \frac{C_{\mathrm{A}}}{C_{\mathrm{A0}}} = -\ln\,(1 - X_{\mathrm{A}}) = (k_1 + k_2 C_{\mathrm{C}})t = kt \qquad (13\text{-}37d)$$

where $k = k_1 + k_2 C_{\mathrm{C}}$. In this case, a series of runs are made with various concentrations of the catalyst and a plot is made of $k = (k_1 + k_2 C_{\mathrm{C}})$ versus C_{C}. The slope of this straight line is k_2, and the intercept is k_1.

The Autocatalytic Reactions. When one of the products of a reaction acts as a catalyst, it is called an autocatalytic reaction. The simplest reaction is given by

$$\mathrm{A} + \mathrm{B} \xrightarrow{k} \mathrm{B} + \mathrm{B}$$

for which the rate equation is

$$-r_{\mathrm{A}} = -\frac{dC_{\mathrm{A}}}{dt} = kC_{\mathrm{A}} C_{\mathrm{B}} \qquad (13\text{-}38)$$

Since, when A is consumed, the total moles of A and B remain unchanged, at any time t, the following relation holds:

$$C_0 = C_A + C_B = C_{A0} + C_{B0} = \text{const.}$$

Then the rate equation becomes

$$-r_A = -\frac{dC_A}{dt} = kC_A(C_0 - C_A) \qquad (13\text{-}38a)$$

Integration with the use of partial fractions yields the solution

$$\ln \frac{C_{A0}(C_0 - C_A)}{C_A(C_0 - C_{A0})} = \ln \frac{C_B/C_{B0}}{C_A/C_{A0}} = C_0kt \qquad (13\text{-}38b)$$

In terms of the initial reaction ratio $M = C_{B0}/C_{A0}$ and the fractional conversion X_A of A, the solution is

$$\ln \frac{M + X_A}{M(1 - X_A)} = C_{A0}(M + 1)kt = (C_{A0} + C_{B0})kt \qquad (13\text{-}38c)$$

Reactions in Series. A typical example of reactions in series is

$$A \xrightarrow{k_1} B \xrightarrow{k_2} C \qquad (13\text{-}39)$$

The rate equations are

$$-r_A = -\frac{dC_A}{dt} = k_1C_A \qquad (13\text{-}39a)$$

$$r_B = \frac{dC_B}{dt} = k_1C_A - k_2C_B \qquad (13\text{-}39b)$$

$$r_C = \frac{dC_C}{dt} = k_2C_B \qquad (13\text{-}39c)$$

and the initial conditions are

$$C_A = C_{A0} \qquad C_B = 0 \qquad C_C = 0 \qquad \text{at } t = 0$$

Concentration of A by integration is

$$-\ln \frac{C_A}{C_{A0}} = k_1t \qquad \text{or} \qquad C_A = C_{A0}e^{-k_1t} \qquad (13\text{-}39d)$$

Using the relation of Eq. (13-39*d*) in (13-39*b*), one obtains

$$\frac{dC_B}{dt} + k_2C_B = k_1C_{A0}e^{-k_1t} \qquad (13\text{-}39e)$$

which can be integrated by the method of *integrating factor* and using initial condition $C_{B0} = 0$ at $t = 0$ find the constant of integration to yield

$$C_B = C_{A0}k_1\left(\frac{e^{-k_1t}}{k_2 - k_1} + \frac{e^{-k_2t}}{k_1 - k_2}\right) \qquad (13\text{-}39f)$$

Since the total number of moles do not change, $C_{A0} = C_A + C_B + C_C$. Using this, one obtains

$$C_C = C_{A0}\left(1 + \frac{k_2}{k_1 - k_2}e^{-k_1t} + \frac{k_1}{k_2 - k_1}e^{-k_2t}\right) \qquad (13\text{-}39g)$$

The maximum concentration of B occurs at

$$t_{max} = \ln\frac{k_2/k_1}{k_2 - k_1} \qquad (13\text{-}39h)$$

and the maximum concentration of B is given by

$$\frac{C_{Bmax}}{C_{A0}} = \left(\frac{k_1}{k_2}\right)^{k_2/(k_2 - k_1)} \qquad (13\text{-}39i)$$

First-Order Reversible Reactions

Reaction: $$A \underset{k_2}{\overset{k_1}{\rightleftharpoons}} B \qquad (13\text{-}40)$$

Rate equations: $$\frac{dC_B}{dt} = -\frac{dC_A}{dt} = k_1C_A - k_2C_B \qquad (13\text{-}40a)$$

or in terms of X_A (if $M = C_{B0}/C_{A0}$),

$$C_{A0}\left(\frac{dX_A}{dt}\right) = k_1(C_{A0} - C_{A0}X_A) - k_2(C_{A0}M + C_{A0}X_A) \qquad (13\text{-}40b)$$

The boundary conditions are: at $t = 0$, $M = C_{B0}/C_{A0}$, $X_A = 0$. At equilibrium,

$$\frac{dC_A}{dt} = 0 \qquad \text{or} \qquad \frac{dX_A}{dt} = 0 \tag{13-40c}$$

Therefore, at equilibrium condition, K_e is given by

$$K_e = \frac{C_{Be}}{C_{Ae}} = \frac{M + X_{Ae}}{1 - X_{Ae}} \tag{13-40d}$$

where equilibrium constant

$$K_e = \frac{k_1}{k_2} \tag{13-40e}$$

If the above equations are combined, the rate equation in terms of the equilibrium conversion is

$$\frac{dX_A}{dt} = \frac{k_1(M + 1)}{M + X_{Ae}}(X_{Ae} - X_A) \tag{13-40f}$$

If the concentrations are measured in terms of X_{Ae}, the equilibrium conversion, integration gives

$$-\ln\left(1 - \frac{X_A}{X_{Ae}}\right) = -\ln\frac{C_A - C_{Ae}}{C_{A0} - C_{Ae}} = \frac{M + 1}{M + X_{Ae}}k_1 t \tag{13-41}$$

and a plot of $-\ln(1 - X_A/X_{Ae})$ versus t will be a straight line.

Second-Order Reversible Reactions

Bimolecular-type second-order reactions are as follows:

$$A + B \underset{k_2}{\overset{k_1}{\rightleftharpoons}} R + S \tag{13-42}$$

$$2A \underset{k_2}{\overset{k_1}{\rightleftharpoons}} R + S \tag{13-42a}$$

$$2A \underset{k_2}{\overset{k_1}{\rightleftharpoons}} 2R \tag{13-42b}$$

$$A + B \underset{k_2}{\overset{k_1}{\rightleftharpoons}} 2R \tag{13-42c}$$

With the restrictions that $C_{A0} = C_{B0}$, $C_{R0} = C_{S0} = 0$ in Eq. (13-42), the rate equation becomes

$$-r = -\frac{dC_A}{dt} = C_{A0}\frac{dX_A}{dt} = k_1C_A^2 - k_2C_R^2 \tag{13-42d}$$

$$= k_1C_{A0}^2(1 - X_A)^2 - \frac{k_2}{k_1}X_A^2 \tag{13-42e}$$

where the fractional conversion $X_A = 0$ at $t = 0$ and $dX_A/dt = 0$ at equilibrium, and therefore the final solution is

$$\ln\frac{X_{Ae} - (2X_{Ae} - 1)X_A}{X_{Ae} - X_A} = 2k_1C_{A0}\left(\frac{1}{X_{Ae}} - 1\right)t \tag{13-43}$$

All other reversible second-order rate equations (13-42) have the same solution with the boundary conditions assumed in the above solution.

Reversible Reactions in General

For orders other than 1 or 2, the integration of the rate equation is difficult. Therefore, the differential method of analysis should be used to search the form of the rate equation. Sometimes a complex equation of the type below fits the data well.

$$-r_A = -\frac{dC_A}{dt} = k_1\frac{C_A}{1 + k_2C_A} \tag{13-44}$$

Taking reciprocals, one obtains

$$\frac{1}{-r_A} = \frac{1 + k_2C_A}{k_1C_A} = \frac{1}{k_1}\frac{1}{C_A} + \frac{k_2}{k_1} \tag{13-44a}$$

and a plot of $1/-r_A$ versus $1/C_A$ would be a straight line with a slope equal to $1/k_1$ and an intercept of k_2/k_1.

Another method of analysis can be obtained in the above case by multiplying each side of Eq. (13-44*a*) by k_1/k_2 and solving for $-r_A$ to yield another form as

$$-r_A = \frac{k_1}{k_2} - \frac{1}{k_2}\frac{-r_A}{C_A} \tag{13-45}$$

on the basis of which, $-r_A$ versus $-r_A/C_A$ will be a linear plot, and values of k_1 and k_2 can be determined from the slope $-1/k_2$ and intercept k_1/k_2.

Example 13-7. For the irreversible thermal dissociation of paraldehyde at 260°C and constant volume the following data were obtained.

Time, h	0	1	2	3	4	∞
P total, mmHg	100	173	218	248	266	300

Determine the order of the reaction and the rate constant.

Solution. Paraldehyde decomposes according to the equation

$$(CH_3CHO)_3 \longrightarrow 3CH_3CHO$$

If the reaction is first order, the rate equation is given by

$$r_A = -\frac{dC_A}{dt} = k_1 C_A$$

Assuming the reactant and product obey the ideal-gas law, the concentration is given by

$$C_A = \frac{n_A}{V} = \frac{p_A}{RT}$$

Therefore, in terms of the partial pressure p_A,

$$-\frac{dp_A}{dt} = k_1 p_A$$

$$\frac{dp_A}{p_A} = -k_1 dt$$

$$\ln p_A = -k_1 t + C \text{ (a constant)}$$

At $t = 0$, $p_A = p_{A0}$, $C = \ln p_{A0}$. The solution of the differential equation is

$$\ln \frac{p_A}{p_{A0}} = -k_1 t$$

Thus, the graph of $\ln (p_A/p_{A0})$ versus t should be a straight line. Now

$$(CH_3CHO)_3 \longrightarrow 3CH_3CHO$$

If n_A is the number of moles of paraldehyde at time t and n_{A0} the moles of paraldehyde at $t = 0$, then moles of acetaldehyde $= 3(n_{A0} - n_A)$ at time t. Thus total moles at time $t = n_T = n_A + 3(n_{A0} - n_A)$. Assuming the gas law applies, we can get (since $n = pV/RT$, and V and T are constant)

$$P_T = p_A + 3(p_{A0} - p_A)$$

where P_T = total pressure
p_A = partial pressure of A (paraldehyde) at time t
p_{A0} = partial pressure of A at time $t = 0$
$p_A = \frac{1}{2}(3p_{A0} - P_T)$

Using the above relation for P_T, Table 13-3 is prepared.

TABLE 13-3 Calculations of p_A/p_{A0}

Time, h	P_T, mmHg at 0°C	p_A	p_A/p_{A0}
0	$100 = p_{A0}$	100	1.0
1	173	63.5	0.635
2	218	41.0	0.410
3	248	26	0.26
4	266	17	0.17
	300	0	0.000

A plot of p_A/p_{A0} versus t on semilog paper (Fig. 13-2) gives a straight line. Therefore the assumed first order for the reaction is correct. Slope of the straight line is $-k_1/2.3$. Thus

$$\frac{k_1}{2.3} = -(\text{slope})$$

and

$$k_1 = -\frac{2.3(0+1)}{0-5.2} = 0.442\ \text{h}^{-1}$$

Example 13-8. The reaction $2NOCl \longrightarrow 2NO + Cl_2$ is studied at 200°C. The concentration of NOCl initially consisting of NOCl only changes as follows:

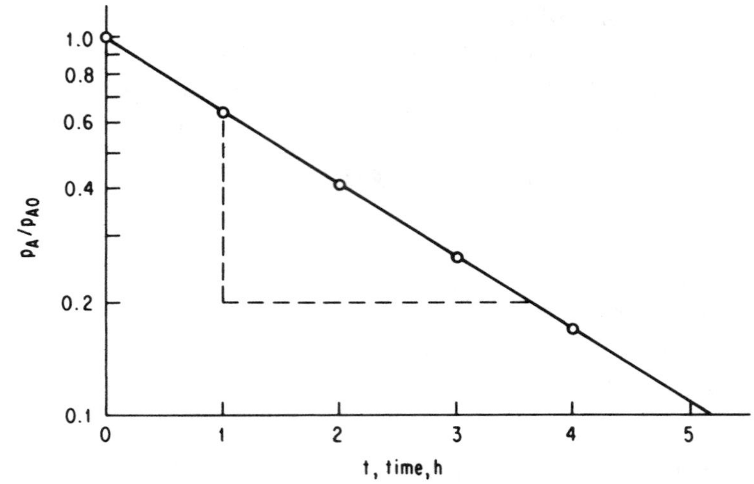

Figure 13-2. Plot of p_A/p_{A0} versus t (Example 13-7).

t, s	0	200	300	500
N_{NOCl}, g·mol/L	0.02	0.0159	0.0144	0.0121

Find the order of the reaction and the rate constant.

Solution. If the reaction is second order, the rate equation is

$$-\frac{dC_A}{dt} = k_1 C_A^2$$

which on integration and with boundary condition $C_A = C_{A0}$ at $t = 0$ yields the solution.

$$\frac{1}{C_A} - \frac{1}{C_{A0}} = k_1 t$$

Thus a plot of $1/C_A$ versus t will be a straight line of slope k_1. Prepare a table as follows:

t	0	200	300	500
$1/C_A$	50	62.3	69.44	82.64

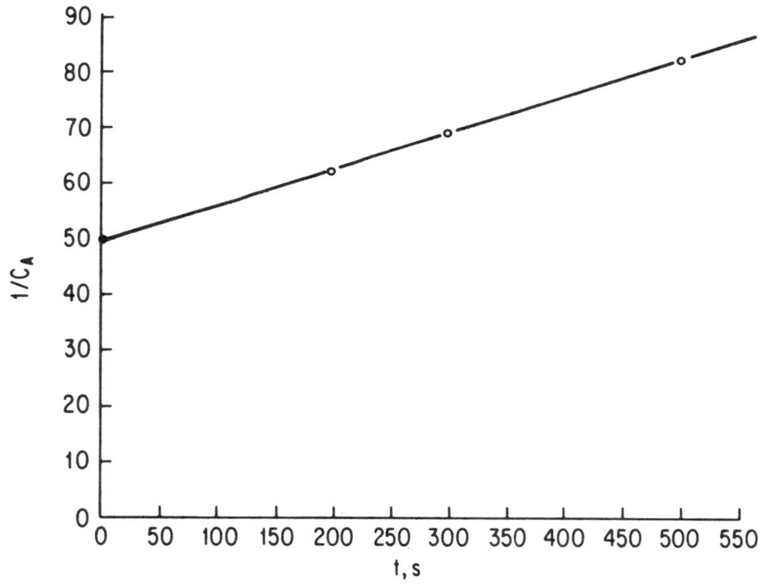

Figure 13-3. Plot of $1/C_A$ versus t (Example 13-8).

$1/C_A$ versus t is plotted in Fig. 13-3. The graph is a straight line. Therefore, the assumed order of the reaction is correct. The slope of the line is

$$k_1 = \frac{86 - 50}{550 - 0} = \frac{36}{550} = 0.0655 \text{ L/g} \cdot \text{mol} \cdot \text{s}$$

Example 13-9. The hydrolysis of methyl acetate is an autocatalytic reaction and is first order with respect to methyl acetate and first order with respect to acetic acid. The reaction is elementary, bimolecular, and can be considered irreversible at constant volume for design purposes. The following data are given.

$$\text{Initial concentration of methyl acetate} = 0.5 \text{ g} \cdot \text{mol/L}$$

$$\text{Initial concentration of acetic acid} = 0.05 \text{ g} \cdot \text{mol/L}$$

The conversion in 1 h is 60 percent in a batch reactor. Calculate (**a**) the rate constant and indicate the rate equation, (**b**) the time at which the rate passes through a maximum, and (**c**) based on the above information, the type of optimum reactor system you would specify for the plant to process 200 ft^3/h. What would be the reactor volume in this system?

Solution.

$$CH_3COOCH_3 + H_2O \longrightarrow CH_3COOH + CH_3OH$$

$$A \longrightarrow B$$

a. For autocatalytic reaction, $A + B \longrightarrow B + B$ and

$$-r_A = -\frac{dC_A}{dt} = k_1 C_A C_B$$

Using the solution for the autocatalytic reaction, the value of the initial reactant ratio M can be calculated:

$$\ln \frac{M + X_A}{M(1 - X_A)} = (C_{A0} + C_{B0})k_1 t \qquad (13\text{-}38c)$$

$$M = \frac{C_{B0}}{C_{A0}} = \frac{0.05}{0.5} = 0.1$$

At $t = 1$ h, $X_A = 0.6$,

$$\ln \frac{0.1 + 0.6}{0.1(1 - 0.6)} = (0.5 + 0.05)k_1(1)$$

or
$$k_1 = \frac{\ln(0.7/0.04)}{0.55} = 5.204 \text{ L/g}\cdot\text{mol}\cdot\text{h}$$

The rate equation is

$$-r_A = -\frac{dC_A}{dt} = 5.2 C_A C_B \qquad \text{L/g}\cdot\text{mol}\cdot\text{h}$$

b. Rate is maximum when $C_A = C_B$. But

$$C_A + C_B = C_{A0} + C_{B0} = 0.5 + 0.05 = 0.55$$

Therefore
$$C_A = C_B = \tfrac{1}{2}(0.55) = 0.275$$

and then

$$X_A = \frac{C_{A0} - C_A}{C_{A0}} = \frac{0.5 - 0.275}{0.5} = 0.45$$

Find t by substituting $X_A = 0.45$ and $k_1 = 5.204$ found in part **a** in the integrated equation. Thus

$$\ln \frac{0.1 + 0.45}{0.1(1 - 0.45)} = 0.55(5.204)t$$

From which, $t = 0.8045$ h.

c. For solution of this part, refer to Example 13-12.

Example 13-10. The gas-phase decomposition $A \longrightarrow B + 2C$ is conducted in a constant-volume reactor. Runs 1 to 5 were conducted at 100°C; run 6 was carried out at 110°C. (**a**) From the data given in Table 13-4 below, determine the reaction order and specify the reaction rate, and (**b**) what are the activation energy and frequency factor for this reaction?

TABLE 13-4 Half-Life $t_{1/2}$ as Function of Initial Concentration C_{A0}

Run No.	C_{A0}, g·mol/L	Half-Life $t_{1/2}$, min
1	0.025	4.1
2	0.0133	7.7
3	0.0100	9.8
4	0.050	1.96
5	0.075	1.30
6	0.025	2.0

Solution.

a.

$$t_{1/2} = \frac{2^{n-1} - 1}{k(n-1)} \frac{1}{C_{A0}^{n-1}}$$

$$\ln t_{1/2} = \ln \frac{2^{n-1} - 1}{k(n-1)} + (1-n)\ln C_{A0}$$

The plot of C_{A0} versus $t_{1/2}$ on log-log paper is shown in Fig. 13-4. From the graph the slope of the line = −1.0. Therefore, the order of the reaction is given by $1 - n = -1.0$. Thus the order of the reaction is 2, and specific reaction rate, when $t_{1/2} = 5$ min and $C_{A0} = 0.02$ g·mol/L at 100°C, is

$$k_{100} = \frac{2^{n-1} - 1}{n-1} \frac{1}{C_{A0}^{n-1} t_{1/2}}$$

$$= \frac{2-1}{1} \frac{1}{0.02(5)} = \frac{1}{0.10} = 10 \text{ L/g} \cdot \text{mol} \cdot \text{min}$$

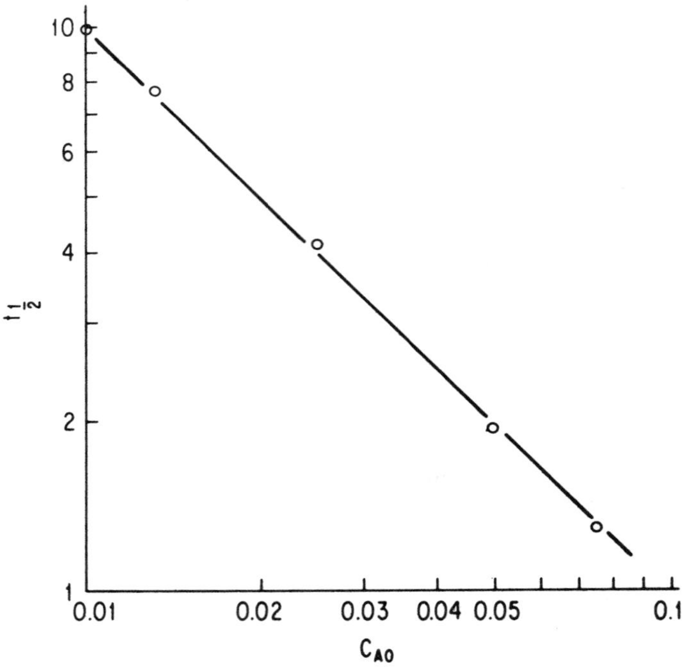

Figure 13-4. Plot of $t_{1/2}$ versus C_{A0} (Example 13-10).

At 110°C, $C_{A0} = 0.025$ g · mol/L, $t_{1/2} = 2$ min, and

$$k_{110} = \frac{2-1}{1}\,\frac{1}{0.025(2)} = 20 \text{ L/g}\cdot\text{mol}\cdot\text{min}$$

b. Activation energy:

$$k_T = \alpha e^{-E/RT}$$

$$k_{100} = \alpha e^{-E/R(373)}$$

and

$$k_{110} = \alpha e^{-E/R(383)}$$

$$\ln\frac{k_{110}}{k_{100}} = -\frac{E}{R(383)} + \frac{E}{R(373)}$$

$$\ln\frac{20}{10} = \ln 2 = \frac{E}{R}\left(\frac{1}{373} - \frac{1}{383}\right)$$

or

$$E = \frac{\ln 2(1.987)}{\frac{1}{373} - \frac{1}{383}} = 19{,}676 \text{ cal/g}\cdot\text{mol}$$

Frequency factor:

At 100°C,

$$\alpha = \frac{k_{100}}{\exp\,[-E/1.987(373)]} = 3.39 \times 10^{12}\ \text{L/g}\cdot\text{mol}\cdot\text{min}$$

At 110°C,

$$\alpha = \frac{k_{110}}{\exp\,[-19{,}676/1.987(383)]} = 3.39 \times 10^{12}\ \text{L/g}\cdot\text{mol}\cdot\text{min}$$

Reactor Design

Reaction equipment in which homogeneous reactions are carried out are of three types: (1) batch reactors, (2) steady-state flow reactors, and (3) unsteady-state flow or semibatch reactors.

Batch Reactors. In a batch reactor, neither the reactants nor the products flow into or leave the system when the reaction is carried out. They are either the constant-volume or constant-pressure (variable volume) reactors. The operation of a batch reactor is unsteady-state. Although the composition throughout the reactor is ideally uniform at a given instant, it changes with time.

Steady-State Flow Reactor. Steady-state flow reactors are of two types: (1) ideal stirred tank or CSTR, also known as mixed reactor or mixed flow reactor; (2) plug-flow reactor, also known as piston flow, ideal tubular, or unmixed flow reactor.

Ideal Stirred Tank Reactor. In this reactor, the contents are well mixed and uniform in concentration throughout. The composition of the exit stream from this reactor is the same as the composition of the fluid in the reactor.

Plug-Flow Reactor. The flow of fluid in a plug-flow reactor is orderly with no backward or forward mixing or diffusion of the fluid elements in the direction of the flow path. There may be lateral mixing of the fluid in a plug-flow reactor. The plug flow is characterized by the fact that the residence time in the reactor is the same for all the elements of the fluid.

In the following treatment, the reactor volume V is the volume of the reaction space or the volume of the fluid.

Mass and Energy Balances

Ideal Batch Reactor. Assuming the reaction A $\longrightarrow$ products, the mole balance on species A in a batch reactor of volume V, where the composition is uniform throughout, results in

$$\frac{dn_A}{dt} = r_A V \tag{13-46}$$

This equation is true for both constant and variable volume. Since the reactant A is disappearing, we write the equation in the form

$$-\frac{dn_A}{dt} = -r_A V \tag{13-47}$$

Now in terms of the conversion X_A, the reaction rate is

$$-\frac{dn_A}{dt} = -\frac{dn_{A0}(1 - X_A)}{dt} = n_{A0}\frac{dX_A}{dt} \tag{13-48}$$

$$n_{A0}\frac{dX_A}{dt} = -r_A V \tag{13-48a}$$

The solution of this equation with the initial condition $t = 0$, $X_A = 0$, is given by

$$t = n_{A0}\int_0^{X_A} \frac{dX_A}{-r_A V} \tag{13-49}$$

which gives a relation showing the time required to obtain a conversion X_A. If the density of the fluid remains constant, one obtains

$$t = C_{A0}\int_0^{X_A} \frac{dX_A}{-r_A} = -\int_{C_{A0}}^{C_A} \frac{dC_A}{-r_A} \tag{13-50}$$

If the volume of the reaction mixture changes proportionately with the conversion, the equation becomes

$$t = n_{A0} \int_0^{X_A} \frac{dX_A}{-r_A V_0 (1 + \epsilon_A X_A)} = C_{A0} \int_0^{X_A} \frac{dX_A}{-r_A (1 + \epsilon_A X_A)} \tag{13-51}$$

where ϵ_A is the fractional change in volume of the system between complete conversion and no conversion of reactant A.

The reaction time t is the measure of the processing rate in a batch reactor.

Flow Reactors. The performance of flow reactors is evaluated in terms of space time and space velocity. These terms are defined as follows.

Space Time. This is the reciprocal of space velocity and is given by

$$\tau = \frac{1}{S} = \text{time required to process one reactor volume of feed measured to specified conditions}$$

$$= \text{time}$$

Space Velocity. The space velocity is related to space time as follows:

$$S = \frac{1}{\tau} = \text{number of reactor volumes of feed at specified conditions which can be treated in unit time}$$

$$= (\text{time})^{-1} \tag{13-52}$$

Note that the values of the space time and space velocity will depend upon the conditions of the stream, viz., temperature, pressure, and state. If the conditions are those of the feed stream, the following relations can be established:

$$\tau = \frac{1}{S} = \frac{C_{A0} V}{F_{A0}}$$

$$\tau = \frac{\text{(moles A feed/volume of feed)(volume of reactor)}}{\text{moles A feed/time}}$$

$$= \frac{V}{v_0} = \frac{\text{reactor volume}}{\text{volumetric feed rate of A}} \tag{13-53}$$

If a standard condition is chosen to express the volumetric feed rate, the following relation holds:

$$\tau' = \frac{1}{S} = \frac{C'_{A0}V}{F_{A0}} = \frac{C'_{A0}}{C_{A0}} \tag{13-54}$$

where the prime denotes the values at standard conditions chosen.

Steady-State Mixed-Flow Reactor or Continuous-Flow Stirred-Tank Reactor. The design equations for a stirred-tank reactor (Fig. 13-5) are established. Mole balance on component A, with the

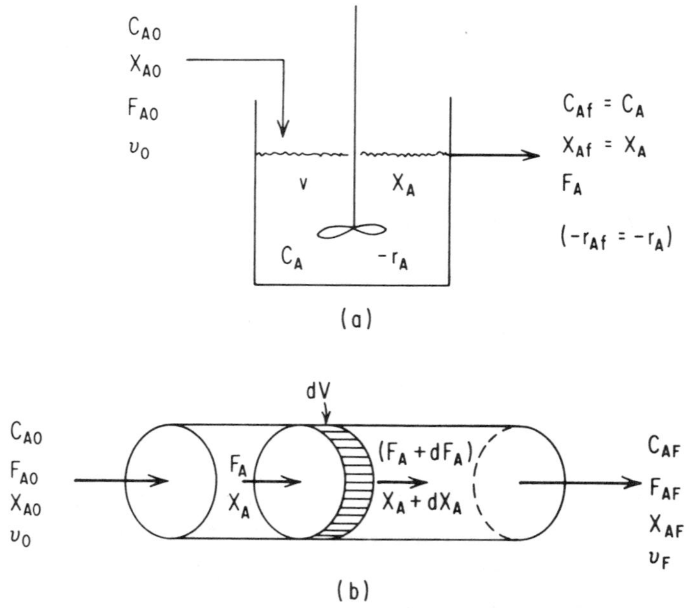

Figure 13-5. (*a*) Steady-state mixed-flow reactor; (*b*) steady-state plug-flow reactor.

reactor as the system gives

$$n_E - n_B = \bar{n}_B = \bar{n}_I \, \Delta t - \bar{n}_O \, \Delta t + n_P \, \Delta t \tag{13-55}$$

where n_B and n_E are moles in the beginning and end of the accounting period Δt. $n_E - n_B = 0$ from steady-state operation. Therefore

$$\bar{n}_I - \bar{n}_O = -\bar{n}_P$$

$\bar{n}_I$ = input of A moles/time $= F_{A0}(1 - X_{A0}) = F_{A0}$ if $X_{A0} = 0$

$\bar{n}_O$ = output of A moles/time $= F_A = F_{A0}(1 - X_A)$

and

$-n_P = -r_A V$ which is the disappearance of A in moles per unit time

$$= \frac{\text{moles A reacting}}{\text{time(volume of fluid)}} \text{(volume of reactor)}$$

With no conversion in the inlet stream, one obtains

$$F_{A0} - F_{A0}(1 - X_A) = -r_A V \tag{13-55a}$$

or

$$F_{A0}X_A = -r_A V \tag{13-55b}$$

Then volume of back-mix reactor is

$$V = \frac{F_{A0}X_A}{-r_A} \tag{13-55c}$$

Also

$$\frac{V}{F_{A0}} = \frac{X_A}{-r_A} = \frac{\tau}{C_{A0}}$$

or

$$\tau = \frac{1}{S} = \frac{V}{v_0} = \frac{VC_{A0}}{F_{A0}} = \frac{C_{A0}X_A}{-r_A} \tag{13-55d}$$

where X_A and r_A are evaluated at the exit stream conditions. If $X_{A0} \neq 0$, i.e., the feed is partially converted, then

$$\frac{V}{F_{A0}} = \frac{X_{Af} - X_{Ai}}{(-r_A)_f} \tag{13-55e}$$

where f and i denote the exit and inlet conditions, respectively,

or
$$\tau = \frac{VC_{A0}}{F_{A0}} = \frac{C_{A0}(X_{Af} - X_{Ai})}{(-r_A)_f}$$

For the special case, when the density is constant,

$$\frac{V}{F_{A0}} = \frac{X_A}{-r_A} = \frac{C_{A0} - C_A}{C_{A0}(-r_A)} \tag{13-56}$$

or
$$\tau = \frac{V}{v} = \frac{C_{A0} - C_A}{-r_A} \tag{13-56a}$$

Steady-State Plug-Flow Reactor. Referring to Fig. 13-5*b*, the mass balance on a differential element gives

$$M_E - M_B = \overline{M}_I\ \Delta t - \overline{M}_O\ \Delta t + \Sigma M_P\ \Delta t \tag{13-57}$$

$$M_E - M_B = 0 \qquad \text{since steady state}$$

$$-(M_I - M_O) = +\Sigma M_P$$

$$M_I = F_A \qquad M_O = F_A + dF_A \qquad -\Sigma M_P = -r_A\ dV$$

and
$$dF_A = +r_A\ dV \tag{13-57a}$$

also
$$dF_A = d[F_{A0}(1 - X_A)] = -F_{A0}\ dX_A \tag{13-57b}$$

From Eqs. (13-57*a*) and (13-57*b*)

$$F_{A0}\ dX_A = -r_A\ dV \tag{13-57c}$$

from which, after separation of variables and integration,

$$V = F_{A0} \int_0^{X_A} \frac{dX_A}{-r_A} \tag{13-58}$$

which gives the plug-flow-reactor volume for the conversion X and then

$$\tau = \frac{V}{v_0} = C_{A0} \int_0^{X_A} \frac{dX_A}{-r_A} \tag{13-59}$$

If the feed on which the conversion is based is partially converted at the entrance to the reactor,

$$\frac{V}{F_{A0}} = \frac{V}{C_{A0}v_0} = \int_{X_{Ai}}^{X_{Af}} \frac{dX_A}{-r_A} \tag{13-60}$$

or

$$\tau = \frac{V}{v_0} = C_{A0} \int_{X_{Ai}}^{X_{Af}} \frac{dX_A}{-r_A} \tag{13-61}$$

For the special case when density is constant,

$$\frac{V}{F_{A0}} = \frac{\tau}{C_{A0}} = \int_0^{X_{Af}} \frac{dX_A}{-r_A} = -\frac{1}{C_{A0}} \int_{C_{A0}}^{C_{Af}} \frac{dC_A}{-r_A} \tag{13-62}$$

or

$$\tau = \frac{V}{v_0} = C_{A0} \int_0^{X_{Af}} \frac{dX_A}{-r_A} = -\int_{C_{A0}}^{C_{Af}} \frac{dC_A}{-r_A} \tag{13-63}$$

Example 13-11. A reaction A $\longrightarrow$ R is to be carried out in a batch reactor. The rates of the reaction as a function of C_A are given in Table 13-5.

TABLE 13-5 The Rates of Reaction as a Function of C_A

C_A, mol/L	$-r_A$, mol/L·min	$\frac{1}{-r_A}$
0.1	0.100	10.000
0.2	0.300	3.34
0.3	0.500	2.00
0.4	0.600	1.67
0.5	0.500	2.00
0.6	0.250	4.00
0.7	0.100	10.00
0.8	0.060	16.67
1.0	0.050	20.00
1.3	0.045	22.20
2.0	0.042	23.81

How long a batch must be reacted to reach a concentration of 0.3 mol/L from initial concentration of C_{A0} = 1.3 mol/L? The reaction is liquid phase.

Solution. Since the reaction is liquid phase, the density can be considered constant, and then t is given by

$$t = -\int_{C_{A0}}^{C_A} \frac{dC_A}{-r_A} \tag{13-64}$$

We plot $1/-r_A$ versus C_A and obtain the value of the integral (Fig. 13-6) $-(dC_A/-r_A)$ between C_A = 0.3 and C_{A0} = 1.3 mol/L. The area under the

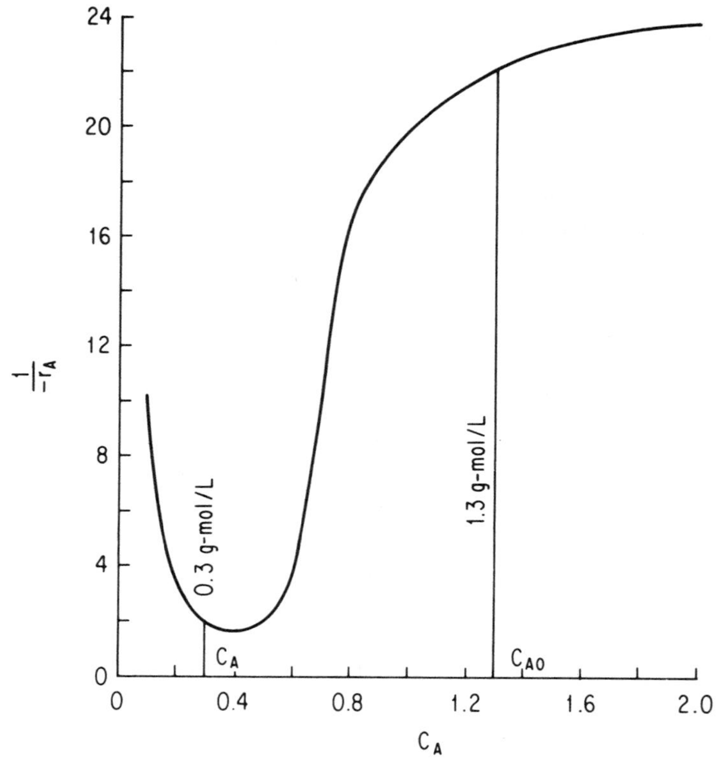

Figure 13-6. Plot of $1/-r_A$ versus C_A (Example 13-11).

curve is found by Simpson's rule as

$$\text{Area} = -\frac{0.1}{3}[2 + 4(1.67) + 2(2) + 4(4) + 2(10) + 4(16.7) + 2(18.5) + 4(20) + 2(20.9) + 4(21.7) + 22.22]$$

$$= -12.8$$

$$t = -(-12.8) = 12.8 \text{ min}$$

Example 13-12. Hydrolysis of CH_3COOCH_3 is an autocatalytic reaction and has a specific reaction rate constant $k = 5.2$ L/g · mol · h. The initial concentration of CH_3COOCH_3 is 0.5 g · mol/L, and the initial concentration of CH_3COOH is 0.05 g · mol/L. The reaction is first order with respect to both methyl acetate and acetic acid. On the basis of the above information, what type of optimum reactor system would you specify for the plant to process 200 L/h of feed? What would be the reactor volume in this system?

Solution. Since the reaction is liquid phase, the density changes are negligible.

$$CH_3COOCH_3 + H_2O \longrightarrow CH_3COOH + CH_3OH$$

$$A \longrightarrow B$$

For a continuous back-mix reactor, the reactor volume is given by

$$V = \frac{F_{A0}X_A}{-r_A} = \frac{C_{A0} - C_A}{C_{A0}(-r_A)}F_{A0} = \frac{v_0 C_{A0} X_A}{-r_A}$$

$$F_{A0} = 200(0.5) = 100 \text{ g} \cdot \text{mol/h}$$

Assume that the conversion required is 90 percent.

For a back-mix reactor, the reactor volume is the total area bounded by the rectangle covering the conversions X_A and 0 multiplied by $v_0 C_{A0}$ or F_{A0} (Fig. 13-7*a*), whereas for a plug-flow reactor, the volume is area under the curve only multiplied by $v_0 C_{A0}$ or F_{A0} (Fig. 13-7*b*). Calculate points from Table 13-6 to plot the curve.

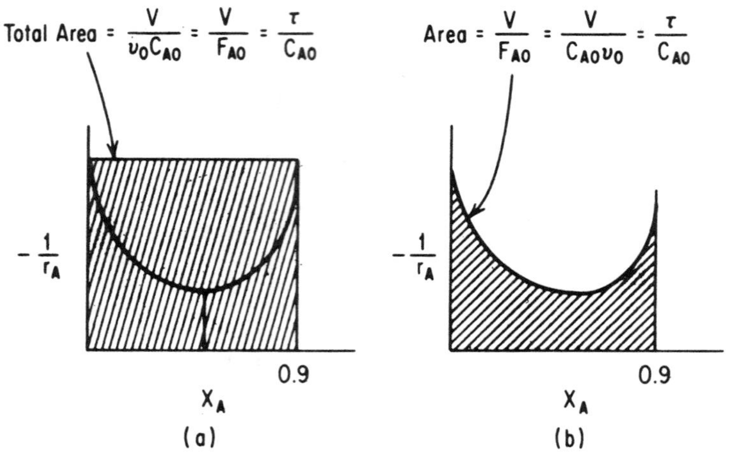

Figure 13-7. Methods of calculation of volumes: (*a*) mix-flow reactor; (*b*) plug-flow reactor.

TABLE 13-6 Calculation of Rates and Conversions

C_A	C_B	$-r_A = 5.2C_AC_B$	$\frac{1}{-r_A}$	$X_A = \frac{C_{A0} - C_A}{C_{A0}}$
0.5(C_{A0})	0.05	0.13	7.69	0
0.4	0.15	0.312	3.20	0.2
0.3	0.25	0.39	2.56	0.4
0.2	0.35	0.364	2.75	0.6
0.1	0.45	0.234	4.27	0.8
0.05	0.50	0.130	7.69	0.9

For minimum reactor volume, use back mix for first portion and then plug flow to complete (Fig. 13-8*a*). The rate is maximum when $X_A = 0.45$ or $C_A = 0.275$. Assuming $X_{Af} = 0.9$,

$$\text{Volume of back-mix reactor} = F_{A0}\left(\frac{X_A}{-r_A}\right)$$

$$= 100(2.54)(0.45)$$

$$= 114.3 \text{ L}$$

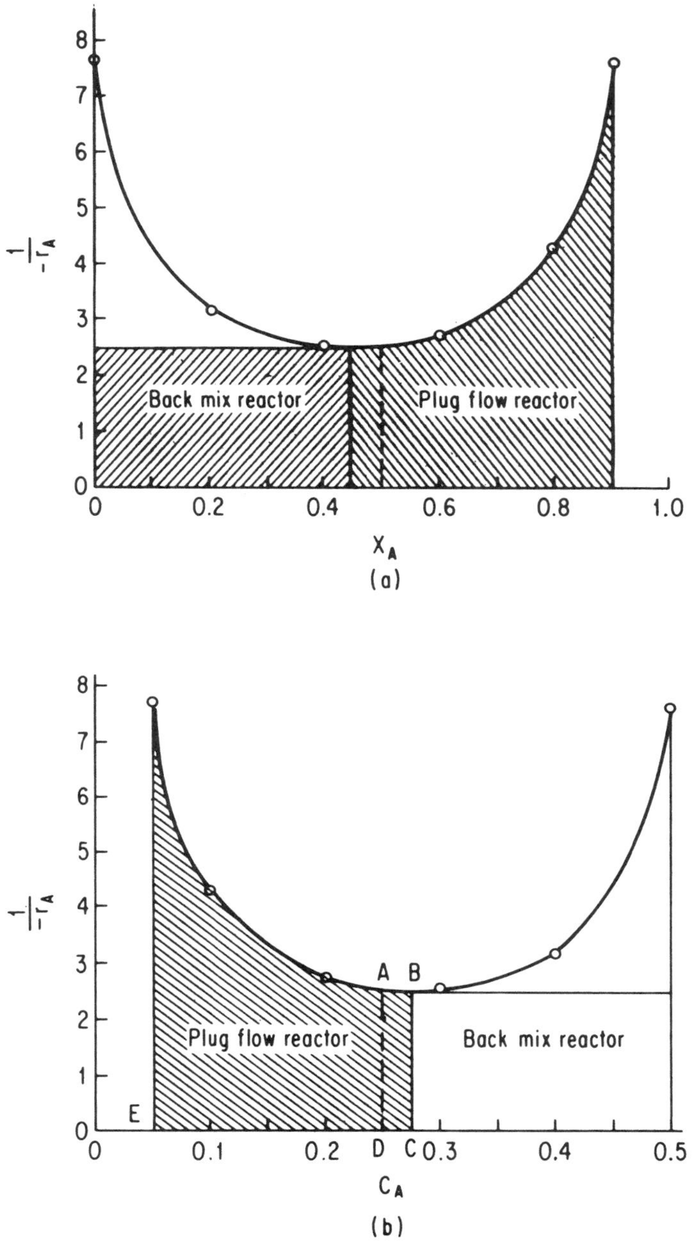

Figure 13-8. Solution of Example 13-12 (*a*) in terms of conversion, (*b*) in terms of concentration.

Plug-Flow Reactor. Use trapezoidal rule from 0.45 to 0.5 and Simpson's rule from 0.50 to 0.9.

$$A = \tfrac{1}{2}(a + b)h = \tfrac{1}{2}(0.05)(2.54 + 2.5) = 0.126$$

$$= \frac{0.05}{3}[10.23 + 4(2.64 + 2.95 + 3.75 + 5.25) + 2(2.75 + 3.3 + 4.27)]$$

$$= \frac{0.05}{3}(89.23) = 1.4872$$

$$\text{Total } A = 0.126 + 1.4872 = 1.6132$$

$$V = F_{A0}(1.6132) = 100(1.6132) = 161.3$$

Volume of back-mix reactor = 114.3 L

Volume of plug-flow reactor = 161.3 L

The problem can be solved directly in terms of concentrations instead of conversions as done above (Fig. 13-8*b*). For $X_{Af} = 0.9$, $C_{Af} = 0.05$ mol/L.

Volume of Back-Mix Reactor.

$$\text{Space time } \tau = \frac{C_{A0}X_A}{-r_A} = \frac{C_{A0} - C_A}{-r_A} = \frac{V}{v}$$

$$= \frac{C_{A0} - C_A}{-r_A}$$

$$= (0.5 - 0.275)(2.54) = 0.5715 \text{ h}$$

$$V = v\tau = 0.5715(200) = 114.3 \text{ L}$$

Volume of Plug-Flow Reactor. This requires the determination of area under the curve from $C_A = 0.05$ to $C_A = 0.275$. Determine the area *ABCD* by trapezoidal rule and the area *ADEF* by Simpson's rule.

$$\text{Area } ABCD = \tfrac{1}{2}(0.025)(2.54 + 2.5) = 0.063 = \tau_{ABCD}$$

$$\text{Area } ADEF = \frac{0.025}{3}[7.69 + 2.54 + 4(5.15 + 3.8 + 3.04 + 2.62) + 2(4.27 + 3.4 + 2.75)]$$

$$= 0.746 \text{ h} = \tau$$

$$\text{Plug flow} = 0.063 + 0.746 = 0.809 \text{ h}$$

Volume of plug-flow reactor = 0.809(200) = 161.8 L

Mixed-Flow Reactors in Series. Consider n reactors of equal volume. With steady state and no density change, a material balance on the reactor i for component A gives

$$\tau = \frac{C_0 V}{F_0} = \frac{V}{v} = \frac{C_0(X_i - X_{i-1})}{-r_A} \tag{13-65}$$

Since, $\Delta\rho = 0$, i.e., $\epsilon = 0$, the above can be written in terms of the concentrations as

$$\tau = \frac{C_{Ai-1} - C_{Ai}}{kC_{Ai}} \tag{13-66}$$

where $-r_A = kC_{Ai}$ is the specific reaction rate for first-order system, or

$$\frac{C_{Ai-1}}{C_{Ai}} = 1 + k\tau_i \tag{13-67}$$

Since the space-time (or mean residence time t) is the same for all the reactors,

$$\frac{C_{A0}}{C_{An}} = \frac{1}{1 - X_{An}} = \frac{C_{A0}}{C_{A1}}\frac{C_{A1}}{C_{A2}}\cdots\frac{C_{An-1}}{C_{An}} = (1 + k\tau)^n \tag{13-68}$$

or

$$C_{An} = \frac{C_{A0}}{(1+kt)^n} \tag{13-69}$$

For the system as a whole by rearrangement,

$$\tau_n = n\tau = \frac{n}{k}\left[\left(\frac{C_{A0}}{C_{An}}\right)^{1/n} - 1\right] \tag{13-70}$$

Notice that the above equation reduces to the plug-flow equation

$$\tau = \frac{1}{k}\ln\frac{C_{A0}}{C_A} \tag{13-71}$$

For reactions other than first order, explicit solution for C_{Ai} in terms of C_{A0} is quite complicated. For a small number of stages, numerical solutions are available. For plug flow (second-order rate equation),

$$\frac{C_{A0}}{C_A} = 1 + C_{A0}k\tau_p \qquad (13\text{-}72)$$

where τ_p is the space time for plug flow.

Comparisons of the performance of a series of n equal-size mixed reactors with a plug-flow reactor for first- and second-order reactions are available in the form of graphs.[2]

Energy Balances

Chemical reaction rate is a strong function of temperature. When heat is evolved or absorbed in a reaction or when heat is added or removed from the reaction system, there is a necessity to account for the temperature effect on the reaction rate. Therefore, an energy balance besides a mass balance is required. The following example illustrates the method of solution of problems requiring energy balance.

Example 13-13. The liquid phase hydrolysis of dilute aqueous anhydride solution can be treated as first order and irreversible. A batch reactor for carrying out the hydrolysis is charged with 200 liters of anhydride solution at 25°C. The initial concentration of anhydride is 2.16×10^{-4} g · mol/cm^3, and other data are as follows:

Density of solution $\rho = 1.05$ g/cm^3

Specific heat of solution = 0.9 cal/g · °C   assume const.

Heat of reaction = −50,000 cal/g · mol   assume const.

The rate of reaction is given by the equation

$$r_A = kC_A$$

where $k = 18.11 \times 10^6 \exp(-11{,}000/RT)$
C = g · mol/cm^3
T = K

What time is required for a conversion of 70 percent when the reactor is operated adiabatically? What time is required for the same conversion if the reactor were operated isothermally?

Solution. For a batch reactor, time θ is given by

$$\theta = C_0 \int_0^{X_F} \frac{dX_A}{-r_A}$$

$$-r_A = 18.11 \times 10^6[\exp(-11{,}000/1.987T)]C_0(1 - X_A)$$

Establish energy balance. Since the reaction is adiabatic, $Q_t = 0$ and we can get

$$M_t C_p \, dT = F_{A0}(-\Delta H_R) \, dX_A$$

or $$200(1000)(1.05)(0.9) \, dT = 200(1000)(2.16 \times 10^{-4})(50{,}000) \, dX_A$$

or $$dT = 11.43 \, dX_A$$

By integration, $T = 11.43X_A + I$, where I is integration constant. When $X_A = 0$, $T = 298°\text{K}$, $I = 298$, and therefore

$$T = 11.43X_A + 298$$

Using this expression for T, prepare Table 13-7 for various values of X_A from 0 to 0.7.

TABLE 13-7 Calculation of Rates of Reaction

X_A	T, K	$-r_A \times 10^4$	$\frac{1}{-r_A}C_0$
0	298.0	0.3345	6.457
0.1	299.14	0.3591	6.684
0.2	300.29	0.3854	7.005
0.3	301.43	0.4133	7.467
0.4	302.57	0.4429	8.129
0.5	303.72	0.4746	9.101
0.6	304.86	0.5081	10.63
0.7	306.00	0.5437	13.24

Integration by the Simpson rule gives time for adiabatic reaction:

$$\tfrac{1}{3}h[6.457 + 4(7.76) + 13.24] = \tfrac{1}{3}(0.35)[6.457 + 4(7.76) + 13.24]$$

$$= 5.92 \text{ min}$$

For isothermal reaction at 25°C,

$$-r_A = 18.11 \times 10^6[\exp(-11{,}000/298R)]C_0(1 - X_A)$$

Substitution in the expression for θ gives

$$\theta = \int_0^{X_{AF}} \frac{C_0\, dX_A}{C_0(1 - X_A)(18.11 \times 10^6)\exp[-11{,}000/1.987(298)]}$$

$$= +6.456 \int_0^{X_{AF}} \frac{dX_A}{1 - X_A}$$

$$= -6.456 \ln[(1 - X_A)]_0^{X_{AF}}$$

$$= -6.456 [\ln(0.3) - \ln(1)] = 7.77 \text{ min}$$

In most cases, the solutions will not be as simple as in the above problem. When the expression for T cannot be integrated analytically, numerical and graphical techniques will have to be employed.

References

1. S. M. Walas, *Reaction Kinetics for Chemical Engrs.*, McGraw-Hill, New York, 1959, p. 35.
2. O. Levenspiel, *Chemical Reaction Engineering*, 2d ed., Wiley, New York, 1972, pp. 136–137.

Additional Reading.

J. M. Smith, *Chemical Engineering Kinetics*, 3d ed., McGraw-Hill, New York, 1981.

Chapter

14

Process Control

A number of process design aspects other than those discussed in the previous chapters have received some attention from time to time in the P.E. examinations. These are: process control, plant safety, explosion protection, waste treatment and disposal, water and energy conservation, and plant environmental dust and pollution control. Treatment of all of these subjects is beyond the limited space of this book. Therefore, only the subject of process control is very briefly dealt with in this chapter.

Control Systems

A control system or scheme is characterized by an output variable (e.g., temperature) that is automatically controlled through the manipulation of inputs (input variables). In an open-loop control system, the inputs to the process are regulated independently without using the controlled output variable to adjust the inputs. In a closed-loop or feedback control system, the output controlled variable is used to adjust the inputs to the process. In a feed-forward control, the measurement of one input variable is used to adjust another input variable.

The method of cascade control is frequently used to decrease the process upsets. It involves the use of the output of a primary controller to adjust the set point of a secondary controller and is commonly used in the feedback control. The ratio and selector controls are two other control modes which require two or more interconnected instruments.

Block Diagrams

In control-system analysis, the block diagrams are used to show the functional relationship betwen the various parts of the system. Each part is represented by a rectangle or box with one input and one output. The "transfer function" (or its symbol), which is the mathematical relationship between output (response function) and input (forcing function), is written inside the box (Fig. 14-1*a*). The blocks are connected by arrows which indicate the flow of the information in the system. The outputs and inputs are considered as signals.

The comparison of the signals is shown by circles with signs (+, −, ×, ÷) written outside them (Fig. 14-1*b*). The symbols to denote addition, +; multiplication, ×; or division, ÷; may be written inside the circles. These circles are termed the comparators. Branching of a signal into more than one direction indicates the flow of the signal without any modification in it.

Laplace Transforms in Control-System Analysis

The analysis of a control system begins with the determination of the transfer functions for the various parts of the system

Figure 14-1. Components of a control system block diagram: (*a*) block diagram of transfer function; (*b*) comparator.

shown in the block diagram by the application of either the mass, energy, or force balance to each part of the system. The solutions of the developed linear differential equations are simplified with the use of the Laplace transforms, which make it possible to transform the differential equations into algebraic equations. The transformation replaces the independent time variable t by the complex variable s.

The Laplace transform of the time function $f(t)$ is defined by the operation ($\mathcal{L}$ is the Laplace operator)

$$\mathcal{L}[f(t)] = \int_0^\infty f(t)\, e^{-st}\, dt = F(s) \tag{14-1}$$

where $F(s)$ is a function of the complex variable s, a parameter which is constant with respect to the integration process. $F(s)$ is called the *Laplace transform of f(t)* as a function of the complex variable s. The following two useful theorems[1] concerning the Laplace transforms of functions can be easily established:

Theorem 1: $\mathcal{L}[c_1 f_1(t) \pm c_2 f_2(t)] = c_1 \mathcal{L} f_1(t) \pm c_2 \mathcal{L} f_2(t)$ (14-1*a*)

Theorem 2: $\mathcal{L} f'(t) = s\mathcal{L} f(t) - f^{(1)}(0) = sF(s) - f(0)$ (14-1*b*)

and in general, for the nth derivate,

$$\mathcal{L}F^n(t) = s^n F(s) - s^{n-1} f(0) - s^{n-2} f^{(1)}(0) - \cdots - f^{n-1}(0) \tag{14-1c}$$

where $f^{(i)}(0)$ is the ith derivate of $f(t)$ with respect to t and evaluated for $t = 0$.

In the control-system analysis, it is usually not necessary to perform the integration of Eq. (14-1). Instead, use is made of the available tables[1,2a] of the Laplace transforms.

The initial- and final-value theorems are useful in checking the transfer functions for systems of known initial and final values. The final-value theorem is also useful in obtaining the steady-state error or offset of a control system. These theorems are

Initial-value theorem: $\lim_{t \to 0} f(t) = \lim_{s \to \infty} sF(s)$ (14-1*d*)

Final-value theorem: $\lim_{t \to \infty} f(t) = \lim_{s \to 0} sF(s)$ (14-1*e*)

Transfer Functions in Terms of Laplace Transforms

In terms of the Laplace transforms, the transfer function is the ratio of the Laplace transform of the output (response variable) to the Laplace transform of the input (the forcing or disturbing variable). The convention to represent the transfer function in the block diagram is $KG(s)$, where $G(s)$ represents the dynamic portion of the transfer function and K is the gain of the element. In the notation of the block diagram, (s) may be omitted from $G(s)$. A block diagram using this notation is shown in Fig. 14-2.

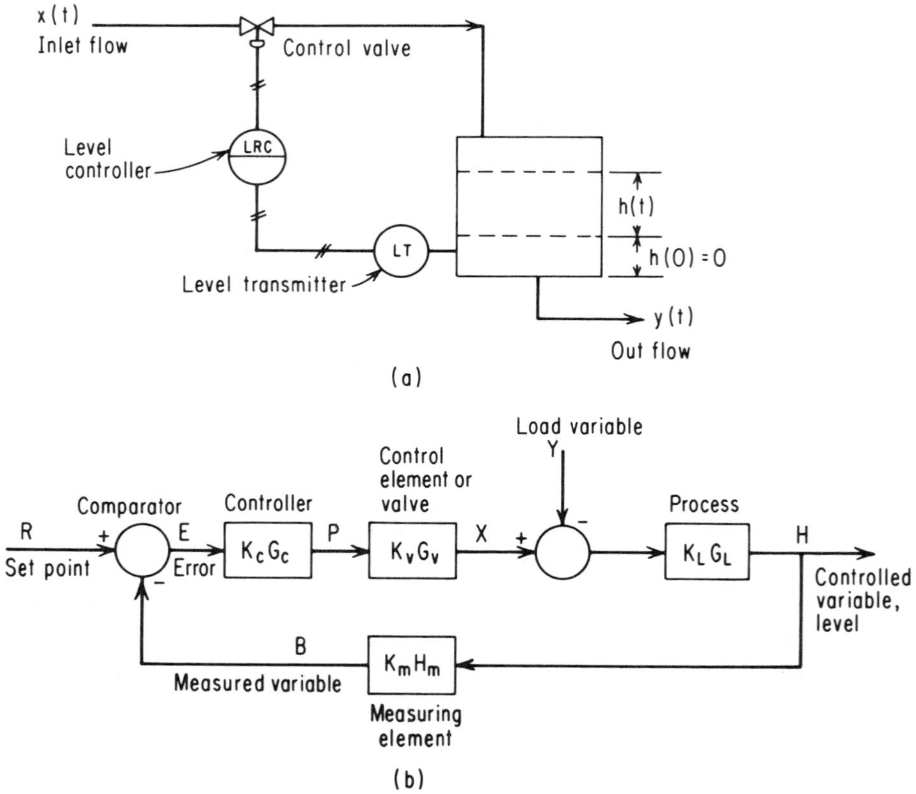

Figure 14-2. (*a*) Control system for liquid filling system; (*b*) block diagram for liquid filling system (Example 14-1).

The evaluation of the transfer functions is illustrated by Example 14-1.

Example 14-1. Obtain the transfer function of the transient variation in the liquid height for the tank filling system in Fig. 14-2*a*.

Solution. By making unsteady-state mass balance with the assumption of a constant density, one obtains

$$x(t) = y(t) + A\frac{dh\,(t)}{dt}$$

where A is the cross-sectional area of the tank. Laplace transformation gives

$$X(s) = Y(s) + A[sH\,(s) - h(0)]$$

For simplification, let $h(0) = 0$. Therefore

$$X(s) = Y(s) + AsH(s)$$

Then the transfer function for the level system is given by

$$\frac{\text{Output}}{\text{Input}} = \frac{H(s)}{X(s) - Y(s)} = \frac{1}{As} = K_L G_L(s)$$

$1/As$ is the transfer function of the tank-filling system. It can be expressed as $K_L G_L(s)$ where $K_L = 1/A$ and $G_L(s)$ or $G_L = 1/s$. In the block diagram of Fig. 14-2*b*, this transfer function is shown as $K_L G_L$.

Overall System Transfer Functions

By combining the transfer functions of the individual system elements, an overall system transfer function is obtained for the analysis of the transient response of the control system as a whole. The overall transfer function for a system can be obtained by the following two simple rules[2]:

1. If there are several transfer functions in a series in a loop, the overall transfer function for the loop is the product of the individual transfer functions in the series. For example, in Fig.

14-2*b* the overall transfer function $X(s)/E(s)$ is given by

$$\frac{X(s)}{E(s)} = (K_c G_c)(K_v G_v)$$

2. In a single-loop feedback system, the overall transfer function relating any two variables Y and X is given by the equation

$$\text{Overall transfer function} = \frac{Y}{X} = \frac{\pi_f}{l \pm \pi_\ell} \qquad (14\text{-}2)$$

where π_f is the product of individual transfer functions between the locations of the signals Y and X and π_ℓ is the product of the individual transfer functions in the loop.

The plus sign is to be used in the denominator when the feedback is negative and the minus sign when the feedback is positive. This rule can also be used to reduce a multiloop system to a single-loop system. Example 14-2 illustrates the method of obtaining an overall transfer function.

Example 14-2. For the system shown in Fig. 14-3, obtain the overall transfer functions $C(s)/U(s)$ and $C(s)/R(s)$.

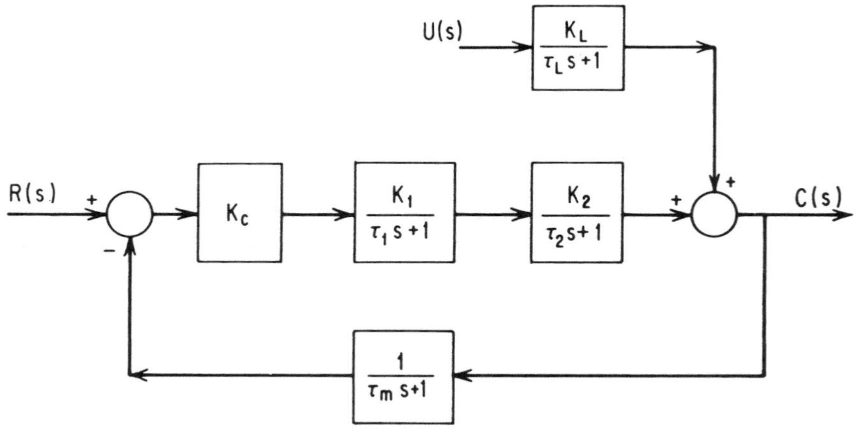

Figure 14-3. Control system (Example 14-2).

Solution.

a. For transfer function C/U,

$$\pi_f = \frac{K_L}{\tau_L s + 1} \qquad \pi_1 = \frac{K_c K_1 K_2}{(\tau_1 s + 1)(\tau_2 s + 1)(\tau_m s + 1)}$$

and

$$\frac{C}{U} = \frac{K_L/(\tau_L s + 1)}{1 + [K_c K_1 K_2/(\tau_1 s + 1)(\tau_2 s + 1)(\tau_m s + 1)]}$$

$$= \frac{K_L(\tau_1 s + 1)(\tau_2 s + 1)(\tau_m s + 1)}{(\tau_L s + 1)[(\tau_1 s + 1)(\tau_2 s + 1)(\tau_m s + 1) + K_c K_1 K_2]}$$

b. For the transfer function between C and R,

$$\pi_f = K_c \frac{K_1}{\tau_1 s + 1} \frac{K_2}{\tau_2 s + 1}$$

and

$$\pi_\ell = \frac{K_c K_1 K_2}{(\tau_1 s + 1)(\tau_2 s + 1)(\tau_m s + 1)}$$

Then

$$\frac{C}{R} = \frac{K_C[K_1/(\tau_1 s + 1)][K_2/(\tau_2 s + 1)]}{1 + K_c K_1 K_2/(\tau_1 s + 1)(\tau_2 s + 1)(\tau_m s + 1)}$$

which simplifies to

$$\frac{C}{R} = \frac{K_c K_1 K_2(\tau_m s + 1)}{(\tau_1 s + 1)(\tau_2 s + 1)(\tau_m s + 1) + K_c K_1 K_2}$$

Control Actions

Various types of controller actions are possible.

On-Off or Bang-Bang Controller. In this case, when the measured variable is below the set point, the controller is on with a maximum output. However the controller is off and the output is zero when the measured variable is above the set point. Because of the either-on-or-off mode of the operation, the on-off control action is inherently cyclic in nature. The on-off control is also called a *two-position control*.

Proportional and Floating Controllers. It is preferable to obtain noncyclic steady operation when disturbances are absent. For this, the change in the controlled variable should be some continuous function of the error. Different types of the proportional and floating actions are given in Table 14-1.

TABLE 14-1 Proportional and Floating Control Actions

Controller Type	Functional Relationship Between Controller Output and Error	Transfer Function
Proportional	$p = K_c e$	$\frac{P}{E} = K_c$
Integral	$p = \frac{1}{\tau_i}\int_0^t e\,dt$	$\frac{P}{E} = \frac{1}{\tau_i s}$
Proportional + integral	$p = K_c\left(e + \frac{1}{\tau_i}\int_0^t e\,dt\right)$	$\frac{P}{E} = K_c\left(1 + \frac{1}{\tau_i s}\right)$
Proportional + derivative	$p = K_c\left(e + \tau_D\frac{de}{dt}\right)$	$\frac{P}{E} = K_c(1 + \tau_D s)$
Proportional + integral + derivative	$p = K_c\left(e + \frac{1}{\tau_i}\int_0^t e\,dt + \tau_D\frac{de}{dt}\right)$	$\frac{P}{E} = K_c\left(1 + \frac{1}{\tau_i s} + \tau_D s\right)$

where p = fractional change in controller output
e = fractional change in error
K_C = controller gain
τ_i = integral time
τ_D = derivative time

In a proportional control, the controlled variable eventually becomes constant at a value which differs from the desired set point resulting in a steady-state error or *off-set*. There is no offset with the integral control since the controller output keeps changing until the error is reduced to zero. The integral action, however, has the disadvantage of the oscillatory response.

A derivative action is added to the proportional integral control to speed up the response so that the system returns to the original steady-state value rapidly with little or no oscillation.

Proportional Band

This term is used to express the proportional control action and is defined as the percentage of the maximum range of the input variable required to cause 100 percent change in the controller output or

$$\%\text{ proportional band} = 100\,\frac{\Delta e}{\Delta e_{\text{max}}} \tag{14-3}$$

where Δe_{max} is the maximum range of the error signal and Δe is the change in the error that gives a maximum change in the output of the controller. The proportional band is also called the proportional band width and is often expressed as the percentage of the chart width.

For a proportional pneumatic controller, the maximum output is 3 to 15 psig or $\Delta P_{\text{max}} = 12$ psi, which is related to error signal by the equation

$$\Delta P_{\text{max}} = K_C\,\Delta e \tag{14-4}$$

Combining Eqs. (14-3) and (14-4) gives

$$\%\text{ proportional band} = \frac{100}{K_C}\,\frac{\Delta P_{\text{max}}}{\Delta e_{\text{max}}}$$

where K_C is the gain of the controller.

Some pneumatic controllers are calibrated in the sensitivity units or psi/inch of the pen travel. With a 4-in chart and 3 to 15 psig controller output range, the sensitivity is given by the relation

$$S = \frac{15 - 3}{4} = 3\text{ psi/in travel} \tag{14-5}$$

Example 14-3. The liquid level at the bottom of a distillation column is controlled with a pneumatic proportional controller by throttling a control valve located in the bottoms discharge line. The level may vary from 0.15 to 1 m from the bottom tangent line of the column. With the controller set point held constant, the output pressure of the controller varies from 100 (valve fully closed) to 20 kPa (valve fully open) as the level increases from 0.3 to 0.9 m from the bottom tangent line. (**a**) Find the percent proportional band and the gain of the controller. (**b**) If the proportional band is changed to 80 percent, find the gain and the change in level necessary to cause the valve to go from fully open to a fully closed position.

Solution.

a. Proportional band

$$= 100\frac{\Delta e}{\Delta e_{\max}} \qquad (14\text{-}3)$$

$$= \frac{\text{span of control variable for fully closed position of control valve}}{\text{full span or range}}$$

$$= \frac{0.9 - 0.3}{1 - 0.15}100 = 70.6 \text{ percent}$$

$$\text{Gain of controller} = \frac{\Delta P}{\Delta e} = \frac{\text{change in controller output pressure}}{\text{corresponding change in control variable}}$$

$$= \frac{100 - 20}{0.9 - 0.3} = 133.3 \text{ kPa/m}$$

b. If proportional band is 80 percent,

$$\frac{\Delta e}{\Delta e_{\max}} = 0.8$$

$$\Delta e = 0.8\Delta e_{\max} = 0.8(1 - 0.15) = 0.68 \text{ m}$$

$$\text{Gain} = \frac{\Delta e}{\Delta e_{\max}} = \frac{100 - 20}{0.68} = 117.65 \text{ kPa/m}$$

Hence, the change in the level necessary to cause the value to go from a fully open to a fully closed position is 0.68 m, and the gain of the controller is 117.65 kPa/m.

Transient Response of Control Systems

The transient response of a control system to either the change in the set point or the load variable is obtained with the use of some specified input. The commonly used inputs with their Laplace transforms are given in Table 14-2.

TABLE 14-2 Forcing Functions and Their Transforms

Input	**Function, $f(t)$**	Laplace Transform
Step	$aU(t)$	a/s
Ramp	at	a/s^2
Impulse	$a[\delta(t)]$	a
Sinusoidal	$\sin \omega t$	$\omega/(s^2 + \omega^2)$

The overall transfer function is multiplied by the Laplace transform of the specified input. An inverse Laplace transform of the resulting expression gives the response function in the time domain.

An important consideration in the closed-loop control system is the stability of the system. If the response function in the time domain $y(t)$ is bounded as $t \to \infty$ for all bounded inputs, the system is said to be stable; if $y(t) \to \infty$ as $t \to \infty$, the system is unstable.

First- and Second-Order Systems

If the transfer function of a system has a first-order denominator, the system is said to be first order. Only one parameter, called the time constant τ, is required to characterize the dynamic behavior of a first-order system. A second-order system, however, requires two parameters to describe its dynamic behavior. The denominator of the transfer function of a second-order system can be expressed in the quadratic form as

$$s^2 + 2\zeta\omega s + \omega^2 = 0 \qquad (14\text{-}6)$$

where ζ = damping coefficient
ω = radian frequency of oscillation = $\sqrt{1 - \zeta^2}/\tau$
τ = period of oscillation

The roots of Eq. (14-6) will be real or complex depending upon the value of the parameter ζ. The nature of the roots will in turn determine the response function. For three possible types of roots the system response will be as given in Table 14-3.

TABLE 14-3 Responses of a Second-Order System

ζ	Nature of Roots	Response
< 1	Complex	Oscillatory or underdamped
$= 1$	Real and equal	Critically damped
> 1	Real	Nonoscillatory or overdamped

Example 14-4. The block diagram of a control system having a first-order process, a measurement lag, and containing a two-mode (proportional-derivative) control action is shown in Fig. 14-4. (**a**) Find an expression for the period of oscillation τ and the damping coefficient for the closed-loop response. Assume a regulatory operation. (**b**) Compute and compare the offset for a step change of the load when $\tau_1 =$ 2 min, $\tau_m = 20$ s, $\xi = 0.8$, and when (*i*) $\tau_D = 0$, (*ii*) $\tau_D = 2$ s.

Solution. For regular operation, $R(s) = 0$ and from Eq. (14-2),

$$\frac{C(s)}{U(s)} = \frac{\text{product of transfer functions between } C \text{ and } U}{1 + \text{product of all transfer functions in the closed loop}}$$

$$= \frac{1/(\tau_1 s + 1)}{1 + [K(1 + \tau_D s)/(\tau_m s + 1)(\tau_1 s + 1)]}$$

$$= \frac{\tau_m s + 1}{\tau_m \tau_1 s^2 + (\tau_m + \tau_1 + K\tau_D)s + (1 + K)}$$

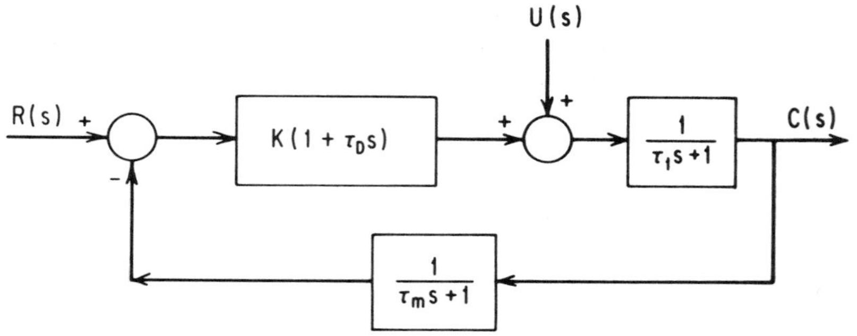

Figure 14-4. System block diagram (Example 14-4).

a. The denominator (called the *characteristic equation*) of the above equation is expressed in the form of Eq. (14-6)

$$\tau_m\tau_1 s^2 + (\tau_m + \tau_1 + K\tau_D)s + 1 + K = 0$$

or

$$s^2 + \frac{\tau_m + \tau_1 + K\tau_D}{\tau_m\tau_1}s + \frac{1+K}{\tau_m\tau_1} = 0$$

By a comparison of the above equation with Eq. (14-6), one obtains

$$2\zeta\omega = \frac{\tau_m + \tau_1 + K\tau_D}{\tau_m\tau_1}$$

and

$$\omega^2 = \frac{1+K}{\tau_m\tau_1}$$

Hence

$$\omega = \frac{1}{\tau} = \sqrt{\frac{1+K}{\tau_m\tau_1}}$$

and

$$\zeta = \frac{\tau_m + \tau_1 + K\tau_D}{2\omega\tau_m\tau_1} = \frac{\tau_m + \tau_1 + K\tau_D}{2\sqrt{(1+K)(\tau_m\tau_1)}}$$

Since the disturbance is a step of size a, $U(s) = a/s$ and

$$C(s) = \frac{\tau_m s + 1}{\tau_m\tau_1 s^2 + (\tau_m + \tau_1 + K\tau_D)s + (1+K)}\frac{a}{s}$$

By the final-value theorem [Eq. (14-1e)], the offset is given by

$$\text{Offset} = \underset{s\to 0}{sC(s)} = \frac{(\tau_m s + 1)a}{\tau_m\tau_1 s^2 + (\tau_m + \tau_1 + K\tau_D)s + (1+K)} = \frac{a}{1+K}$$

b(i). For $\tau_1 = 2$ min $= 120$ s, $\tau_m = 20$ s, $\tau_D = 0$, $\zeta = 0.8$,

$$\zeta = \frac{\tau_m + \tau_1 + K\tau_D}{2\sqrt{(1+K)(\tau_m\tau_1)}}$$

$$0.8 = \frac{20 + 120 + 0}{2\sqrt{(1+K)(20)(120)}} = \frac{70}{\sqrt{(1+K)(2400)}}$$

Therefore, $(1+K)(2400) = 4900/0.64 = 7656.25$ and $K = 2.19$.

$$\omega = \frac{1}{\tau} = \sqrt{\frac{1+K}{\tau_m\tau_1}} = \sqrt{\frac{1+2.19}{20(120)}} = 0.0365$$

$$\text{Offset} = \frac{a}{1+K} = \frac{a}{1+2.19} = \frac{a}{3.19} = 0.313a$$

b(ii). For $\tau_1 = 120$ s, $\tau_m = 20$ s, $\tau_D = 2$ s, $\zeta = 0.8$

$$\zeta = 0.8 = \frac{\tau_m + \tau_1 + K\tau_D}{2\sqrt{(1+K)(\tau_m\tau_1)}}$$

$$= \frac{20 + 120 + 2K}{2\sqrt{(1+K)(20)(120)}} = \frac{70 + K}{\sqrt{(1+K)(2400)}}$$

Hence, squaring both sides and simplifying gives the quadratic equation

$$K^2 - 1396K + 3364 = 0$$

from which

$$K = \frac{1396 \pm \sqrt{1396^2 - 4(3364)}}{2}$$

which gives $K = 1393.6$ or 2.414. When $K = 1393.6$,

$$\omega = \frac{1}{\tau} = \sqrt{\frac{1+K}{\tau_m\tau_1}} = \sqrt{\frac{1 + 1393.6}{20(120)}} = 0.762$$

Then $$\text{Offset} = \frac{a}{1+K} = \frac{a}{1+1393.6} = 0.0007a$$

When $K = 2.414$,

$$\omega = \frac{1}{\tau} = \sqrt{\frac{1+K}{\tau_m\tau_1}} = \sqrt{\frac{1 + 2.414}{20(120)}} = 0.0377$$

and $$\text{Offset} = \frac{a}{1+K} = \frac{a}{1+2.414} = 0.293a$$

Stability of the Higher-Order (>2) Systems

The denominator of the closed-loop transfer function for a higher-order system is a polynomial larger than second degree. For a control system to be stable, the roots of the characteristic equation must have negative real parts. For a further review of topics such as Routh's test, root locus method, frequency response analysis, Bode diagrams, etc., other texts[2,3] should be consulted.

References

1. Ray Wylie, *Advanced Engineering Mathematics*, 4th ed., McGraw-Hill, New York, 1975, pp. 268–269, 295.
2. D. R. Coughanowr and L. B. Koppel, *Process Systems Analysis and Control*, McGraw-Hill, New York, 1965: (**a**) pp. 16–17; (**b**) p. 134.

Additional Reading

3. P. Harriott, *Process Control*, McGraw-Hill, New York, 1967.

Chapter

15

Engineering Economics

There is a special emphasis on economics in the P.E. examinations. In the final part of the examination, the candidates are required to answer one economics question in each of the morning and afternoon sessions; the economics problems constitute 25 percent of the total number of problems to be solved.

Time Value of Money

When money is borrowed by individuals, corporations, or public service organizations, they are required to pay compensation for the use of the borrowed money. The sum of money loaned is called *principal* and the compensation paid by the borrower or earned by the lender is termed *interest*.

Rate of Interest

This is the ratio of the amount of interest payment to the principal per unit of the interest period. It might also be termed as the rate of return or yield on the productive investment of capital. It is denoted by i and expressed as percent per interest period. Thus

$$i = \frac{\text{(total interest paid)(100)}}{\text{(principal)(number of interest periods)}}$$

Time-Scale Presentation

Economic disbursements on a time scale are represented as follows:

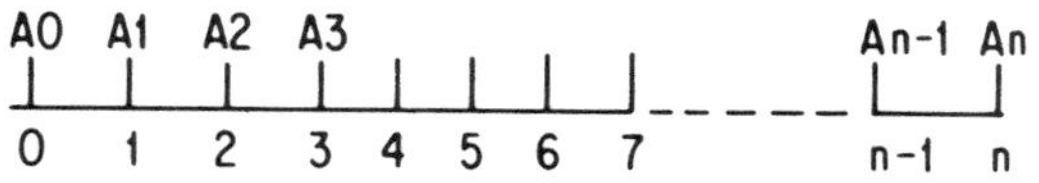

Here 0 to n indicate the time periods, *usually in years* and A_0, $A_1, \ldots, A_n$ are disbursements. The zero on the time scale is the beginning of the year, and 1 is the end of one year (or first period) or beginning of the second year (or second period) and so on.

Simple Interest

When the interest earned on an investment is not reinvested with the original investment to form new interest-earning capital, the interest is called *simple interest*. Thus

$$I' = Pin \tag{15-1}$$

$$F = P(1 + in) \tag{15-2}$$

where I' = interest
i = interest rate as fraction per interest period
n = number of interest periods
P = present worth
F = future worth

Example 15-1. What will be the future worth after 14 months if a sum of $100 is invested today at a sample interest rate of 10% per year?

Solution. From Eq. (15-2), $F = P(1 + in)$

$$P = 100 \qquad i = 0.1 \qquad n = \tfrac{14}{12} = 1.1667$$

Hence
$$F = \$100[1 + 0.1(1.1667)] = \$111.67$$

Compound Interest

When the interest earned on an investment is reinvested along with the original investment to earn interest, the interest earned is called *compound interest*. The compounding of interest is represented on a time scale as follows:

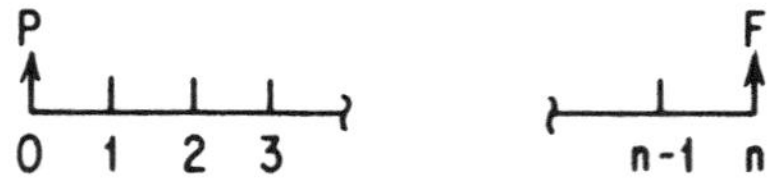

Here F and P are the future and present worths, respectively. Given P, F is found by

$$F = P(1 + i)^n = P(F/P, i, n) \tag{15-3}$$

Given F, P is found by

$$P = \frac{F}{(1 + i)^n} = F(P/F, i, n) \tag{15-4}$$

The terms $(F/P, i, n)$ and $(P/F, i, n)$ are the functional symbols to represent the factors $(1 + i)^n$ and $1/(1 + i)^n$, respectively. Interest tables with these and other factors are available in engineering economics texts.[1a,2a] The term $(F/P, i, n)$ is called the *single-payment future worth factor* and the term $(P/F, i, n)$ is called the single-payment present worth factor. Unless otherwise specified, the interest rate i is taken as percent per year.

Example 15-2. How many years* will it take to double an investment if the interest is 10% per year?

*Rule of 72. This rule indicates that the number of years N over which a future value is double its present worth at the annual compound interest rate i is given by $N = 72/i$. In other words, $(1 + i)^n$ doubles about every N years. This is a rule of thumb.

Solution. From Eq. (15-3)

$$F = 2P = P(F/P, 10\%, n)$$

or

$$(F/P, 10\%, n) = 2$$

From the interest table,[1a,2a] $n = 7.3$ years approximately.

Nominal and Effective Annual Interest Rates

The nominal interest rate expresses the interest rate as percent per year. However, in many instances the compounding of interest is done more than once a year. For example, "8% compounded semiannually" means 4% interest is paid on the original investment at the end of the six months and it is then added to the investment to earn interest for the next six months. To convert the nominal interest rate into an effective annual interest rate (as the name signifies, the compounding is done annually in terms of an effective annual interest rate i), the following formula is used:

$$i = \left(1 + \frac{j}{m}\right)^m - 1 \tag{15-5}$$

where j is the nominal interest rate, fraction per period, and m is the number of the interest periods or compoundings per year. The relation between the present worth and future worth is given by

$$F = P\left(1 + \frac{j}{m}\right)^{mn} \tag{15-6}$$

where m and j are defined above and n is the number of years.

Example 15-3. A savings bank offers loans on two choices of interest: (**a**) 10 percent compounded every month, and (**b**) 12 percent semiannually. Which option would give a lower debt?

Solution. Effective annual interest rate, $i = (1 + j/m)^m - 1$

a. $j = 0.1 \qquad m = 12$

$$i = \left(1 + \frac{0.1}{12}\right)^{12} - 1 = 10.47 \text{ percent per year}$$

b. $j = 0.12,\ m = 2$

$$i = \left(1 + \frac{0.12}{2}\right)^{2} - 1 = 12.36 \text{ percent per year}$$

The option (**a**) would give a lower debt since the effective interest to be paid would be smaller.

Example 15-4. What will be the equivalent amount after 10 years, if \$2000 is deposited today at an interest rate of 10% compounded semiannually?

Solution. There are two ways of solving this problem.
a. Calculate the effective annual interest rate:

$$i = \left(1 + \frac{j}{m}\right)^{m} - 1 \qquad j = 0.1 \qquad m = 2$$

$$= \left(1 + \frac{0.1}{2}\right)^{2} - 1 = 0.1025$$

Future worth $= F = P(1 + i)^n$

$$= \$2000(1 + 0.1025)^{10} = \$5306.60$$

b. Equation (15-6) can be directly used:

$$j = 0.1 \qquad m = 2 \qquad n = 10$$

$$F = P\left(1 + \frac{j}{m}\right)^{mn}$$

$$= \$2000\left(1 + \frac{0.1}{2}\right)^{2(10)} = \$5306.60$$

Annuity

An annuity is a series of uniform end-of-period payments, such that each period is equal and the first payment is made after the first interest period. On the time scale, it looks like the following diagram (A = uniform series of payment).

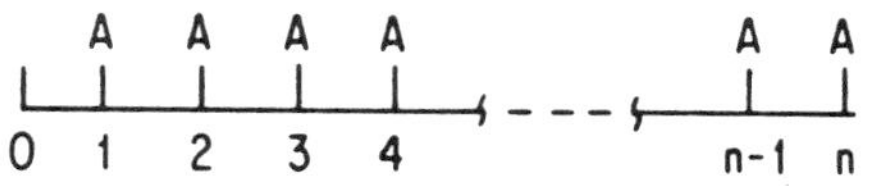

Various formulas may be developed for the annuity payment. Future worth of a uniform series payment is

$$F = A\frac{(1+i)^n - 1}{i} = A(F/A, i, n) \tag{15-7}$$

Annuity for a future worth is

$$A = F\frac{i}{(1+i)^n - 1} = F(A/F, i, n) \tag{15-7a}$$

Present worth of an annuity is

$$P = A\frac{(1+i)^n - 1}{i(1+i)^n} = A(P/A, i, n) \tag{15-7b}$$

Example 15-5. If I start saving in a bank at a rate of $1000 per year at the end of each year from now, how much will I accumulate at the end of 10 years if the interest rate is 10% per year?

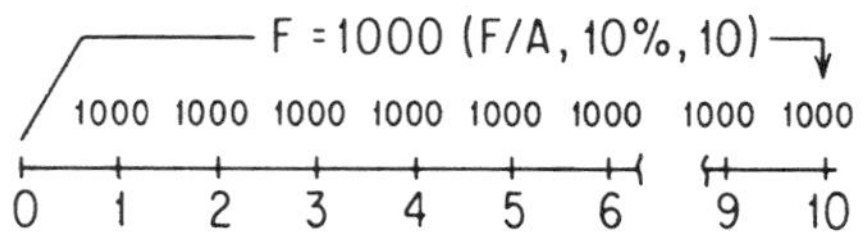

Solution.

$$F = 1000(F/A, 10\%, 10) = 1000(15.937)^* = \$15{,}937$$

*In solving problems, use of the interest tables[1a,2a,3] is made. They are not reproduced in this book. The calculations may also be done with the help of a handheld calculator and interest factors.

Example 15-6. If I start saving in a bank at the rate of $1000 per year with the first savings deposit made today, how much will I accumulate at the end of 10 years if I make a total of 10 equal payments (including the first one) and the interest rate is 10% per year?

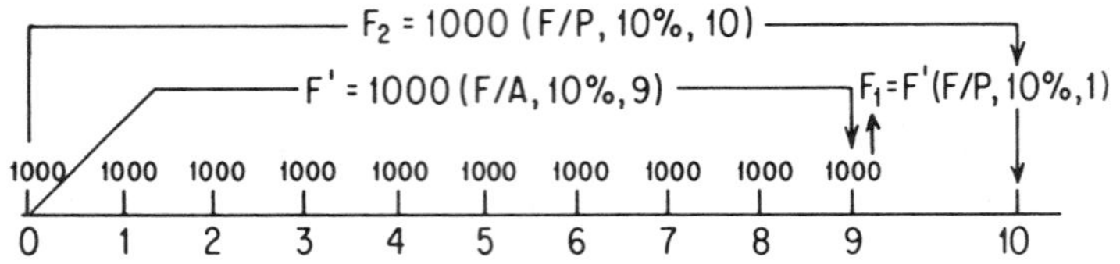

Solution. Note that the annuity formula applies to the end-of-period payment only. Hence, the first payment is not to be included in the annuity calculation. Also, a total of 10 payments, including the first one, is made, so the payment ends at the end of nine years. The future worth of the annuity made for nine years will carry the interest for one more year to calculate the future worth at the end of 10 years:

$$1000(F/A, i, n)(F/P, i, n) + 1000(F/P, i, n)$$
$$= 1000(F/A, 10\%, 9)(F/P, 10\%, 1) + 1000(F/P, 10\%, 10)$$
$$= 1000(13.579)(1.1) + 1000(2.594)$$
$$= \$17,530.90$$

Example 15-7. How much would I accumulate in a bank at the instant of the last payment if I made 10 annual payments of $1000 each with the first deposit made today at an interest rate of 10% per year?

Solution.

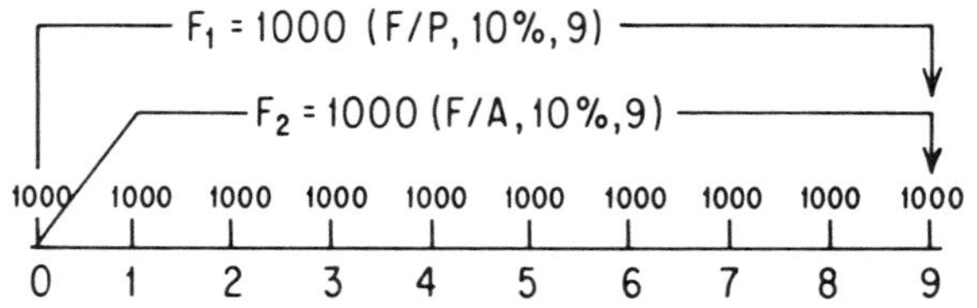

$$F = 1000(F/A, i, n) + 1000(F/P, i, n)$$
$$= 1000(F/A, 10\%, 9) + 1000(F/P, 10\%, 9)$$
$$= 1000(13.579) + 1000(2.358) = \$15,937$$

Sinking-Fund Factor

A sinking fund is an interest-earning fund established to accumulate a desired amount for withdrawal in the future through deposition of a uniform series of payments. The sinking-fund factor is needed to calculate the number of the end-of-period uniform payments necessary to accumulate the desired sinking fund.

$$A = F\frac{i}{(1+i)^n - 1} = F(A/F, i, n) \tag{15-8}$$

The expression $i/[(1+i)^n - 1] = (A/F, i, n)$ is called the sinking-fund factor.

Example 15-8. A man wants to deposit in a savings bank a certain amount of money annually by the end of the current year and for a total of 30 years. How much should he deposit annually so that he will have \$30,000 by the end of 30 years if the interest rate is 10%?

Solution.

$$A(F/A, 10\%, 30) = 30{,}000$$

From the interest tables,

$$(F/A, 10\%, 30) = 164.494$$

Therefore $$A(164.494) = 30{,}000$$

from which $$A = \$182.38$$

Capital-Recovery Factor

This factor is used to calculate the amount of the end-of-period payment necessary to accumulate a given present value when the number of interest periods and the interest rate are known. Substitution of the value of F [Eq. (15-3)] in Eq. (15-8) gives

$$A = P\frac{i(1+i)^n}{(1+i)^n - 1} = P(A/P, i, n) \tag{15-9}$$

The expression $i(1+i)^n/[(1+i)^n - 1]$ is the capital-recovery factor and is abbreviated by $(A/P, i, n)$. The capital-recovery factor is used to calculate the capital-recovery depreciation. It is given by

$$\text{Capital-recovery factor} = (A/P, i, n) = \frac{\text{annual return}}{\text{capital invested}} \qquad (15\text{-}9a)$$

Example 15-9. A savings bank is offering house mortgage loans at $8\frac{1}{2}\%$ interest compounded monthly. What will be the monthly payment for a loan of \$43,000 for a 30-year mortgage if the first installment is due after one month from the date of signing the deed? What will be the effective annual interest rate?

Solution.

$$\text{Effective annual interest rate} = \left(1 + \frac{j}{m}\right)^n - 1$$

$$= \left(1 + \frac{0.085}{12}\right)^{12} - 1 = 8.84\%/\text{year}$$

$$\text{Number of interest periods} = 30(12) = 360$$

$$i = \text{monthly interest rate} = \frac{0.085}{12} = 0.007083$$

$$43{,}000 = A(P/A, i = 0.7083\%, n = 360)$$

$$43{,}000 = A\frac{(1+i)^n - 1}{i(1+i)^n}$$

$$= A\frac{(1.007083)^{360} - 1}{0.007083(1.007083)^{360}}$$

$$= A(130.05844)$$

Therefore $\qquad A = \$330.62$ per month

Relation between Capital-Recovery Factor and Sinking-Fund Factor

$$\text{Capital-recovery factor} = \frac{i}{(1+i)^n - 1} + i$$

$$= \text{sinking-fund factor} + \text{interest rate} \qquad (15\text{-}9b)$$

From Eqs. (15-8) and (15-9), two other relations follow:

$$F = A\frac{(1+i)^n - 1}{i} = A(F/A, i, n) \tag{15-9c}$$

and

$$P = A\frac{(1+i)^n - 1}{i(1+i)^n} = A(P/A, i, n) \tag{15-9d}$$

$[(1 + i)^n - 1]/i$ and $[(1 + i)^n - 1]/i(1 + i)^n$ are called the uniform series compound-amount factor and uniform series present-worth factor, respectively.

Arithmetic-Gradient Conversion Factor

If successive payments made at equal intervals of time differ by a constant amount, the series of payments is termed *uniform-gradient series*. The payments increasing by constant gradients are represented on the time scale as follows:

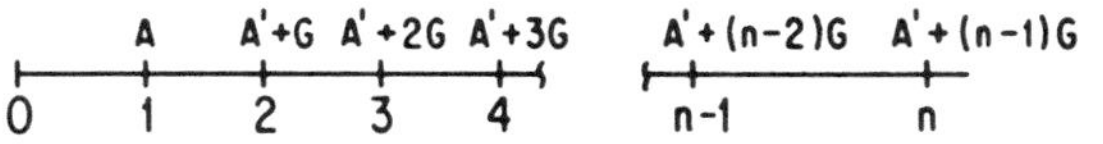

As the time scale shows, the annual payment increases uniformly by G. This is converted to the uniform annuity by the following formula:

$$A = A' + G(A/G, i, n) \tag{15-10}$$

In the above equation, $G(A/G, i, n)$ is an equivalent uniform annuity payment. If the payments decrease by a constant gradient, the sign of $G(A/G, i, n)$ in Eq. (15-10) will be minus.

The composite annuity A in Eq. (15-10) can be used to determine the present or future worths. The present worth may be directly computed by the following formula:

$$P = A'(P/A', i, n) + G(P/G, i, n) \tag{15-11}$$

Note that whereas the annuity counts after the first year, the gradient starts at the end of the second year.

Gradient Series Factors

These are as given below.

Gradient present worth $= P_G$

$$= \frac{G}{i}\left[\frac{(1+i)^n - 1}{i} - n\right]\frac{1}{(1+i)^n}$$

$$= G(P/G, i, n) \qquad (15\text{-}11a)$$

$$\text{Gradient future worth } F_G = \frac{G}{i}\left[\frac{(1+i)^n - 1}{i} - n\right]$$

$$= G(F/G, i, n) \qquad (15\text{-}11b)$$

Gradient equivalent annuity A_G

$$= \frac{G}{i}\left[\frac{(1+i)^n - 1}{i} - n\right]\frac{i}{(1+i)^n - 1}$$

$$= G(A/G, i, n) \qquad (15\text{-}11c)$$

Example 15-10. For my new car purchased today, I have to pay five year-end installments of $1000 each for the next 5 years. For the first year, the maintenance will be covered by the dealer. For the second year, the maintenance will cost me $100. For the remaining years, the maintenance will be $200, $300, and $400. How much should I set aside in a bank to meet the installments and maintenance expenses for the next 5 years if the interest rate is 8%?

Solution.

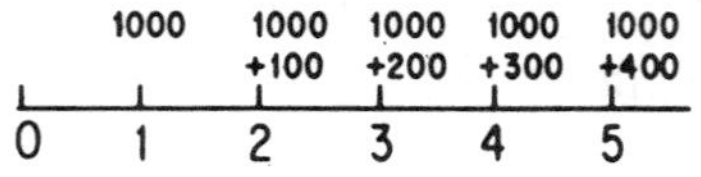

$$
\begin{aligned}
P &= A(P/A, 8\%, 5) \\
&= [A' + G(A/G, 8\%, 5)](P/A, 8\%, 5) \\
&= [1000 + 100(1.8465)](3.993) \\
&= \$4730.31
\end{aligned}
$$

An alternative method is

$$
\begin{aligned}
P &= 1000(P/A, 8\%, 5) + 100(P/F, 8\%, 2) + 200(P/F, 8\%, 3) \\
&\qquad + 300(P/F, 8\%, 4) + 400(P/F, 8\%, 5) \\
&= 1000(3.993) + 100(0.8573) + 200(0.7938) + 300(0.7350) \\
&\qquad + 400(0.6806) \\
&= \$4730.23
\end{aligned}
$$

The alternative method takes a longer time to compute and should be used when the gradient conversion table is not available or the number of years is small.

Example 15-11. One year from now, my daughter will be entering college. To start with, I have to provide her with \$5000. For the remaining four years I will have to provide her with \$4000, \$3000, \$2000, and \$1000, respectively. How much should I set aside in a bank now so that I can continue to provide her with the required funds? Assume 5% interest rate per year.

Solution. Here

$$G = 1000$$

$$
\begin{aligned}
A &= A' - G(A/G, 5\%, 5) = 5000 - 1000(1.9025) \\
&= \$3097.50
\end{aligned}
$$

$$
\begin{aligned}
P &= \text{present worth} = A(P/A, 5\%, 5) = 3097.50(4.329) \\
&= \$13{,}409.08
\end{aligned}
$$

Alternatively,

$$
\begin{aligned}
P = {} & 5000(P/F, 5\%, 1) + 4000(P/F, 5\%, 2) + 3000(P/F, 5\%, 3) \\
&\qquad + 2000(P/F, 5\%, 4) + 1000(P/F, 5\%, 5)
\end{aligned}
$$

$$P = 5000(0.9524) + 4000(0.9070) + 3000(0.8638) + 2000(0.8227) + 1000(0.7835)$$

$$= \$13{,}410$$

Sinking Fund with Beginning-of-Period Deposit

In the case of certain sinking funds, a fund is established at the beginning of an interest period and the first deposit is made at the same time. The second payment is then made at the beginning of the second interest period. For n deposits in the fund and n interest periods, the deposits of the sinking fund can be represented on the time scale as follows:

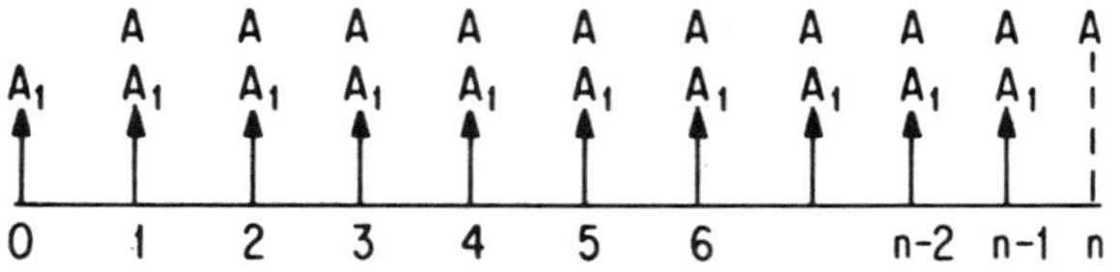

Notice that the final deposit is made one interest period before the terminal date.

The series of uniform beginning-of-period deposits can be converted into a series of uniform end-of-period deposits by the following relations:

$$A_1 = F(A/F, i, n)(P/F, i, 1) \qquad (15\text{-}12a)$$

$$F = A_1(F/A, i, n)(F/P, i, 1) \qquad (15\text{-}12b)$$

$$P = A_1(F/P, i, 1)(P/A, i, n) \qquad (15\text{-}12c)$$

where A_1 is the beginning-of-period uniform payment and A is the end-of-period uniform payment.

Example 15-12. Solve Example 15-6 using the relations of Eq. (15-12).

Solution. Here $n = 10$, $i = 10$ percent:

$$F = A_1(F/A_1, 10\%, 10)(F/P, 10\%, 1)$$
$$= 1000(15.937)(1.1)$$
$$= \$17{,}530.70$$

Amount of Sinking Fund at Intermediate Period

Sometimes the amount of a sinking fund at an intermediate period is required. If n is the number of the total annuities, r the number of payments already made, P the principal, and F_n the future worth of the n payments, the following relationships apply. Value of n payments at the end of rth year:

$$P_r = F_n[P/F, i, (n - r)] \tag{15-13a}$$

Value of r payments made at the end of rth year:

$$F_r = A(F/A, i, r) \tag{15-13b}$$

Value of $n - r$ deposits to be made at the end of rth year:

$$U_r = F_n(P/F, i, n) - A(F/A, i, r)$$
$$= A(F/A, i, n)[(P/F, i, n) - A(F/A, i, r)] \tag{15-13c}$$
$$= P(A/P, i,n)\{(F/A,i,n)[(P/F,i,n-r) -(F/A,i,r)]\} \tag{15-13d}$$

Example 15-13. A \$20,000 mortgage is being amortized by means of 20 uniform annual payments at an interest rate of 8%. The mortgage can be paid off at any time by a lump sum equaling the unpaid balance and a prepayment charge (a penalty for paying off the loan before the due date) on the unpaid balance at 2%. What amount has to be paid after 8 payments have been made?

Solution. $n = 20$, $i = 8$ percent, $r = 8$, $n - r = 12$, $P = \$20{,}000$. Value of 12 payments to be made is $U_r = U_8$.

$$U_r = P(A/P, \text{i, n})[(F/A, i, n)(P/F, i, n - r) - (F/A, i, r)]$$

$$U_r = 20{,}000(A/P, 8\%, 20)[(F/A, 8\%, 20)(P/F, 8\%, 12) - (F/A, 8\%, 8)]$$

$$= 20{,}000(0.10185)[(45.762)(0.3971) - 10.637]$$

$$= \$15{,}348.98$$

$$\text{Prepayment charge} = \$15{,}348.98(0.02) = \$306.98$$

$$\text{Total payment required} = \$15{,}348.98 + \$306.98 = \$15{,}655.96$$

Equivalent Groups of Money Values

The principle of the equivalence of money based on a given interest rate on the time scale applies not only to the individual sums of money but also to the groups of money sums. Sometimes it is required to translate a group of the money sums to a single equivalent sum or to another group of money sums at another date. The following examples will illustrate the procedure involved.

Example 15-14. A owes B the sum of $5000 due December 31, 1984, $3000 due December 31, 1986. By mutual consent A would discharge the debt by making a payment of $7000 on December 31, 1987 and a balance payment on December 31, 1988 with interest at 8%. What is the amount of the final payment?

Solution. Let the final payment be X.

First Plan	Agreed Revised Plan
$5000 on 12/31/84	$7000 on 12/31/87
$3000 on 12/31/86	X on 12/31/88

The two groups of the money sums are equivalent, based on an interest of 8%. Their total values are also equal at any point in time. Choose a convenient standard date for the evaluation of the two groups of money sums. Let this be December 31, 1988. Then since the two groups are equivalent, the following equality can be written as

$$7000(F/P, 8\%, 1) + X = 5000(F/P, 8\%, 4) + 3000(F/P, 8\%, 2)$$

or
$$7000(1.08) + X = 5000(1.3605) + 3000(1.1664)$$

from which
$$X = \$2741.70$$

Calculation of Interest or Investment Rate

When equivalent money sums are known and it is required to calculate the interest rate on which their equivalence is based, the calculations involve trial and error and, in addition, many times an approximate interpolation. The following example illustrates the procedure.

Example 15-15. A person purchased a plot of land for \$20,000 in January 1972. Taxes on the property were as follows: \$300 in 1973, \$300 in 1974, \$350 in 1975, 1976 each, \$400 in 1977 and 1978 each and \$500 in 1979. The owner paid the charges up to and including 1976 but not subsequently. The owner sold the lot at the end of 1979 for \$40,000, paid the back taxes at 9% interest compounded annually, and also paid a commission of 5% to the real estate broker. What rate of return did the owner realize on the investment?

Solution. Net receipt of the investor at the end of 1979:

1977 taxes = $400(F/P, 9\%, 2) = 400(1.1881)$	= \$ 475.24
1978 taxes = $400(F/P, 9\%, 1) = 400(1.09)$	= 436.00
1979 taxes = 500	= 500.00
Total equivalent taxes	= \$ 1,411.24
Commission 5% of \$40,000	= 2,000.00
Total disbursements at end of 1979	= \$ 3,411.24
Selling price	= 40,000.00
Therefore, net receipt at end of 1979	= \$36,588.76

Thus the following two groups of equivalent money sums can be obtained:

Disbursements	Date	Net Receipts on 12/31/79
\$20,000	1/1/73	\$36,588.76
300	12/1/73	
300	12/31/74	
350	12/31/75	
350	12/31/76	

Let $i\%$ be the return on investment. Then the equivalence of the two groups of the money sums can be written as

$$20{,}000(F/P, i, 7) + 300(F/P, i, 6) + 300(F/P, i, 5) + 350(F/P, i, 4) + 350(F/P, i, 3) = 36{,}588.76$$

Since the value of $i\%$ is not known, the solution is by trial and error.

Trial 1. Assume $i = 9$ percent. Then

$$\text{LHS} = 20{,}000(1.8280) + 300(1.6771) + 300(1.5386) + 350(1.4116) + 350(1.2950) = \$38{,}472.02$$

Trial 2. Assume $i = 8$ percent. Then

$$\text{LHS} = 20{,}000(1.7138) + 300(1.5869) + 300(1.4693) + 350(1.3605) + 350(1.2597) = \$36{,}109.93$$

Then, by a straight-line interpolation,

$$i = \frac{36{,}588.76 - 36{,}109.93}{38{,}472.02 - 36{,}109.93}(9 - 8) + 8 = 8.2 \text{ percent}$$

Continuous Compounding of Interest

The compound amount formula is given by Eq. (15-6) as

$$F = P\left(1 + \frac{j}{m}\right)^{mn}$$

If m (number of the interest periods per year) is increased without limit, m becomes very large and approaches infinity, and j/m becomes very small and approaches zero. This condition is called *continuous compounding* and is calculated by

$$F = P\left(1 + \frac{j}{m}\right)^{mn}_{m\to\infty} = Pe^{jn} \qquad (15\text{-}14)$$

This is the continuous-compounding single-payment formula. The continuous-compounding single-payment present-worth formula is

$$P = Fe^{-jn} \qquad (15\text{-}14a)$$

Tables of the continuous-compound amount factors (e^{jn} and e^{-jn}, respectively) are available.[1c,2b] Effective continuous compounding rate becomes

$$i = \lim_{m\to\infty}\left(1 + \frac{j}{m}\right)^{m} - 1 = e^{j} - 1 \qquad (15\text{-}14b)$$

Capitalized Cost

The sum of the first cost and present worth of perpetual disbursements is called the *capitalized cost*. It is used to study the long life assets such as the *railways, dams, tunnels*, and *similar* structures, which provide extended service.

The capitalized cost can be expressed by the following relation:

$$\text{Capitalized cost} = P + \frac{A}{i} + \frac{(P - L)(A/F, i, n)}{i} \qquad (15\text{-}15)$$

where P = first cost or initial investment

A = difference between annual receipts and disbursement or simple disbursement when there is no annual receipt

i = interest rate, fraction per year

L = salvage value

The last term of the expression [Eq. (15-15)] of the capitalized cost vanishes when the service life is very large (50 years or more).

Example 15-16. The initial cost of a dam is \$25,000,000. The annual maintenance cost is \$200,000. If the interest rate is 10% per year, determine the capitalized cost.

Solution. The service life of a dam is large and $(A/F, i, n) = 0$.

$$\text{Capitalized cost} = P + \frac{A}{i}$$

$$= \$25{,}000{,}000 + \frac{\$200{,}000}{0.1} = \$27{,}000{,}000$$

Example 15-17. How much is to be deposited now as fixed deposit in a bank so that you would receive an annual year-end payment of \$10,000 forever? Assume interest rate is 8%.

Solution.

$$P = A(P/A, i, n = \infty) = A\frac{(1+i)^n - 1}{i(1+i)^n} = \underset{n=\infty}{A}\ \frac{1 - 1/(1+i)^n}{i}$$

$$= \frac{A}{i} = \frac{10{,}000}{0.08} = \$125{,}000$$

Depreciation[1b,2b]

Depreciation means a decrease in the value of an asset. In the economic analysis, it may refer to the market value or value to the owner. It may also mean the systematic allocation of the cost of an asset over its useful life. Two major types of depreciation used in economics are described next.

Depreciation for Tax Computation. This is an annual fractional loss of fixed investment allowed by the government to compute the taxable income. This is beyond the control of the owner.

$$\text{Taxable income} = R - dI_F \tag{15-16}$$

$$\text{Income tax} = (R - dI_F)t \tag{15-17}$$

where R = gross income (when cost of production excludes depreciation), \$/year
d = depreciation, fraction \$/\$ · year
I_F = fixed investment, \$
t = tax rate \$/\$

Depreciation for Recovery of Fixed Investment. This is an annual fraction of the fixed investment set aside by an owner or corporation to ensure that the original investment will be completely recovered in a given number of years. This depreciation and the tax are deducted from the gross income to give the net profit. The capital-recovery depreciation charge is eI_F.

$$P_N = R - eI_F - (R - dI_F)t \tag{15-18}$$

where P_N = net profit, \$/year
R = gross income when the cost of production excludes depreciation, \$/year
e = depreciation for recovery of investment, \$/\$ · year
d = depreciation allowed by government, \$/\$ · year
I_F = fixed investment, \$
t = tax rate, \$/\$ earned (for economic evaluation, use $t = 0.5$)

Depreciation Calculations

The computation of the depreciation involves a consideration of (1) the cost of the asset, (2) its useful life, and (3) the salvage value of the asset at the end of its useful life.

Salvage Value. Salvage value or future worth of an asset to be realized at its disposal has to be estimated and may be:

1. Positive if the used item can be resold
2. Zero if the used item is disposed of at no cost or if the salvage value is only a small portion of the cost of the asset
3. Negative if the resale value of the asset is less than the cost of its disposal

Two other definitions are important.

Book Value. The book value B is given by

$$B = \text{purchase cost} - \text{depreciation charge}$$

Sunk Cost. Past costs generally do not affect the present or future and are therefore disregarded. These costs are called *sunk costs*. A sunk cost is given by

$$\text{Sunk cost } S = \text{book value} - \text{actual value}$$

Some of the more important methods of depreciation are discussed next.

Straight-Line Method. In this method, a constant depreciation charge is made every year. Thus

$$\text{Annual depreciation} = \frac{P - L}{n} \tag{15-19}$$

where P = initial investment or cost
L = salvage value
n = economic life of the asset

$$\text{Depreciation reserve at end of } r \text{ years} = \frac{r}{n}(P - L) \tag{15-20}$$

$$\text{Book value at the end of } r \text{ years} = P - \frac{r}{n}(P - L) \tag{15-21}$$

where r is the number of years the asset is in use since its purchase.

Sum-of-the-Digits Depreciation. This method provides a larger depreciation in the early years of the life of an asset than in the later years. The sum of the digits depreciation (for any year r) is

$$\frac{\text{Digit representing remaining years of life}}{\text{Sum of the digits for entire life}}(P-L)$$

$$=\frac{n-r+1}{(1+2+3+\cdots+n)}(P-L) \tag{15-22}$$

$$=\frac{2(n-r+1)}{n(n+1)}(P-L) \tag{15-22a}$$

Book value at the end of r years is given by

$$B=(P-L)\frac{n-r}{n}\frac{n-r+1}{n+1}+L$$

$$=P-\frac{(P-L)r}{n(n+1)}[2n-(r-1)] \tag{15-23}$$

and the depreciation reserve at the end of r years is given by

$$\frac{2[r+(r+1)+(r+2)+\cdots+n]\,(P-L)}{n(n+1)}$$

$$=\left[1-\frac{(n-r)(n-r+1)}{n(n+1)}\right](P-L) \tag{15-24}$$

Declining-Balance Depreciation. This method also allows one to have an accelerated depreciation rate in the early part of an asset's life.

$$\text{Rate of depreciation}=1-\left(\frac{L}{P}\right)^{1/n} \tag{15-25}$$

which requires L to be positive to be realistic. Hence, the declining-balance method cannot be applied when the salvage value of an asset is zero.

$$\text{Depreciation charge in } r\text{th year} = P\left(\frac{L}{P}\right)^{(r-1)/n}\left[1 - \left(\frac{L}{P}\right)^{1/n}\right] \tag{15-25a}$$

$$\text{Book value at the end of the } r\text{th year} = P\left(\frac{L}{P}\right)^{r/n} \tag{15-25b}$$

$$\text{Depreciation reserve at end of } r\text{th year} = P\left[1 - \left(\frac{L}{P}\right)^{r/n}\right] \tag{15-25c}$$

Double-Declining-Balance Method. Another declining-balance method allowed by the income tax code is to use a depreciation rate which is two times the straight-line depreciation rate. (Rates of 1.25 or 1.5 times that of the straight-line method are allowed under certain circumstances.) The double-declining-balance depreciation is

$$\frac{200 \text{ percent}}{n} \tag{15-26}$$

The double-declining-balance depreciation charge in rth year is

$$\frac{2P}{n}\left(1 - \frac{2}{n}\right)^{r-1} \tag{15-26a}$$

The book value at the end of the rth year is

$$P\left(1 - \frac{2}{n}\right)^{r} \tag{15-26b}$$

and the depreciation reserve at the end of the rth year is

$$P\left[1 - \left(1 - \frac{2}{n}\right)^{r}\right] \tag{15-26c}$$

When the double-declining-balance method is used and the salvage value at the end of the economic life of the asset is expected to be less than the book value predicted by the double-declining-balance method, a switch is made to the straight-line depreciation to the advantage of the taxpayer. If r_s is the year at the end of which switch is made, the depreciation by straight-line method for the rest of the economic life span is calculated by the following formula.

$$\text{Annual depreciation} = \frac{P\left(1 - \frac{2}{n}\right)^{r_s} - L}{n - r_s} \qquad (15\text{-}27)$$

Sinking-Fund Depreciation. The sinking fund assumes a fund in which periodic uniform deposits are made annually at some assumed interest rate. The future worth of the annuity equals the cost of the asset minus the salvage value by the end of the depreciation life of the asset.

$$\text{Annual depreciation} = (P - L)(A/F, i, n) \qquad (15\text{-}28)$$

The book value at the end of the rth year is

$$P - (P - L)(A/F, i, n)(F/A, i, r) \qquad (15\text{-}29)$$

Depreciation reserve at the end of the rth year is

$$(P - L)(A/F, i, n)(F/A, i, r) \qquad (15\text{-}30)$$

Capital-Recovery Depreciation. Annual capital-recovery depreciation is given by

$$(P - L)(A/P, i, n) + Li \qquad (15\text{-}31)$$

Note that the capital-recovery depreciation should be used to compare the time value of the alternatives. The formula for the capital-recovery depreciation is derived as follows:

$$\begin{aligned} P(A/P, i, n) - L(A/F, i, n) &= P(A/P, i, n) - L[(A/P, i, n) - i] \\ &= (P - L)(A/P, i, n) + Li \end{aligned}$$

Annual capital-recovery depreciation is the annual sinking-fund depreciation plus the interest on initial investment, or

$$(P - L)(A/P, i, n) + Li = (P - L)(A/F, i, n) + iP \qquad (15\text{-}32)$$

Example 15-18. The installed cost of a refrigeration unit is \$250,000. It has a salvage value of \$20,000 after an economic life of 10 years. Using the straight depreciation method, calculate the annual depreciation, book value, and depreciation reserve at the end of 5 years.

Solution. Annual depreciation equals

$$\frac{P-L}{n} = \frac{250{,}000 - 20{,}000}{10} = \$23{,}000/\text{year}$$

The depreciation reserve at the end of 5 years is

$$r\left(\frac{P-L}{n}\right) = 5(23{,}000) = \$115{,}000$$

and the book value is

$$250{,}000 - 115{,}000 = \$135{,}000$$

Example 15-19. Compare the present worth for the depreciation using the straight-line, double-declining-balance with a switch to the straight line at the end of the fourth year and the sum of the digit depreciations for an asset valued at \$14,000 at zero time with \$2000 salvage value after 6 years and with an interest rate of 12% per year.

Solution. Straight-line depreciation is calculated as follows. The annual depreciation is

$$\frac{P-L}{n} = \frac{14{,}000 - 2000}{6} = \$2000/\text{year}$$

and present worth is

$$2000(P/A, 12\%, 6) = 2000(4.111) = \$8222$$

Double-declining-balance depreciation is

Annual depreciation: $\dfrac{2P}{n}\left(1 - \dfrac{2}{n}\right)^{r-1}$ during rth year

First year depreciation: $\frac{2(14{,}000)}{6}\left(1-\frac{2}{6}\right)^{(1-1)} = \$4666.67$

Second year depreciation: $\frac{2(14{,}000)}{6}\left(1-\frac{2}{6}\right)^{(2-1)} = \$3111.11$

Third year depreciation: $\frac{2(14{,}000)}{6}\left(1-\frac{2}{6}\right)^{(3-1)} = \$2074.08$

Fourth year depreciation: $\frac{2(14{,}000)}{6}\left(1-\frac{2}{6}\right)^{(4-1)} = \$1382.72$

Depreciation for fifth and sixth years:

$$\frac{P\left(1-\frac{2}{n}\right)^{r_s} - L}{n - r_s} = \frac{14{,}000\left(1-\frac{2}{6}\right)^4 - 2000}{6-4} = \$382.72/\text{year}$$

Present worth:

$$4666.67(P/F, 12\%, 1) + 3111.11(P/F, 12\%, 2) + 2074.08(P/F, 12\%, 3) + 1382.72(P/F, 12\%, 4) + 382.72(P/F, 12\%, 5) + 382.72(P/F, 12\%, 6) = \$9413$$

Sum-of-digit depreciation follows.

Annual depreciation: $\frac{2(n-r+1)(P-L)}{n(n+1)} = \frac{2(6-r+1)(14{,}000-2000)}{6(7)}$

$= (7-r)(571.43)$

Annual depreciations calculated are given in Table 15-1.

TABLE 15-1 Annual Depreciations

r	1	2	3	4	5	6
Depreciation	3428.57	2857.14	2285.71	1714.29	1142.86	571.43
$(P/F, 12\%, r)$	0.8929	0.7972	0.7118	0.6355	0.5674	0.5066

Present worth = 3428.57(0.8929) + 2857.14(0.7972) + 2285.71(0.7118)
+ 1714.29(0.6355) + 1142.86(0.5674) + 571.43(0.5066)
= $8993

Cost Comparison of Alternatives[4,5,6]

Cost comparison of different alternatives are made in terms of the equivalent money values based on a selected interest rate and at a selected point in time.

Annual Cost Method. In this method, the initial investment, if any, is spread over the economic life by multiplication with the capital-recovery factor $(A/P, i, n)$ or by other specified depreciation methods. The annual cost thus obtained is added to the other annual costs to obtain the total annual cost. When the alternatives involve different life spans, the comparisons are to be made through the lowest common multiple of the life spans of the alternatives. For example, if one alternative has a 4-year life span and another has a 10-year life span, the comparisons must be made for 20 years. However, in certain cases it is permissible to compare the alternatives with different lives on the basis of their own service lives, e.g., when the alternatives are compared by assuming that each asset will be replaced at the end of its useful life by an identical asset. Also, the annual costs based on the different lives of the assets are used when the least common multiple of the lives of the alternatives is unusually high (e.g., alternatives with 7- and 11-year lives have a least common multiple of 77 years) and exceeds a reasonable analysis period* for an economic study.

Example 15-20. Consider the following alternatives of a project. The first alternative is the base case, or do-nothing, case. The interest rate is 10%.

Alternative	1	2	3
Initial investment	0	$300,000	$250,000
Annual operating cost	$200,000	150,000	120,000
Economic life	24 years	12 years	8 years
Salvage value	0	0	0

*Industries with stable technologies generally use 10 to 20 years, while government agencies use 50 years or more.

On the basis of the annual cost, determine the most economic alternative.

Solution. The study period is the lowest common multiple of the estimated lives and is 24 years.

Alternative 1. Annual cost = \$200,000

Alternative 2.

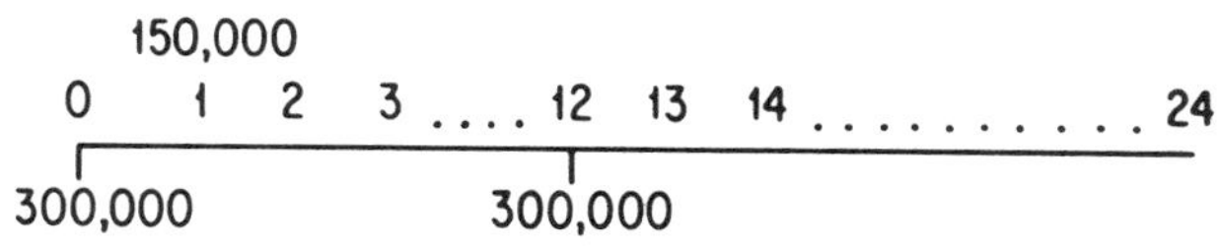

Equivalent annual cost:

$$150{,}000 + [300{,}000 + 300{,}000(P/F, 10\%, 12)](A/P, 10\%, 24)$$

$$= 150{,}000 + [300{,}000 + (300{,}000)(0.3186)](0.1113) = \$194{,}028$$

Alternatively, equivalent annual cost:

$$150{,}000 + (300{,}000)(A/P, 10\%, 12) = 150{,}000 + 300{,}000(0.14676)$$

$$= \$194{,}028$$

This holds because the expense pattern is the same in the second life span.

Alternative 3.

Equivalent annual cost:

$$120{,}000 + (250{,}000)(A/P, 10\%, 8) = 120{,}000 + 250{,}000(0.18744)$$

$$= \$166{,}860$$

Conclusion. Alternative 3 is the most economical alternative.

Calculation of Annual Costs. Annual cost is the sum of (1) the depreciation charge for the year, (2) the annual interest charge, and (3) annual operating cost. Thus

$$\text{Annual cost} = (P - L)(A/P, i, n) + Li + \text{OC}$$

where OC signifies the annual operating costs such as maintenance. The above relation is based on the capital-recovery depreciation which takes into account the compounding of the interest. An approximate annual cost of the capital recovery (CR) is based on a straight-line depreciation and an average interest and is given by

$$\text{Approximate CR} = \frac{P - L}{n} + (P - L)\frac{i}{2}\left(\frac{n + 1}{n}\right) + Li$$

and approximate annual cost is given by

$$\frac{P - L}{n} + (P - L)\frac{i}{2}\left(\frac{n + 1}{n}\right) + Li + \text{OC}$$

Example 15-21. A company is considering the purchase of two fans. If the interest rate is 9 percent, which fan should be bought? Assume identical replacement of an asset and continuing need for A and B fans.

	Fan *A*	**Fan *B***
First cost	\$10,000	\$8,000
Salvage value	2,000	1,500
Annual maintenance cost	500	300
Useful life	12	9

Solution. Identical equipment and continued requirements are assumed.

Fan A.

Annual cost: $(P - L)(A/P, i, n) + Li + \text{OC}$

$$= (10{,}000 - 2000)(A/P, 9\%, 12) + 2000(0.09) + 500$$

$$= 8000(0.13965) + 180 + 500 = \$1797.2.$$

Fan B.

Annual cost: $(8000 - 1500)(A/P, 9\%, 9) + 1500(0.09) + 300$

$$= 6500(0.16680) + 1500(0.09) + 300 = \$1519.20$$

Since the annual cost of fan B is less, fan B is to be purchased.

Example 15-22. In Example 15-21, compare the annual cost by assuming the straight-line depreciation and simple interest.

Solution.

Fan A.

Annual cost: $\frac{P-L}{n} + (P-L)\frac{i}{2}\frac{n+1}{n} + Li + \text{OC}$

$$= \frac{10{,}000 - 2000}{12} + \frac{(10{,}000 - 2000)(0.09)(12+1)}{2(12)} + 2000(0.09) + 500$$

$$= \frac{8000}{12} + \frac{8000(0.09)(13)}{24} + 2000(0.09) + 500 = \$1736.67$$

Fan B.

Annual cost:

$$\frac{8000 - 1500}{9} + \frac{(8000 - 1500)(0.09)(9+1)}{2(9)} + 1500(0.09) + 300$$

$$= \frac{6500}{9} + \frac{6500(0.09)(10)}{18} + 1500(0.09) + 300 = \$1482.22$$

The annual cost of fan *B* is less than that of fan *A* by the approximate method of annual cost calculation also.

Present-Worth (PW) Method. In this method, all the uniform annual costs are converted to the present worth by multiplication with the series-payment present-worth factor $(P/A, i, n)$. Any nonuniform annual cost may be converted to its present worth by multiplication with the single-payment present-worth factor $(P/F, i, n)$. For the alternatives with different life spans, the comparisons are to be made through the least common multiple of the life spans of the alternative assets. However, the present-worth comparisons can also be made by (1) the study-period method which requires the selection of a time period (It assumes the best possible replacement.) and (2) the service-period method. This is used when the operational requirement of an asset is limited.

Example 15-23. Solve Example 15-20 by the present-worth method.

Solution. Study period = 8(3) = 12(2) = 24 years

Alternative 1. Present worth:

$$200{,}000(P/A, 10\%, 24) = 200{,}000(8.985) = \$1{,}797{,}000$$

Alternative 2. Present worth:

$$300{,}000 + 150{,}000(P/A, 10\%, 24) + 300{,}000(P/F, 10\%, 12)$$
$$= 300{,}000 + 150{,}000(8.985) + 300{,}000(0.3186)$$
$$= \$1{,}743{,}330$$

Alternative 3.

Present worth:

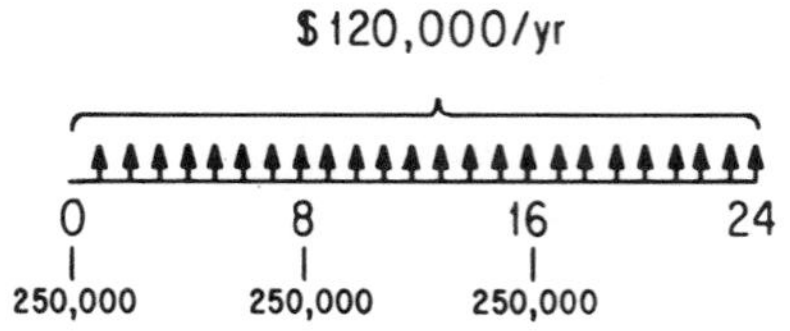

$$250{,}000 + 120{,}000(P/A, 10\%, 24) + 250{,}000(P/F, 10\%, 8)$$
$$+ 250{,}000(P/F, 10\%, 16)$$
$$= + 250{,}000 + 120{,}000(8.985) + 250{,}000(0.4665) + 250{,}000(0.2176)$$
$$= \$1{,}499{,}225$$

Conclusion. Alternative 3 is the most economical.

Capitalized-Cost Method of Comparison of Alternatives. This method assumes a perpetual service life of an alternative or perpetual service through an infinite series of replacement of an alternative having a finite service life. The capitalized cost equals initial investment plus present worth of all perpetual costs. Thus

$$\text{Capitalized cost} = P + \frac{A}{i} + \frac{(P - L)(A/F, i, n)}{i} \qquad (15\text{-}33)$$

where P = initial investment and/or present worth of any future investment

A = perpetual annual cost or perpetual equivalent annual cost of a future one-time cost
L = salvage value

Note: When n is large (> 50 years), the last term may be neglected.

Example 15-24. Compare the capitalized cost of the following alternatives on the basis of 5% interest rate.

Alternative A. Install one machine with an initial investment of \$500,000 and maintenance cost of \$10,000, 10 years from now and every 10th year thereafter.

Alternative B. Install one machine with half the capacity of the machine in Alternative A but with an initial investment of \$300,000 and a maintenance cost of \$7000, 10 years from now and every tenth year thereafter, plus \$1000 per year forever.

Install the second machine with duplicate capacity with an initial investment of \$300,000 on 20th year after the first is installed plus \$1000 per year forever starting from the same year as the second machine is installed.

Solution. The study period is infinite.
Alternative A.

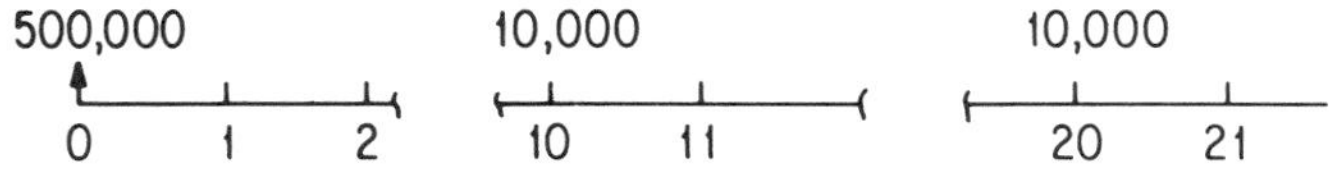

$$P = \text{initial investment} = \$500{,}000$$

Equivalent annual cost each of the first 10 years for the cost of \$10,000 incurred in the tenth year from now is given by

$$F(A/F, 5\%, 10) = 10{,}000(0.07951) = \$795.10/\text{year}$$

For the next 10 years and every year thereafter, this cost of \$795.10/year repeats. Hence, perpetual equivalent annual cost A = \$795.10/year. So, capitalized cost equals

$$P + \frac{A}{i} = 500{,}000 + \frac{795.1}{0.05} = \$515{,}902$$

Alternative B.

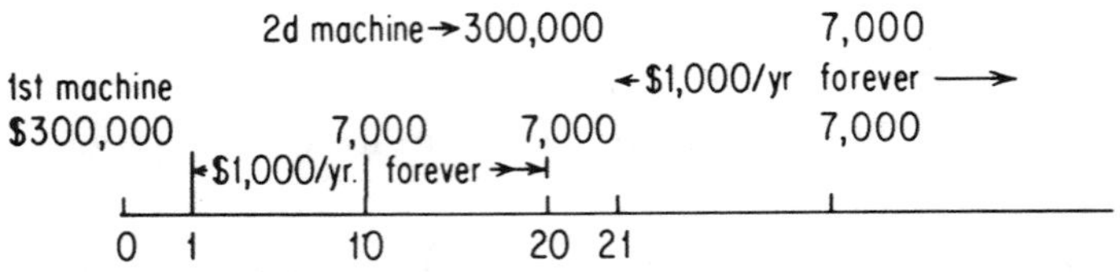

For the first machine,

$$P = \$300{,}000$$

$$A = 1000 + 7000(A/F, 5\%, 10)$$

$$= 1000 + 7000(0.07951) = \$1556.57$$

and capitalized cost equals

$$P + \frac{A}{i} = 300{,}000 + \frac{1556.57}{0.05} = \$331{,}131.4$$

For the second machine, the capitalized cost in the 20th year is

$$\$331{,}131.4$$

and the capitalized cost at time zero is

$$\$331{,}131.4(P/F, 5\%, 20) = \$331{,}131.4(0.37689) = \$124{,}800.11$$

The capitalized cost for both the machines is

$$331{,}131.4 + 124{,}800.11 = \$455{,}931.51$$

Rate-of-Return Method. Rate-of-return (RR) method, also known as *internal rate of return* (IRR) and *return on investment* (ROI), is the most important method of comparing alternatives. Rate of return provides a percentage figure of the relative yield from different ways of using the capital. The basis of the rate-of-return calculations are the present worth and annual worth. The rate of return is calculated by equating to zero the annual or present worths of cash flows involving the receipts and disbursements and by determining the interest rate which would establish the equality. Although the use of either the present worth or annual worth as the basis of calculation is correct, the rate of return is generally formulated in terms of the present worths. The calculation of i usually involves a trial-and-error procedure. If no value of i from the interest tables satisfies the equality, the correct value is determined by a linear interpolation. The relation to be established is to find i so that

$$\text{PW(receipts)} - \text{PW(disbursements)} = 0 \qquad (15\text{-}34)$$

Example 15-25. A company is considering installation of two devices as follows.

	Device *A*	**Device *B***	
Initial cost	\$2000	\$1800	
Maintenance, year · \$	50	60	
Salvage value	0	0	
Annual savings, \$	600	600	1st year but declining by \$30 per year in subsequent years
Useful life, years	6	6	

By the method of rate of return, which device should be purchased?

Solution. PW(benefits) – PW(costs) = 0. i is to be found.

Device A.

PW of costs: $2000 + 50(P/A, i, 6)$

PW of benefits: $600(P/A, i, 6)$

Therefore $2000 + 50(P/A, i, 6) = 600(P/A, i, 6)$

or $2000 - 550(P/A, i, 6) = 0$

Note: When computing the RR by trial and error, a question often arises as to the value of i with which to begin the trials. To narrow down the calculations, divide the average annuity A by P to get A/P. For the given number of years n, find for the average A/P calculated as above the nearest interest rate from the interest tables and use this for the first trial.

In this example,

$$\text{Average } A = 600 - 50 = \$550$$

$$A/P = \frac{550}{2000} = 0.275 \qquad n = 6$$

For $n = 6$ and $A/P = 0.275$, i is between 15 and 18 percent.

Let

$i = 15\%$ LHS $= 2000 - 550(P/A, 15\%, 6) = 2000 - 550(3.784)$
$= -81.2$

$i = 16\%$ LHS $= 2000 - 550(P/A, 16\%, 6) = -26.75$

$i = 18\%$ LHS $= 2000 - 550(P/A, 18\%, 6) = 2000 - 550(3.498)$
$= 76.1$

By interpolation

$$i = 16 + 2\,\frac{0 - (-26.75)}{76.1 - (-26.75)} = 16.5 \text{ percent}$$

Device B.

Note that in this case, the savings per year involve a uniform gradient series.

PW of costs: $1800 + 60(P/A, i, 6)$

PW of benefits: $600(P/A, i, 6) - 30(P/G, i, 6)$

Thus $1800 + 50(P/A, i, 6) - 600(P/A, i, 6) + 30(P/G, i, 6) = 0$

or $1800 - 550(P/A, i, 6) + 30(P/G, i, 6) = 0$

$i = 20\%$ LHS $= 1800 - 550(P/A, 20\%, 6) + 30(P/G, 20\%, 6)$

$= 1800 - 550(3.326) + 30(6.581) = 168.13$

$i = 15\%$ LHS $= 1800 - 550(P/A, 15\%, 6) + 30(P/G, 15\%, 6)$

$= 1800 - 550(3.784) + 30(7.937) = -43.09$

By interpolation $i = 15 + 5\,\dfrac{0 + 43.09}{168.13 + 43.09} = 16.02$ percent

The rate of return of device A is more.

Maximizing the rate of return is not, however, the proper criterion to choose an alternative. For the proper application of RR (i.e., ROI) method, an incremental analysis which examines the differences in the alternatives is required. This is discussed in the next section.

Incremental Analysis Through Rate of Return by Compound Interest Method. By this method, the rate of return is the interest rate i that balances the present worth of all the annual profits with the invested capital.

$$(A/P, i, n) = \frac{\text{annual profit (return)}}{\text{invested capital}} \tag{15-35}$$

When the alternatives are compared by the rate-of-return method, the ratio of the incremental annual profit to the incremental investment is computed to determine the rate of return.

Example 15-26. A corporation is contemplating the installation of a batch organic chemical plant to manufacture one of four competitive products. The following are the estimated figures of the capital investments and net returns.

Product	*A*	*B*	*C*	*D*
Required capital investment, $	20×10^6	30×10^6	35×10^6	50×10^6
Net return, $/year	5×10^6	6.5×10^6	8×10^6	10×10^6

Each of the products would be based on a proven design and technology, and hence the risk factor is minimum and a minimum interest-rate-of-return of 10% would be acceptable. Which of the four products is the most economically attractive? Assume project life of 10 years.

Solution.

Step 1. Tabulate the data in the order of increasing capital investment. This is already done by the presentation of the problem.

Step 2. Check the interest rate of return of the first alternative—the one with the least investment.

$$(A/P, i, n) = \frac{\text{annual return}}{\text{capital invested}}$$

$$(A/P, i, n) = \frac{5 \times 10^6}{20 \times 10^6} = 0.25$$

From the interest table,

$$i > 20\% \quad [A/P, 20\%, 10) \doteq 0.24 \quad \text{and} \quad (A/P, 25\%, 10) \doteq 0.28]$$

Hence, the first alternative is an acceptable one because the minimum acceptable rate of return is 10%.

Step 3. Compute the incremental rate of return. The calculations are given in tabular form below.

Comparison between Products	Incremental Return, \$/year (*a*)	Incremental Investment, \$ (*b*)	(*A*/*P*, *i* = ?, *n*) = *a*/*b*	For *n* = 10, Calculated *i* from Interest Tables, %	Remark
A and *B*	1.5×10^6	10×10^6	0.15	~ 8 (< 10)	Reject *B* Select *A*
A and *C*	3×10^6	15×10^6	0.20	~ 15 (> 10)	Reject *A* Select *C*
C and *D*	2×10^6	15×10^6	0.1333	~ 6 (< 10)	Reject *D*
		Final selection: Product C			

Instead of estimating the incremental rate of return, if the rate of return is determined from the absolute amount of investment and the corresponding net return, a wrong conclusion may be drawn as can be observed from the following table.

Product	Capital Invested	Net Return, \$/year	(*A*/*P*, *i*, 10)	*i*, %
A	20×10^6	5×10^6	0.2500	21.3
B	30×10^6	6.5×10^6	0.2167	17.2
C	35×10^6	8×10^6	0.2286	18.7
D	50×10^6	10×10^6	0.200	15.1

The wrong conclusions that may be drawn from the above table are:

1. Product A offers the highest interest rate of return, and therefore is the best investment. This is a wrong conclusion because it prevents additional investment that would still meet the minimum interest-rate-of-return.
2. Product D requires highest investment and has a 15.1% rate of return which is above the minimum requirement (10%). This is therefore the best investment. This is a wrong conclusion, since additional investment above $\$35 \times 10^6$ would earn 6% interest rate of return as shown earlier.

Benefit-to-Cost (*B/C*) Ratio Method. The B/C ratio of a scheme is calculated by

$$B/C \text{ ratio} = \frac{\text{annual savings or benefits}}{\text{equivalent annual cost}} \tag{15-36}$$

where

$$\text{Equivalent annual cost} = (\text{initial investment})(A/P, i, n) + \text{other annual cost (such as maintenance, etc.)}$$

To justify a project, the B/C ratio must be greater than unity.

Example 15-27. The federal government is considering four sites to construct a dam on a river to save the public from the damage caused by floods. Apart from this benefit, the water from the dam will be used for fire protection and recreation. The following are the estimates of the capital investment, annual cost, and benefits of the different sites.

Site No.	Required Initial Investment	Annual Benefit from Flood Saving, Fire Protection, and Recreation	Annual Operation and Maintenance
1	\$1,500,000	\$300,000	\$30,000
2	2,000,000	350,000	40,000
3	2,500,000	450,000	60,000
4	3,500,000	500,000	80,000

Assuming a 50-year life of the dam and 5% interest rate, determine the most economical site.

Solution.

Step 1. Determine the equivalent annual cost of each site as given below.

Site No.	Equivalent Annual Cost, \$/year
1	1,500,000($A/P, i, n$) + 30,000 = 1,500,000(0.05478) + 30,000 = 112,170
2	2,000,000($A/P, i, n$) + 40,000 = 2,000,000(0.05478) + 40,000 = 149,560
3	2,500,000($A/P, i, n$) + 60,000 = 2,500,000(0.05478) + 60,000 = 196,950
4	3,500,000($A/P, i, n$) + 80,000 = 3,500,000(0.05478) + 80,000 = 271,730

Step 2. Tabulate in the order of increasing equivalent annual cost and check the individual benefit-to-cost (B/C) ratio.

Site No.	Equivalent Annual Cost C, \$/year	Annual Benefit B, \$/year	B/C Ratio	Remarks
1	112,170	300,000	2.67	
2	149,560	350,000	2.34	All sites justified
3	196,950	450,000	2.28	since B/C is > 1
4	271,730	500,000	1.84	in each case

Step 3. Compute the incremental annual cost and incremental annual benefit and estimate the incremental B/C ratio.

Comparison between	Incremental Annual Cost C, \$/year	Incremental Annual Benefit B, \$/year	B/C Ratio	Remarks
Sites 1 and 2	149,560 − 112,170 = 37,390	350,000 − 300,000 = 50,000	1.34	Select Site 2 over 1
Sites 2 and 3	196,950 − 149,560 = 47,390	450,000 − 350,000 = 100,000	2.11	Select Site 3 over 2
Sites 3 and 4	271,730 − 196,950 = 74,780	500,000 − 450,000 = 50,000	0.67 (< 1)	Select Site 3 over 4
		Final selection: Site 3		

After-Tax Comparisons

After-tax comparisons can be made by either the annual cost or by the present-worth and the rate-of-return methods. The following definitions will be useful in following the after-tax methods.

Gross Profit. Generally, the cost of production does not include depreciation. Under this accounting procedure, gross profit is calculated by

$$R = S - C \tag{15-37}$$

where R = gross profit, \$/year
S = sales revenue, \$/year
C = cost of production excluding depreciation \$/year

Net Profit. P_N is related to gross profit by

$$P_N = R - (R - dI_F)t - ei_F \tag{15-37a}$$

where I_F = fixed investment, \$
t = tax rate, \$/\$
e = capital-recovery depreciation, \$/\$ · year
d = government allowed depreciation

when $\qquad d = e \qquad P_N = (R - dI_F)(1 - t) \qquad$ (15-37b)

Sometimes the cost of production includes the depreciation for the sake of simplicity. Under this accounting procedure, the gross profit, which is different from the gross profit defined in Eq. (15-37), is given by

$$R' = S - C' \tag{15-37c}$$

where C' is the cost of production including depreciation. Obviously

$$C' = C + dI_F \tag{15-37d}$$

so $\qquad R' = S - C - dI_F = R - dI_F \qquad$ (15-37e)

When $d = e$, from Eqs. (15-37*b*) and (15-37*e*),

$$\text{Net profit } P_N = (R - dI_F)(1 - t) = R'(1 - t) \qquad (15\text{-}37f)$$

One method of the rate-of-return analysis is given earlier. A simplified method commonly used in the process design is given below.

Return on original investment is given by

$$\text{ROI} = \frac{\text{net profit in a year}}{\text{total investment}}$$

$$= \frac{(R - dI_F)t - eI_F}{I_F + I_w} \qquad \text{where } I_w \text{ is working capital}$$

$$= \frac{(R - dI_F)(1 - t)}{I_F + I_w} \qquad \text{when } e = d$$

$$= \frac{R'(1 - t)}{I_F + I_w} \qquad (15\text{-}38)$$

Generally, the cost reduction alternatives affect the cost of production and not the sales revenue, and no working capital is considered. Thus

$$\text{ROI} = \frac{(R - dI_F)(1 - t)}{I_F} = \frac{[S - (C + dI_F)](1 - t)}{I_F} \qquad (15\text{-}39)$$

The challenge of a higher investment to cut down the cost of production of a defending scheme requiring lower investment is evaluated by estimating the rate of return on the incremental investment as follows:

$$\frac{[(C + dI_F)_1 - (C + dI_F)_2](1 - t)}{(I_F)_2 - (I_F)_1} \qquad (15\text{-}40)$$

where subscripts 1 and 2 refer to alternatives 1 and 2, respectively. The required minimum rates of return depend on the company practice.

Payout Time Method or Payback Method. This method is used in two ways:

1. Taxes and depreciation are ignored. This is the most commonly used form. The payout time T is given by

$$T = \frac{\text{fixed investment}}{\text{gross annual income}} = \frac{I_F}{R} \tag{15-41}$$

2. Taxes and depreciation are accounted for. Then T is

$$T = \frac{I_F}{R - (R - dI_F)t} = \frac{I_F}{R'(1 - t) + dI_F} \tag{15-42}$$

In general, the payout time method should not be used to compare alternatives having different economic life spans.

Venture Profit Method.[2]

$$\begin{aligned} V &= R - (R - dI_F)t - eI_F - i_m(I_F + I_w) \\ &= (1 - t)R - (e - dt + i_m)I_F - i_m I_w \end{aligned} \tag{15-43}$$

when $e = d$ = constant.

$$V = (R - dI_F)(1 - t) - i_m(I_F + I_w) \tag{15-44}$$

where i_m is the minimum acceptable rate of return and V is the venture profit, \$/year. Other terms are defined earlier. The above equation assumes a zero salvage value.

Cash Flows in Annual Disbursement Only. Methods of economic analysis in cases when the cash flows are in annual disbursement only are illustrated by the following example.

Example 15-28. The following four diameters were considered to find the optimum pipe diameter of a pipeline system.

Pipe Diameter	4 in	6 in	8 in	10 in
A. Initial investment	\$12,000	\$13,000	\$16,000	\$18,000
B. Annual pumping and maintenance cost	1200	700	400	200
C. Annual property taxes and insurance	360	450	480	540

Assume a 10-year life and a zero salvage value for the pipeline. Minimum attractive rate of return is 15 percent after the income tax. The income tax is at 50% of the incremental annual savings. Estimate the total equivalent annual cost and select the most optimum diameter.

Solution. The equivalent annual cost is

$$C + (I_F - L)(A/P, i, n) + Li + [S - (C + dI_F)]t$$

Step 1. Calculate C.

Pipe Diameter	**4 in**	**6 in**	**8 in**	**10 in**
Annual pumping and maintenance cost	1200	700	400	200
Annual property taxes and insurance	360	450	480	540
$C =$	1560	1150	880	740

Step 2. Calculate the capital-recovery charge. This includes the second and third terms of the equation.

$$(I_F - L)(A/P, i, n) + Li$$

Here, $i = 15\%$, $n = 10$, $L = 0$.

Pipe Diameter	**4 in**	**6 in**	**8 in**	**10 in**
Capital recovery Charge	12,000 × $(A/P, 15\%, 10) =$ \$2391	13,000 × $(A/P, 15\%, 10) =$ \$2590	16,000 × $(A/P, 15\%, 10) =$ \$3188	18,000 × $(A/P, 15\%, 10) =$ \$3587

Step 3. Calculate the annual income tax. The income tax would be calculated by

$$\text{Relative tax} = \left[(C + dI_F)_{\text{base case}} - (C + dI_F)_{\text{other case}}\right]t$$

Note that $d = \frac{1}{10} = 0.1$, $t = 0.5$.

Pipe Size	4 in	6 in	8 in	10 in
Depreciation charge, dI_F	1200	1300	1600	1800
Annual cost (from Step 1)	1560	1150	880	740
Total deduction for taxable income ($C + dI_F$)	\$2760	\$2450	\$2480	\$2540
Deduction less than 4-in case	0	310	280	220
Relative income tax	0	\$155	\$140	\$110

Step 4. Calculate the equivalent annual cost.

Pipe Diameter	4 in	6 in	8 in	10 in
Annual cost (Step 1)	1560	1150	880	740
Capital-recovery charge (Step 2)	2391	2590	3188	3587
Income tax (Step 3)	0	155	140	110
Total equivalent annual cost	3951	3895	4208	4437
Select 6-in-diameter line				

Example 15-29. In a newly installed plant, the solvent losses are costing the company \$350,000/year. The operating costs are: fuel \$275,000/year, cooling water \$90,000/year. A proven design to reduce the solvent loss by 50% would involve an investment of \$75,000. However, with the new solvent-recovery equipment installed, the operating cost would be: fuel \$350,000/year, cooling water \$100,000/year. Depreciation would be 10% of the investment. Calculate the return on the investment. Assume the other costs remain unchanged. The tax rate is 50%.

Solution.

$$\text{ROI} = \frac{(C + dI_F)_{\text{base case}} - (C + dI_F)_{\text{other case}}}{(I_F)_{\text{other case}} - (I_F)_{\text{base case}}}(1 - t)$$

	Base Case	Proposed Project
Investment, I_F	0	$ 75,000
Annual cost C		
A. Solvent losses	$350,000/year	$175,000/year
B. Fuel	275,000/year	350,000/year
C. Cooling water	90,000/year	100,000/year
	$715,000/year	$625,000/year
Depreciation, dI_F^C	0	7,500/year
$C + dI_F$	$715,000/year	$632,500/year

$$\text{Incremental return} = (C + dI_F)_{\text{base case}} - (C + dI_F)_{\text{proposed project}}$$

$$= 715{,}000 - 632{,}500$$

$$= \$82{,}500$$

$$\text{Incremental investment} = (I_F)_{\text{proposed project}} - (I_F)_{\text{base case}}$$

$$= \$75{,}000$$

$$\text{ROI after tax} = \frac{82{,}500(0.5)(\$/\text{year})}{75{,}000(\$)} = 0.55/\text{year or } 55\%/\text{year}$$

Linear Break-Even Analysis[1b]

The break-even capacity of a plant is the capacity at which the annual revenue is equal to the annual cost of production. In other words, at break-even capacity, the gross profit is zero. Let

Q_B = break-even capacity units/time

S = revenue, \$/time

R = gross profit, \$/time

s = selling price, \$/unit

C = total cost of production, \$/time

c = variable cost, \$/unit

F = fixed cost, \$/time

$$Q = \text{number of units produced per unit time}$$

Then the following equations can be derived.

$$S = sQ$$

$$C = Qc + F$$

$$R = S - C = Q(s - c) - F \tag{15-45}$$

$$Q_B = \frac{F}{s - c} \qquad \text{when } Q = Q_B,\ R = 0$$

$$s - c = \text{contribution}$$

$$\frac{Q - Q_B}{Q_B} = \text{margin of profit} = \frac{R}{F} \tag{15-46}$$

Example 15-30. Production of a 250,000 tons/year capacity plant involves the following costs.

Plant investment = \$5,000,000

Manufacturing cost = \$50/ton

Selling price = \$60/ton

Assuming a reasonable depreciation, calculate the break-even capacity.

Solution. Assume 10% straight-line depreciation.

$$F = \text{fixed cost} = 0.1(5 \times 10^6) = \$500{,}000/\text{year}$$

$$s = \text{selling price} = \$60/\text{ton}$$

$$c = \text{variable cost, } \$50/\text{ton}$$

$$\text{Break-even capacity} = \frac{F}{s - c} = \frac{500{,}000}{60 - 50} = 50{,}000 \text{ tons/year}$$

Analytical Method of Optimization

This method is applied when the dependent variable in the objective function can be expressed as a function of one or more variables. Example 15-31 illustrates this method.

Example 15-31. A flat-roofed and flat-bottomed cylindrical storage tank is to be designed to store a liquid of density ρ lb/ft^3 and capacity V ft^3. The tank is open to atmosphere. The cost of the roof is \$$C_T$/ft^2 and the cost of bottom is \$$C_B$/ft^2. The vertical surface cost is \$$kt$/ft^2 where k is constant and t is the wall thickness of the tank in inches. The wall thickness is calculated by

$$t = \frac{6PD}{SE}$$

where P = internal pressure, psig
D = diameter of tank, ft
S = allowable stress, psi
E = joint efficiency

Find the dimension of the tank for the lowest cost.

Solution. Let H be the height of tank, ft, and C_T the total cost

$$C_T = \pi DHkt + (C_B + C_T)(\tfrac{1}{4}\pi D^2) \qquad (1)$$

$$t = \frac{6PD}{SE} \qquad (2)$$

$$P = \frac{H\ (\text{ft})\ \rho(\text{lb/ft}^3)}{144\ \text{in}^2/\text{ft}^2} = \frac{H\rho}{144}\text{psig} \qquad (3)$$

Also $$\tfrac{1}{4}\pi D^2 H = V \qquad (4)$$

From Eqs. (1) to (4), one derives:

$$C_T = \frac{\pi DVk(0.04167)DV\rho}{(\frac{1}{4}\pi D^2)(\frac{1}{4}\pi D^2)(SE)} + (C_B + C_T)\tfrac{1}{4}\pi D^2$$

$$= \frac{0.2122V^2k\rho}{D^2(SE)} + (0.7854)(C_B + C_T)D^2$$

Employing $dC_T/dD = 0$ for minimum C_T gives

$$-\frac{0.2122(2)V^2k\rho}{D^3(SE)} + 0.7854(2)(C_B + C_T)D = 0$$

from which

$$D = 0.721\left[\frac{V^2k\rho}{(C_B + C_T)SE}\right]^{1/4} \tag{5}$$

Note: $d^2C_T/dD^2 > 0$; hence this diameter gives minimum C_T. Substituting D from Eq. (5) in Eq. (4), H can be determined.

In practice, the calculated diameter as obtained from Eq. (5) is substituted in Eq. (2) to determine the economic thickness. If the calculated thickness is less than a certain minimum allowable thickness (usually $\frac{1}{4}$ in for steel), the minimum allowable thickness is taken and dimensions recalculated.

References

1. J. L. Riggs, *Engineering Economics*, McGraw-Hill, New York, 1977: (*a*) pp. 594–609; (*b*) pp. 46–47; (*c*) p. 160.
2. D. G. Newman, *Engineering Economics Analysis*, Engineering Press, San Jose, 1976: (*a*) pp. 430–459; (*b*) p. 460.
3. *Chemical Engineers Handbook*, 5th ed., R. H. Perry (ed.), McGraw-Hill, New York, 1973, pp. 1-33 to 1-34.
4. J. Happel and D. G. Jordan, *Chemical Process Economics*, 2d ed., Marcel Dekker, New York, 1975, pp. 56–62.
5. E. L. Grant and W. G. Ireson, *Principles of Engineering Economy*, 5th ed., Ronald Press, New York, 1970, pp. 201–227.
6. M. S. Peters and K. D. Timmerhaus, *Plant Design and Economics for Chemical Engineers*, McGraw-Hill, New York, 1980, pp. 302–345.

Additional Reading

1. M., Kurtz, *Engineering Economics for Professional Engineering Examination*, 2d ed., McGraw-Hill, New York, 1975.
2. V. W. Uhl and A. W. Hawkins, *Technical Economics for Chemical Engineers*, A.I.Ch.E., New York, 1971.

Chapter

16

Plant-safety

The purpose of this chapter is to offer a guideline, not a rigid procedure, in the safety aspect of process engineering. The user must exercise his/her judgement as to the appropriateness of a proposed method and deviate if a comparatively safe design is achieved as a result of such deviation. Under any circumstance, however, safety must not be compromised with cost.Theoretical derivations of the formulas have been avoided to get to the point as quickly as possible.

TOXICOLOGY AND INDUSTRIAL HYGIENE

The degree to which a substance is poisonous is termed its toxicity. A toxicant is any chemical substance which when inhaled, ingested, absorbed through the skin or when applied to, injected into or developed within the body, in relatively small amounts may cause, by its chemical reaction, damage to the body structure or disturbance to human function or even death. Most chemicals enter the body through eyes, respiratory tract, digestive tract and skin. Two levels of toxicity are defined: (1) acute "short term" exposure that causes initiation poisoning, (2) chronic-"long period" exposure that causes anemia, leukemia and death.

Threshold Limit Values(TLV) and Exposure Limits

Toxic threshold limit values (TLVs) are the greatest concentrations of a toxicant in air that can be tolerated for a given length of time without any toxic effects. Definitions of commonly used limit values are given below.

The different types of TLV's and exposure limits are as follows:

- **TLV-TWA**
  The time weighted average for a normal eight-hour workday or 40 hour work week to which a worker may be exposed without adverse effect.
- **TLV-STEL**
  The short term exposure limit is the maximum concentration to which a worker may be exposed continuously up to 15 minutes without adverse effect.
- **TLV-C**
  The ceiling limit is the concentration which should not be exceeded at any time.
- **PEL**
  Permissible exposure limit is the maximum exposure allowed by OSHA 29CFR1910.1000.
- **IDLH**
  Immediately dangerous to life or health. It is the maximum concentration from which one could escape within 30 minutes without escape-impairing symptoms or any irreversible health effects.

TLVs are expressed either in parts per million by volume (ppm) or milligram per cubic meter (mg/m^3).

$$1 \text{ ppm} = \frac{PM}{0.08205T} \text{ mg/m}^3 \cong 0.04 \text{ M @ } 25°\text{C and 1 atm}$$

Where: P = pressure in atm,
T = Temperature in K,
M = Molecular mass

Example 16-1. The TLV-TWA of ammonia is 25 ppm. What is the corresponding value in mg/m³ at 1 atm, and 25°C?

Solution.

$$\text{Value in mg/m}^3 = \frac{1 \times 17 \times 25}{0.0805 \times 298} = 17.4$$

Time weighted average of concentration is calculated by the following formula:

$$\mathbf{TWA} = \frac{C_1 t_1 + C_2 t_2 + \text{---} + C_n t_n}{t_1 + t_2 + \text{---} + t_n} \tag{16-1}$$

Where: C_i = Concentration
t_i = Time of Exposure

Example 16-2. Estimate the time weighted concentration for eight hours if a worker is exposed to concentrations of 100 ppm for 1 hr, 90 ppm for 2 hrs and 80 ppm for 5 hrs.

Solution.

$$\text{TWA} = \frac{100 \times 1 + 90 \times 2 + 80 \times 5}{1 + 2 + 5} = 85$$

The effect of more than one toxicant with different TLV-TWAs may be calculated from:

$$\sum_{i=1}^{n} \frac{C_1}{(TLV - TWA)_i} \tag{16-2}$$

If this sum exceeds 1, then the workers are overexposed.

Example 16-3. A processing room contains 20 ppm Heptane (TLV-TWA = 400 ppm), 50 ppm methyl alcohol, (TLV-TWA = 200 ppm), 40 ppm methyl chloride (TLV-TWA = 50 ppm). Determine whether combined effect of these toxicants exceeds the exposure limit.

Solution.

$$\sum_{i=1}^{n} \frac{C_1}{(TLV - TWA)_i}$$

$$= \frac{20}{400} + \frac{50}{200} + \frac{40}{50}$$

$$= \quad 1.1 > 1$$

Hence exposure limit is exceeded.

Estimation of Dilution Air

When volatile substances are emitted in working spaces such as a laboratory, fresh air is ventilated into the working space to maintain average concentration of the volatile matter below TLV-TWA. The quantity of such dilution air in CFM(cubic feet/minute) may be estimated by:

$$\mathbf{CFM} = \frac{(3.87 \times 10^8)W}{M \times TLV \times K} \qquad (16\text{-}3)$$

Where: W = Mass of volatile matter evaporated/time, lb/min

M = Molecular mass of the volatile matter

TLV = Threshold limit value of the volatile substance, ppm

K = Nonideal mixing factor

= 1 for perfect mixing

= 0.1 to 0.5 for most real situations

Noise

The noise power level in absolute decibels (dBA) is given by:

$$\mathbf{Lw = 10\ log_{10}\ (W/W_O)} \qquad (16\text{-}4)$$

Where:

Lw	=	Sound power in dBA
W	=	Intensity of sound wave, Watt/m^2
W_O	=	Threshold of human hearing, 10^{-12} Watt/m^2

If Lw_1, Lw_2, ... are the noise power levels of machine 1, 2, etc., the combined noise power level of the machines is given by:

$$\mathbf{Lw\ (combined) = 10\ log_{10}\ (10^{Lw1/10} + 10^{Lw2/10} + ...)} \qquad (16\text{-}5)$$

Example 16-4. In a workshop which has a noise level of 60 decibel, a new machine with a noise level of 70 decibel is moved in. What will be the combined noise level?

Solution.

$$\begin{aligned} Lw\ (combined) &= 10\ log_{10}\ (10^{60/10} + 10^{70/10}) \\ &= 10\ log_{10}\ (10^6 + 10^7) \\ &= 10\ log_{10}\ [10^6\ (1+10)] \\ &= 10\ log_{10}\ (10^6) + 10\ log_{10}(11) \\ &= 60 + 10.41 \\ &= 70.41\ dBA \end{aligned}$$

The noise pressure level (Lp) in absolute decibels (dBA) is given by:

$$\mathbf{Lp = 20\ log_{10}\ (P/Po)} \qquad (16\text{-}6)$$

Where: P = root-mean-square value of sound pressure in psi
Po = 0.0002 microbar = 3 x 10^{-9} psi

To estimate the noise pressure level (Lp) at a distance greater than that at which it was measured, use the following formula:

Lp (desired distance) = Lp (distance) - 20Log$_{10}$ (desired distance/distance)

Example 16-5. Estimate the noise pressure level at a distance of 100 ft. when the noise pressure level at 3 ft. from the source is 120 dB.

Solution.

$$Lp\ (100\ ft) = Lp\ (3\ ft) - 20\log_{10}(100/3)$$
$$= 120 - 20\log_{10}(100/3)$$
$$= 89.5\ dB$$

The exposure time allowed by OSHA in an environment of noise pressure level L is given by the following formula:

$$\mathbf{T = 16/[2^a]} \tag{16-7}$$

Where:

a = 0.2(L-85)
T = continuous exposure time, hr
L = noise pressure level in dBA

Example 16-6. Estimate the continuous exposure time allowed by OSHA in an environment with noise pressure level of 105 dBA.

Solution.

$$a = 0.2(L-85)$$
$$= 0.2(105-85)$$
$$= 4$$
$$T = 16/2^4 = 1\ hr$$

Example 16-7. A worker is exposed to the following noise levels in an eight hour working period: 50 dBA for 3 hours, 85 dBA for 4 hours and 95 dBA for 1 hour. Determine whether the OSHA exposure limit of the worker is exceeded.

Solution.

NOISE LEVEL	PERMITTED EXPOSURE TIME, HR ,T*	ACTUAL EXPOSURE TIME, HR T_A	TIME FRACTION EXPOSED T_A/T
50	2048	3	0.001
85	8	4	0.500
95	4	1	0.250
			0.750

* T = $16/2^a$, where a = 0.2(L-85)

Since the summation of the time fraction exposed to various noise levels is less than 1, OSHA exposure limit is not exceeded.

Types of noise control in operation and design.
Devices such as insulation barriers,enclosures, silencers, reduced velocity, quieter valve, non-protruding gaskets, smoothing of internal welds, increased thickness of pipe are used to reduce noise.

FIRE AND EXPLOSION

A fire is caused by the presence of sufficient quantity of each of the following:

- Fuel such as gasoline, hydrogen, wood, etc.
- Oxidizer such as oxygen, chlorine, ammonium nitrite
- Ignition source such as heat, mechanical or electrical sparks or electrostatic discharge, flame

Terms Used in Fire and Explosion

- **Autoignition Temperature (AIT)**
  The temperature above which a flammable substance ignites itself by drawing sufficient amount of oxidizer and ignition from the surrounding is called the auto ignition temperature.

- **Flash Point (FP)**
  The lowest temperature at which a flammable substance gives off sufficient vapor to sustain fire only temporarily (hence the name flash) in presence of air is called the flash point.
- **Fire Point**
  The lowest temperature at which a flammable substance gives off sufficient vapor to sustain fire continuously in presence of air is called fire point.
- **Upper Flammability Limit**
  The highest concentration of a flammable fluid in air above which the air-fuel mixture will not burn is called the upper flammability limit.
- **Lower Flammability Limit**
  The lowest concentration of a flammable fluid in air below which the air-fuel mixture will not burn is called the lower flammability limit.
- **Explosion**
  A rapid release of energy through a shock wave is called explosion. When the speed of the shock wave is below the speed of sound, the phenomenon is called deflagration. When it is above the speed of sound, the phenomenon is known as detonation.
- **Minimum Oxygen Concentration (MOC)**
  The MOC is the minimum per cent of oxygen in air plus fuel required to propagate flame. MOC generally varies from 8% (vapor) to 10% (dust). Control points range from 4% (vapor) to 6% (dust).
- **Minimum Ignition Energy (MIE)**
  The minimum energy, expressed in milli Joules, required to start combustion is called minimum ignition energy. It is dependent on pressure and concentration of inert substances.

Inerting

It is the process of lowering the oxygen concentration in a combustible mixture below MOC by adding an inert gas. Three common methods of inerting are: Sweep through purging, pressure purging and vacuum purging.

Sweep Through Purging

The inert gas is added to a vessel at one end and withdrawn at another end, both the addition and withdrawal being at atmospheric pressure.

$$Q_v t = V \ln \frac{(C_1 - C_o)}{(C_2 - C_o)} \qquad (16\text{-}8)$$

Q_v	=	Volumetric flow rate of inert
t	=	Time of purging
V	=	Vessel volume
C_1	=	Initial concentration of oxygen
C_2	=	Final concentration of oxygen
C_o	=	Concentration of oxygen in the inert, usually zero

Pressure Purging

The vessel is pressurized with inert gas and then vented usually to atmospheric pressure. The process is repeated until the oxygen concentration goes below MOC.

Vacuum Purging

The contents of a vessel is withdrawn by a vacuum system until a predetermined vacuum is reached. It is then filled up with the inert gas to the initial pressure. The process is repeated until the oxygen concentration goes below MOC.

The formulas for estimating purging cycles and amount of inert gas for pressure and vacuum purging are:

$$n = \frac{\ln (y_o / y_n)}{\ln (P_H / P_L)} \qquad (16\text{-}9)$$

$$\mathbf{MI} = \mathbf{n\ (P_H\text{-}P_L)(M)(V)/(RT)} \qquad (16\text{-}10)$$

Where:

- n = Number of purging cycles
- y_o = Initial concentration of oxygen
- y_n = Concentration of oxygen after n purge cycles
- P_H = High pressure of the cycle, psia
- P_L = Low pressure of the cycle, psia
- MI = Mass of inert gas required, lb
- M = Molecular mass of inert
- V = Volume of the vessel to be purged, cuft
- R = 10.73, (psia-ft^3/lb mole-°R)
- T = Operating Temperature, °R

(**Note:** The units of y_o and y_n are the same.)

Example 16-8. A 3,000 gallon reactor initially filled with ambient air is to handle dusty solids. It is to be inerted using nitrogen at 95°F. Calculate the mass of nitrogen required if (a) pressure purging is used with 90 psig nitrogen (b) vacuum purging is used. The vacuum system can produce a vacuum of 25 mm Hg Absolute (c) Sweep through purging is used. Assume a safe final concentration of oxygen in the reactor.

Solution.

Since the reactor handles dusty materials, assume final concentration of oxygen as 4%.

y_2 = 4%

y_o = 21%, since initially filled with ambient air

(a) Pressure Purging

$$n = \frac{\ln (21/4)}{\ln (104.7/14.7)} = \frac{1.6582}{1.9632} = 0.84$$

Say one purge.

$$MI = 1(104.7\text{-}14.7)(28)(3000) /(7.48)(10.72)(460\text{+}95)$$
$$= 169.88 \text{ lbs}$$

(b) Vacuum purging

$$P_H = 14.7 \text{ psia}$$
$$P_L = 14.7 \text{ x } 25 / 760 = 0.4836 \text{ psia}$$
$$n = \frac{\ln (21/4)}{\ln (14.7/0.4836)} = \frac{1.6582}{3.4143} = 0.84$$
$$MI = 1(14.7\text{-}.4836)(28)(3000) / (7.48)(10.72)(555)$$
$$= 26.83 \text{ lbs}$$

(c) Sweep Through Purging

$$\text{Volume of inert} = V \ln (C_1/C_2)$$
$$= (3000/7.48)\ln(21/4)$$
$$= 665.1 \text{ cuft}$$

$$\text{Mass of inert} = 665.1 \text{ X } 14.7 \text{ X } 28/(10.72)(555)$$
$$= 46.0 \text{ lbs}$$

OVERPRESSURE PROTECTION OF STORAGE VESSEL AND PROCESS EQUIPMENT

General Requirements

Although proper design, adequate control and instrumentation may safeguard a plant from predictable upset conditions, it is required by various codes and government act to protect all equipment from overpressure due to unpredictable but probable upset conditions. This is usually done by installing a rupture disc or a relief valve or their combination on the equipment to be protected. The sizing of the rupture disc or relief valve requires the estimation of the emergency relief load.

Such relief loads may be handled by one or multiple valves. The set pressure and accumulation limits of pressure relief valves depend on the contingency and number of valves installed to handle the relief load as shown below:

	SINGLE-VALVE		MULTIPLE-VALVE	
	Set Pressure %	Accumulation %	Set Pressure %	Accumulation %
Non fire only				
First valve	100	110	100	110
Additional valve(s)	-	-	105	116
Fire Only				
First valve	100	121	100	121
Additional valve(s)	-	-	105	121
Supplemental valve	-	-	110	121

All values shown above are percentages of maximum allowable working pressure.

Emergency Relief Load Calculation and Sizing Safety Device: Conditions Considered in Emergency Relief

The most common conditions considered in emergency relief load calculation are shown in Table 1. Internal explosion and dust explosion should also be considered. These are, however, beyond the scope of this chapter.

TABLE 1

Item No.	CONDITION	LIQUID RELIEF LOAD	VAPOR RELIEF LOAD
1.	Fire Heat Input		See relief load calculation due to fire
2	Failure of Utility		
	a. Cooling Water		Total incoming vapor and that generated therein.
	b. Fan-failure in air cooled exchanger		80-100% of the vapor in.
	c. Electricity	Estimate by system study.	Estimate by system study.
3.	Reflux pump failure		Total vapor to condensor.
4.	Heat exchanger tube rupture. Consider this condition if the design pressure of the high presure side is greater than 1.5 times the design pressure of the low pressure side.	Estimate flow using flow coefficient =0.7 and flow area =2x cross-section area of one tube. If this exceeds the normal flow, the latter should be used.	Same as liquid load. Considier critical flow.
5.	Overfilling of storage or surge vessel	Maximum liquid pump-in rate.	
6.	Instrument failure such as failure of control valve.	Estimate by system study. Usually determined by maximum flow of control valve.	Same remark as under "Liquid Relief Load".

7.	Operator mistake such as blocking outlet of a positive displacement machine.	Estimate by system study. Usually the normal flow from the machine.	Estimate by system study. Usually the flow from the machine.
8.	Chemical reaction.		Vapor generation due to normal and uncontrolled condition. Consider two-phase flow.
9.	trapped fluid in a heat exchanger.	Relief in gpm (thermal relief)= $= \frac{\beta Q}{500GC}$	
	Where:	β=Coefficient of thermal expansion of blocked fluid per °F =.0001 for water.	
		Q=Maximum heat exchanger duty, Btu/h.	
		G=Sp.Gr. of blocked fluid.	
		C=Sp. Heat of blocked fluid Btu/lb-°F	
10.	Water Hammer.	Equipment may not be protected by a relief device. Use slow closing valve.	
11.	Steam Hammer.		May not be protected by relief device. Use slow-closing valve, or a diffusing element.

Relief Load Calculation Due to External Fire

Relief load calculation due to external fire requires the estimation of the heat transfer area. Table 2 shows the formulae used in computing exposed areas.

To calculate the heat transfer area for fire condition, the total area calculated in Table 2 may be multiplied by a factor shown in Table 3 depending on the agency that governs the operation of the equipment. For regulatory materials (flammable material) in storage tanks, OSHA 1910.106 is mandatory. When the regulated material goes to process equipment, heat transfer area may be calculated by any of the following methods shown in Table 3 unless the insurance company dictates otherwise. Similarly, if the storage tank does not contain any regulated material but is exposable to fire, any of the methods shown in Table 3 may be used.

TABLE 2

Equipment or Equipment Part			Total Area (A_T, Ft.^{2})
1.	Sphere Shell		$A_T = \pi D^2$
2.	Cylindrical Shell		$A_T = \pi DL$
3.	Ends		
	a)	Flat	$A_T = (\pi/4) D^2$
	b)	Conical	$A_T = \left(\pi \frac{D}{2}\right)\sqrt{\left(D^2/4\right)+h^2}$
	c)	2:1 Ellipsoidal	$A_T^* = 1.19 D^2$
	d)	ASME flanged and and dished	$A_T^* = 0.918 D^2$

* Excludes area of straight flange.

Except for the sphere, calculate the total area by adding the areas of the components. Formulas for ends are for one end only.

D = Outside diameter, ft.

L = Tangent to tangent length, ft.

This length may be adjusted per item 3 of Table 3 for vertical tanks.

h = Cone height, ft.

TABLE 3

Estimation of Heat Transfer Area From Total Area

EQUIPMENT	AGENCY GOVERNING THE EQUIPMENT OPERATION		
	NFPA-30 & OSHA 1919.106	**API-520 & API-521 Operating Pressure > 15 psig**	**API-2000 Operating Pressure: 0.5 oz/in² Vacuum Through 15 psig**
1. Sphere	55% of total exposed area.	Area up to the maximum horizontal diameter or up to the height of 25 ft., whichever is greater.	As in API-520/521.
2. Horizontal Tank	75% of total exposed area. If under 200 ft.^{2}, use 100% of total exposed area.	Area equivalent to the average inventory level up to the height of 25 ft.	75% of total exposed area.
3. Vertical Tank	100% of total exposed area for the first 30 ft. Exclude bottom area if the bottom is flat and supported on ground.	Area equivalent to the average inventory level up to the height of 25 ft.	As in OSHA.
4. Process Vessel	-	Area equivalent to the average inventory level up to the height of 25 ft.	-
5. Fractionating Column	-	Area equivalent to liquid level in bottom, and reboiler if part of the column, plus liquid hold up from all trays up to a height of 25 ft.	-

(**Note:** Compressed Gas Association and Chlorine Institute consider total calculated area as the heat transfer area)

After heat transfer area is calculated using the appropriate correction factor in Table 3, the heat from fire is estimated from equations shown in Table 4, corrected by environmental factor, F, shown in Table 5. The vapor generation rate is estimated by dividing the heat load with the latent heat of vaporization. For a multicomponent mixture, an equilibrium flash calculation may be required to estimate the vapor flow and the vapor molecular mass. Tables 6, 7 and 8 are used to calculate the relief area and choose the relief device.

TABLE 4

[Using the area estimated in Tables 2 and 3, the heat load due to fire is calculated by the following equations. A_h = (A_T from Table 2) and corrected per Table 3. See Table 5 for factor, F.]

HEAT TRANSFER AREA, A_h, Ft.^{2}	**Heat Input from Fire, Q, Btu/Hr.**	
	NFPA-30 OSHA 1910.106, API-2000 (Note 5)	**API-521**
$20 < A_h \leq 200$	$Q = 20{,}000\ FA_h$	$Q = 21{,}000\ FA_h^{0.82}$
$200 < A_h \leq 1{,}000$	$Q = 199{,}300\ FA_h^{0.566}$	(Note 1)
$1{,}000 < A_h \leq 2{,}800$	$Q = 963{,}400\ FA_h^{0.338}$	$Q = 34{,}500\ FA_h^{0.82}$
$A_h > 2{,}800$	$Q = 21{,}000\ FA_h^{0.82}$ (Note 3) $= 14.09 \times 10^6 F$ (Note 4)	(Note 2)

Notes:

1) Applies to all heat transfer areas where prompt firefighting efforts and drainage facilities exist.

2) Applies to all heat transfer areas where adequate firefighting and drainage facilities do not exist.

3) For maximum allowable working pressure > 1 psig.

4) For maximum allowable working pressure ≤ 1 psig.

5) API-2000 applies to tanks designed for operating pressures from 1/2 ounce/inch^2 vacuum through 15 psig pressure.

TABLE 5

Environmental factor, F, to be used for equations in Table 4. Use one factor.

	Agency Governing The Equipment Operation			
	NFPA-30	**OSHA 1910.106**	**API-520/521**	**(Note 1) API-2000**
Insulation	0.3	0.3	0.075 x K/t (note 2)	0.075 x K/t (note 2)
Water Spray	0.3 (Note 3)	0.3	1.0	1.0
Insulation and Water Spray	0.15 (Note 3)	0.15	Not Defined	Not Defined
Tank Drainage	0.5 (Note 4)	0.5 (Note 4)	Not Defined	0.5

Notes:

1) Factors given for API-2000 are for unrefrigerated tank. See API-2000 for refrigerated tank.
2) More precise environmental factor is given by $F = k(1660-t_f)/(21000t)$
   k =Conductivity of insulation at 1600°F, Btu-inch/hr. ft^2. °F.
   t =Thickness of insulation, inch. Insulation should resist dislodgement by fire hose streams.
   t_f =Temperature of vessel content at relieving pressure, °F.
3) NFPA requires drainage of the equipment for these factors to apply.
4) For equipment with less than 200 ft^2. wetted surface use this factor as 1.

Sizing of Relief or Safety Valve

Once the relief load is determined, use Tables 6 and 7 to calculate the relief area. Then use Table 8 to select the relief valve.

For calculation of relief load due to thermal expansion of gas caused by external fire, see reference [3]

TABLE 6

Sizing Equation for Gas, Vapor or Steam

CRITICAL FLOW, GAS OR VAPOR

$$A = \frac{W}{CKP_1K_b}\sqrt{\frac{TZ}{M}}$$

A = Relief Area, $inch^2$
W = Flow, lb/h
T = Absolute temperature of inlet vapor, °R, at relief pressure
Z = Compressibility factor
M = Molecular mass of gas or vapor

$$C = 520\sqrt{n\left[\frac{2}{n+1}\right]^{\frac{n+1}{N-1}}}$$

$$n = \frac{MC_p}{MC_p - 1.99}$$

Cp = Specific heat, Btu/lb-°F (If n cannot be estimated assume C = 315)

K = Coefficient of discharge = 0.975, if unknown

P_1 = Set pressure (PSIG) x (1 + accumulation) + barometric pressure

Accumulation

= 0.10 for unfired emergency relief

= 0.21 for fire condition
= 0.33 for piping
K_b = Back pressure correction factor (see Figure 1 & 2)

SUBCRITICAL FLOW, GAS OR VAPOR

$$A = \frac{W}{735FK}\sqrt{\frac{TZ}{MP_1(P_1 - P_2)}}$$

$$F = \sqrt{\left[\frac{n}{n-1}\right]r^{\frac{2}{n}}\left[\frac{1-r^{(n-1)/n}}{1-r}\right]}$$

P_2 = Back pressure, psia.
r = P_2/P_1

Other terms are explained under critical flow formula of gas or vapor:

Note:
Balanced-bellows relief-valves that operate in the subcritical region may be sized using the critical flow formulas.

Saturated Steam, Critical Flow

$$A = \frac{W}{51.5P_1KK_n}$$

$K_n = 1$ when $P_1 \leq 1515$ psia

Other terms are explained under critical flow formula of gas or vapor.

For superheated steam, a correction factor is required; required area is higher than the corresponding area required for saturated steam.

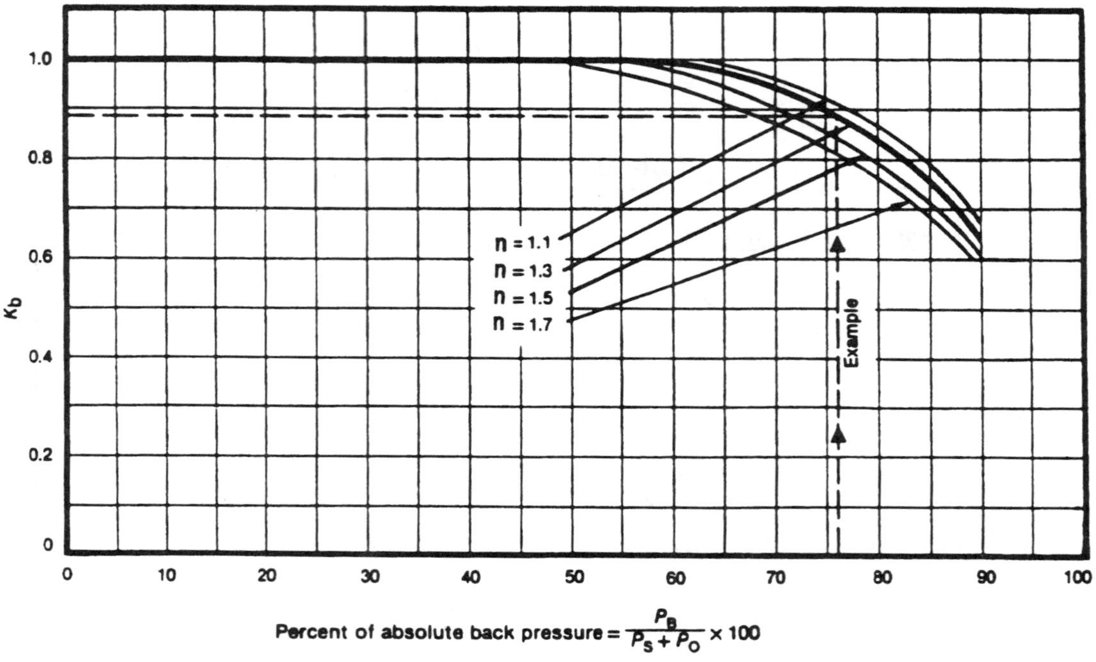

P_B = back pressure, in pounds per square inch absolute.
P_S = set pressure, in pounds per square inch absolute.
P_O = overpressure, in pounds per square inch absolute.

Set pressure (MAWP)	= 100 pounds per square inch gauge.
Overpressure	= 10 pounds per square inch gauge.
Superimposed back pressure (constant)	= 70 pounds per square inch gauge.
Spring set	= 30 pounds per square inch.
Built-up back pressure	= 10 pounds per square inch.
Percent absolute back pressure	$= \frac{(70 + 10 + 14.7)}{(100 + 10 + 14.7)} \times 100 = 76$
K_b (follow dotted line)	= 0.89

Note: This chart is typical and suitable for use only when the make of the valve or the actual critical flow pressure point for the vapor or gas is unknown; otherwise, the valve manufacturer should be consulted for specific data . This correction factor should be used only in the sizing of conventional (non-balanced) pressure relief valves that have their spring setting adjusted to compensate for the superimposed back pressure. It should not be used to size balanced-type valves.

Figure 16-1. Constant Back Pressure Sizing Factor, K_b , for Conventional Safety Relief Valves(Vapros & Gases only). Reprinted, courtesy of the American Petroleum Institute, Washington, D.C.

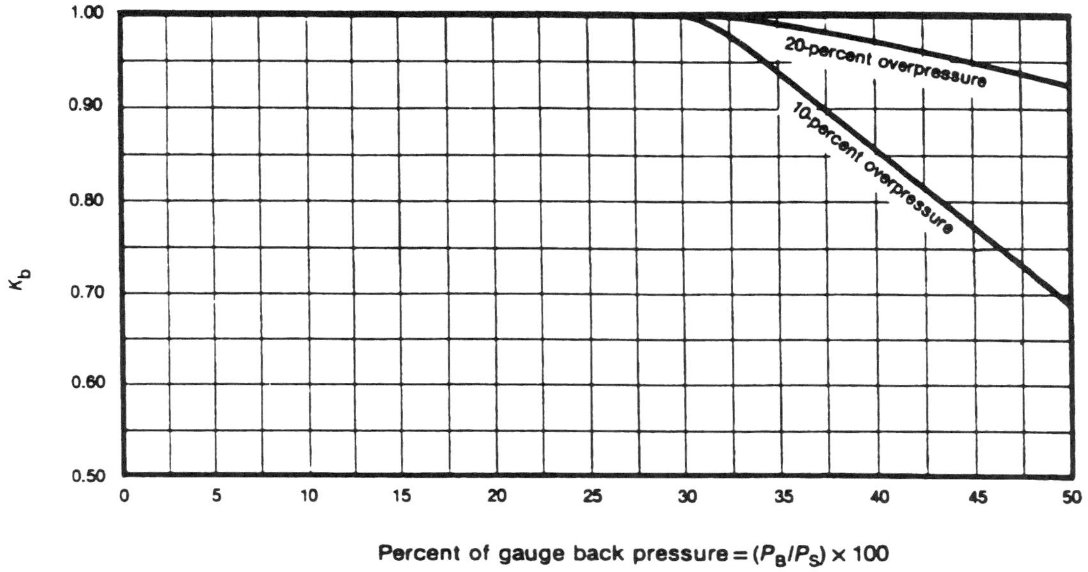

P_B = back pressure, in pounds per square inch gauge.
P_S = set pressure, in pounds per square inch gauge.

Note: The curves above represent a compromise of the values recommended by a number of relief valve manufacturers and may be used when the make of the valve or the actual critical flow pressure point for the vapor or gas is unknown. When the make is known, the manufacturer should be consulted for the correction factor. These curves are for set pressures of 50 pounds per square inch gauge and above. They are limited to back pressure below critical flow pressure for a given set pressure. For subcritical flow back pressures below 50 pounds per square inch gauge, the manufacturer must be consulted for values of K_b.

Figure 16-2. Back Pressure Sizing Factor ,K_b , for Balanced-Bellows Pressure Relief Valves (Vapors and Gases). Reprinted, courtesy of the American Petroleum Institute, Washington,D.C.

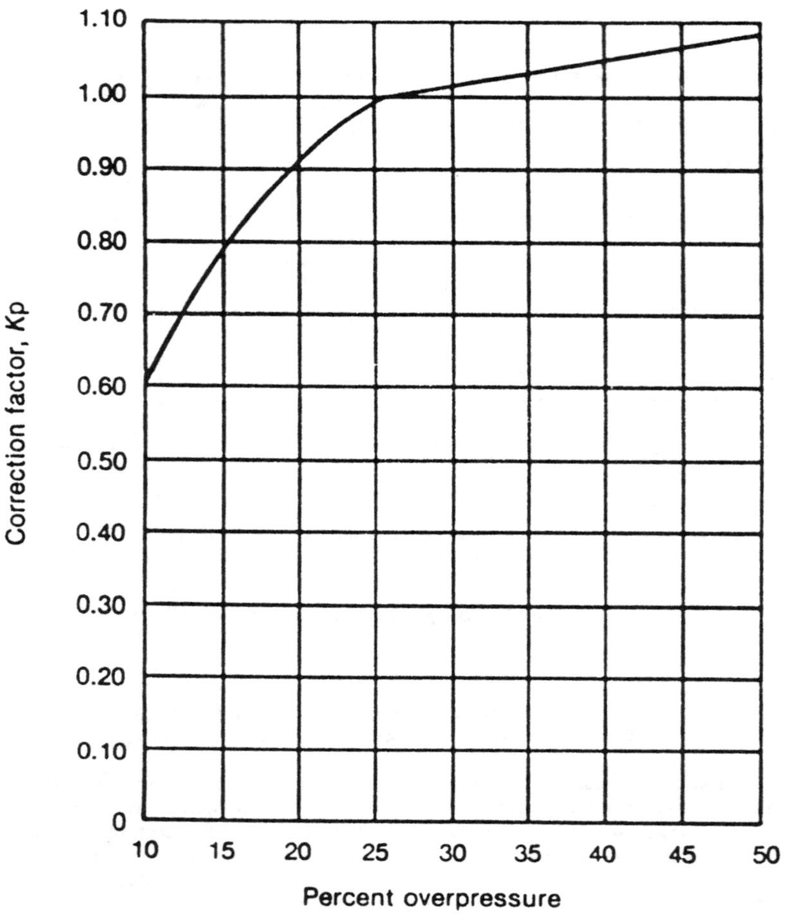

Note: The curve above shows that up to and including 25-percent overpressure, capacity is affected by the change in lift, the change in the orifice discharge coefficient, and the change in overpressure. Above 25 percent, capacity is affected only by the change in overpressure. Valves operating at low overpressures tend to chatter; therefore, overpressures of less than 10 percent should be avoided.

Figure 16-3. Capacity Correction Factors Due to Overpressure for Relief and Safety Relief Valves in Liquid Service. Reprinted, courtesy of the American Petroleum Institute, Washington,D.C.

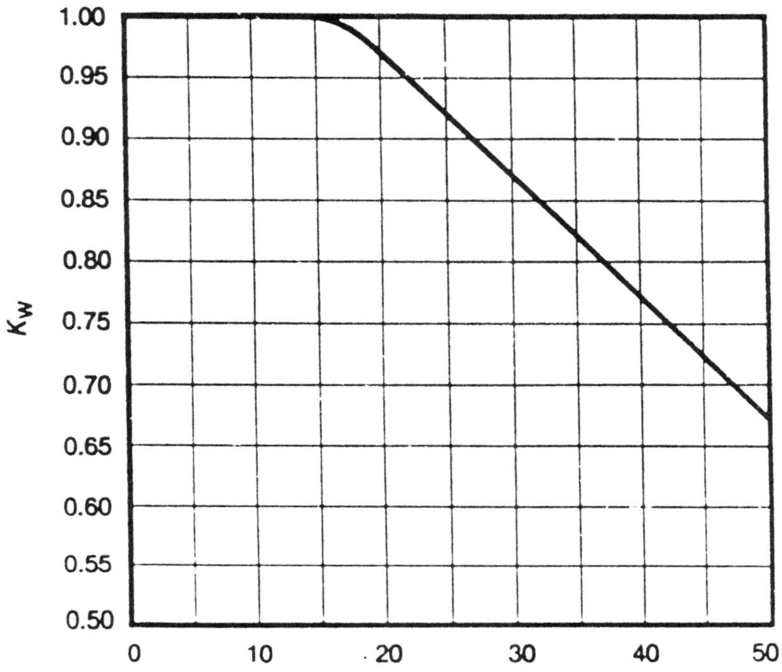

Percent of gauge back pressure = $(P_B/P_S) \times 100$

K_w = correction factor due to back pressure.
P_B = back pressure, in pounds per square inch gauge.
P_S = set pressure, in pounds per square inch gauge.

Note: The curve above represents values recommended by various manufacturers. This curve may be used when the manufacturer is not known. Otherwise, the manufacturer should be consulted for the applicable correction factor.

Figure 16-4. Variable or Constant Back-Pressure Sizing Factor, K_W, for 25 Percent Overpressure on Balanced Bellows Safety Relief Valves (Liquid Only). Reprinted, courtesy of the American Petroleum Institute, Washington,D.C.

When a relief valve is sized for viscous liquid service, it should first be sized as it was for nonviscous-type application so that a preliminary required discharge area, A, can be obtained. From manufacturers' standard orifice sizes, the next larger orifice size should be used in determining the Reynold's number, R, from either of the following relationships:

$$R = \frac{Q(2800G)}{\mu\sqrt{A}}$$

Where:

Q = flow rate at the flowing temperature, in U.S. gallons per minute.
G = specific gravity of the liquid at the flowing temperature referred to water
= 1.00 at 70°F.
μ = absolute viscosity at the flowing temperature, in centipoises.

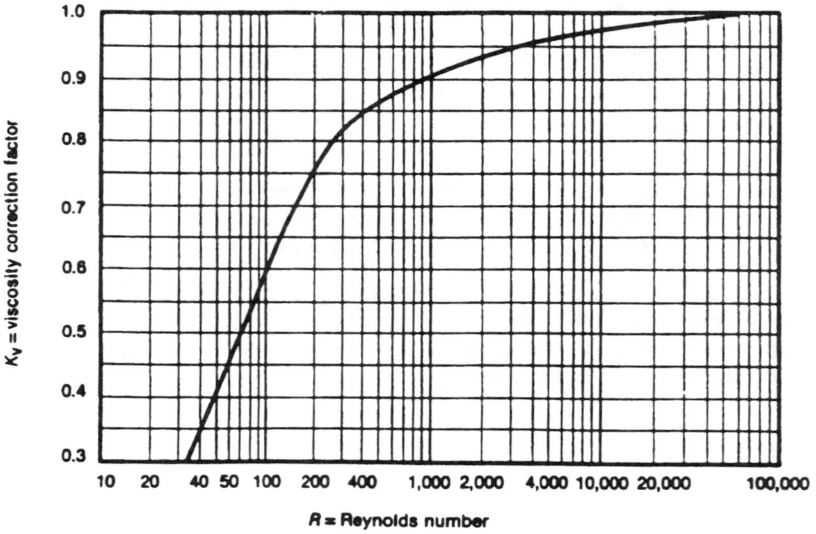

Figure 16-5. Capacity Correction Factor, K_V, Due to Viscosity. Reprinted, courtesy of the American Petroleum Institute, Washington,D.C.

TABLE 7

Sizing Equation for Liquid

1. **When capacity certification is required per ASME Section VIII, Div. I.:**

$$A = \frac{GPM}{24.7\, K_w K_v} \sqrt{\frac{G}{P_1 - P_b}}$$

2. **When Capacity Certification is Not Required:**

$$A = \frac{GPM}{23.56 K_p K_w K_v} \sqrt{\frac{G}{P_1 - P_b}}$$

A	=	Relief Area, inch^2
GPM	=	Flow in gallon/min
G	=	Sp. gr.
K_p	=	Capacity correction factor for overpressure (see Figure 3)
K_w	=	Capacity correction factor for back pressure (see Figure 4)
	=	1 for conventional valve
K_v	=	Viscosity correction factor = 1 for nonviscous fluid (see Figure 5)
P_1	=	Set pressure (psig) x (1 + accumulation) + barometric pressure (psia)
Accumulation		
	=	0.25 when certification is not required, otherwise 10%
	=	0.33 for piping
P_b	=	Back pressure, psia

When a rupture disc is used in series with the relief valve, the combination capacity is to be reduced by 10% per ASME code. In other words, the selected orifice area from Table 8 multiplied by 0.9 should be greater than or equal to the calculated area.

Pressure drop through the inlet line of the relief device should not exceed 3% of the set pressure. The outlet lines should be sized to limit the pressure drop as follows: 10% of the set pressure for the conventional safety relief valve, and 30-50% of the set pressure for balanced valves.

TABLE 8

VALVE	ORIFICE AREA IN^2	VALVE	ORIFICE AREA IN^2
1 D 2	0.110	3 L 4	2.853
1 E 2	0.196	4 M 6	3.60
1½ F 2	0.307	4 N 6	4.34
1½ G 2½	0.503	4 P 6	6.38
1½ H 3	0.785	6 Q 8	11.05
2 J 3	1.287	6 R 8	16.0
3 K 4	1.838	8 T 10	26.0

(The first number of the valve size indicates the nominal size in inch of the inlet, and the number following the letter code indicates the nominal size of the outlet of the relief valve. It is not unusual to have an outlet pressure rating lower than the inlet pressure rating of a relief valve.)

Sizing of Rupture Disc

The diameter of the rupture disc for the critical flow condition is determined by:

$$d = \left(\frac{W}{146P}\right)^{\frac{1}{2}}\left(\frac{T}{M}\right)^{\frac{1}{4}} \tag{16-11}$$

P = bursting pressure in psia. This pressure does not include any accumulation.
d = diameter of rupture disc, inch.

Alternatively, the flow area required for critical flow through a rupture disc may be computed by multiplying the area obtained from the critical flow equations in Table 6 with 1.573. This factor is obtained by replacing the discharge coefficient (0.975) with (0.62) i.e. 0.975/0.62 = 1.573. This area is further corrected for accumulation by multiplying with the ratio of flowing pressure/set pressure, both pressures being in absolute scale. See example 9. The size of the rupture disc for liquid service may be estimated using equation 2 of Table 7 with $K_p = K_w = 1$ and P_1 as the burst pressure. No accumulation is used. A 20% factor of safety should be added to the calculated rupture disc area to account for the structural support (such as vacuum support, etc.). This safety factor also applies when the disc is used in combination with a relief valve.

Example 16-9. A vertical cylindrical storage tank 15' dia x 40' high (tangent to tangent) with flanged-and-dished heads is used to store a flammable liquid of vapor molecular weight 90. Size the relief valve and rupture disc to protect the tank from overpressure due to external fire. Consider OSHA 1910.106 and API-520/521. Assume adequate fire fighting efforts and drainage facilities exist.

Data Available

Vessel diameter: 15'0" O.D.
Vessel height: 40'0" (T-T), flanged and dished ends
Normal liquid level: 35' from bottom tangent line and the bottom tangent line is 7' above grade.
Insulation: 1" (Fire proof), thermal conductivity = 4 Btu-inch/hr.ft.^{2}°F.
Relief valve set pressure: 130 psig, atmospheric discharge.

TABLE 9

Flowing Pressure, psia	152	165	172
Flow Temperature °R	450	500	550
K	312	313	315
Z	1.08	1.08	1.09
Latent heat Btu/lb.	114	112	110

Solution: [See Figure 16-6 for estimation of L]

OSHA

Exposed area:

$=A_h = \pi\ DL + 0.918\ D^2$

(from Table 2)

$=(\pi)(15)(23) + 0.918\ (15)^2$

=1,289.85 ft2

$Q = 963{,}400\ F\ A_h^{0.338}$

(from Table 4)

$= (963{,}400)(.3)(1{,}289.85)^{0.338}$

(F = 0.3 from Table 5)

$=3.253 \times 10^6$ Btu/hr

API-520/521

Exposed area:

$= \pi(15)(18) + 0.918\ (15)^2$

$=\ 1{,}054.35\ ft^2$.

F=0.075 x 4/1 = 0.3

$Q = 21{,}000\ FA_h^{0.82}$

$=21{,}000 \times 0.3 \times (1{,}054.32)^{0.82}$

$=1.898 \times 10^6$ Btu/hr.

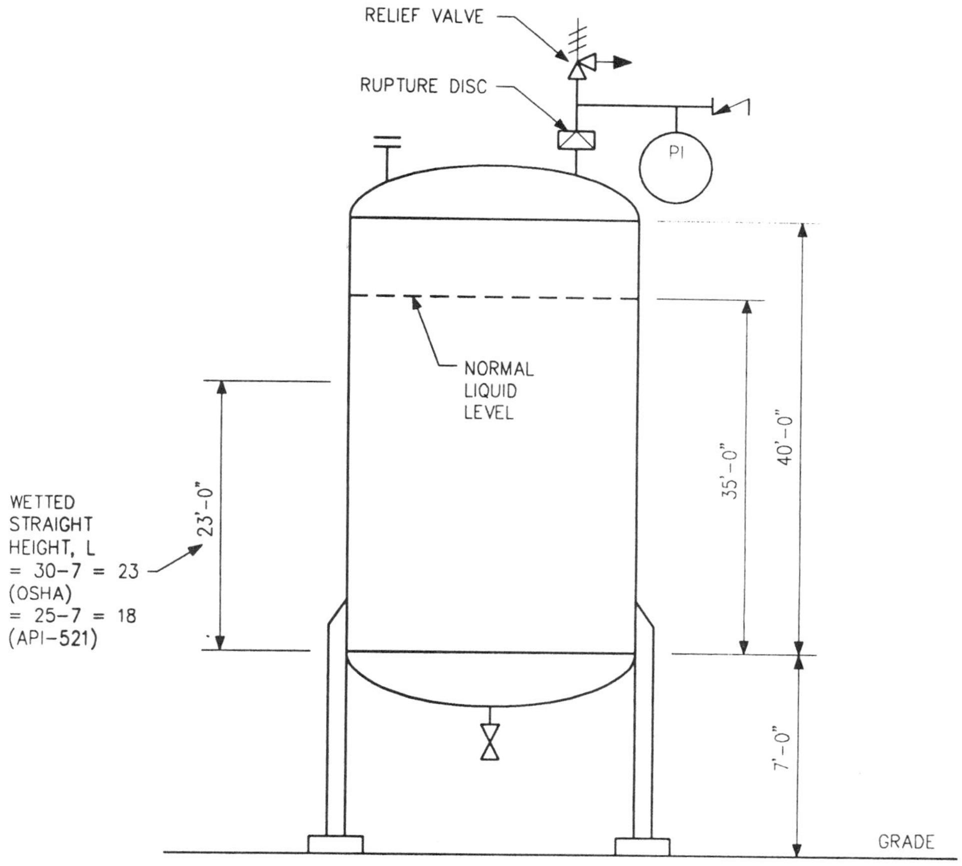

Figure 16-6. Diagram for Example 16-9

OSHA | **API-520/521**

Flowing pressure

P_1=130 (1 + .21) + 14.7 = 172 psia

Corresponding flowing temperature

=550°R and latent heat = 110 Btu/lb

OSHA:

$$W = \frac{Q}{L} = \frac{3.253 \times 10^6}{110} = 29{,}573 \text{ lb/hr}$$

$$A = \frac{W}{KCP_1K_b}\sqrt{\frac{TZ}{M}}$$

$$= \frac{29{,}573}{315 \times .975 \times 172 \times 1}\sqrt{\frac{550 \times 1.09}{90}}$$

$$= 1.445 \text{ in}^2$$

API-520/521:

$$W = \frac{1.898 \times 10^6}{110} = 17{,}250 \text{ lb/hr}$$

$$A = \frac{17{,}250}{315 \times .975 \times 172}\sqrt{\frac{550 \times 1.09}{90}}$$

$$= 0.843$$

Since the valve is used in series with the rupture disc, a 10% allowance must be made. The selection would be: 3K4(OSHA) or 2J3 (API-520/521) [See Table 8]. The rupture disc diameter (OSHA) is given by

$$d = \left[\frac{W}{146P}\right]^{1/2}\left[\frac{T}{M}\right]^{1/4} = \left[\frac{29{,}573}{146 \text{ x } 144.7}\right]^{1/2}\left[\frac{550}{90}\right]^{1/4} = 1.86 \text{ inch}$$

Flow area of rupture disc can also be obtained from the flow area of the relief valve after correcting for the discharge coefficient and pressure.

$$A = (1.445 \text{ x } 1.573) \text{ x } (172/144.7) = 2.702 \text{ in.}^2$$

Diameter of the rupture disc = $[(4/\pi) \text{ x } 2.702]^{1/2}$ = 1.85 in.

Select 3" rupture disc in series with 3K4 relief valve or 2" rupture disc in series with 2J3 relief valve since the rupture disc in series with a relief valve should not be smaller than the inlet size of the relief valve, and 10% allowance must be made for the combination capacity. If the equipment is located in an area not regulated by OSHA, a smaller safety device may be used by adopting API. [Consideration of two phase flow may require higher relief area than what is calculated above.]

Example 16-10. A pump feeds 88 GPM of a fluid at 518 psig through a shell and tube heat exchanger as shown in Figure 16-7. The process flows through the tube side and the coolant flows through the shell side. The tube side is designed for 650 psig at 250°F, and shell side is designed for 100 psig at 350°F. The tubes have inside diameter of 0.62 inch. The heat duty of the exchanger is 240,000 Btu/hr. The coefficient of thermal expansion of water is 0.0001/°F. Size the relief valve for all applicable conditions and choose the final size. Assume appropriate overpressure of the relief valve. Capacity certification of the relief valve is not required.

Solution.

Choose set pressure of relief valve at 100 psig. The relief valve will be located on the shell side.

Blocked Condition

Since cooling water line has two block valves at the inlet and outlet lines, these valves may be kept closed by mistake while hot process fluid runs through the tube side.

$$\textbf{Relief Load} = \frac{\beta Q}{500GC} = \frac{(0.0001)(240{,}000)}{(500)(1)(1)} = 0.048\text{ GPM}$$

$$\textbf{Relief Area A} = \frac{gpm\sqrt{G}}{23.56K_pK_wK_v\sqrt{(P_1 - P_b)}}\ \text{inch}^2$$

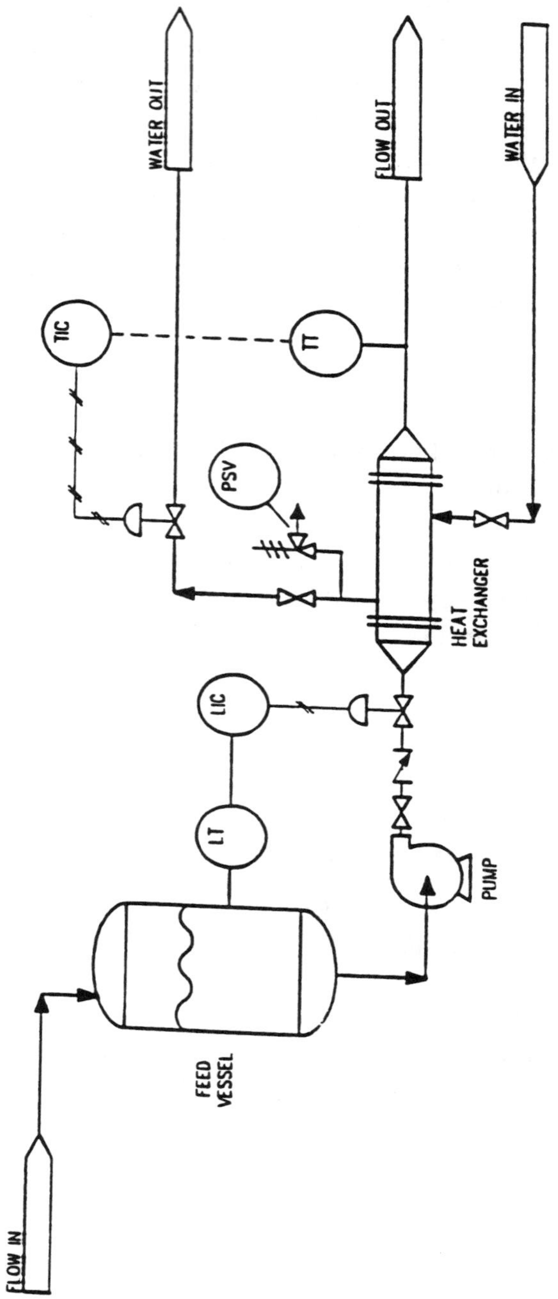

Figure 16-7. Diagram for Example 10

gpm=		relief flow gallons/min = 0.048
G	=	Sp. gr. = $\simeq$ 1
K_p	=	Capacity correction due to overpressure = 1 for 25% overpressure
K_w	=	Capacity correction factor for back pressure. Assume discharge of relief valve goes to a short distance so that back pressure is small. Kw ≈ 1
K_v	=	Viscosity correction factor = 1
P_1	=	Relief pressure = 100 psig x 1.25
P_b	=	Back pressure = 0 psig

$$A = \frac{0.048}{23.56 \times 1\sqrt{(125-0)}} = 0.0002 \text{ sq. inch}$$

Heat Exchanger Tube Rupture

P_1	=	Design pressure of the low pressure side = 100 psig
P_2	=	Design pressure of the high pressure side = 650 psig
		1.5 P_1 = 150 psig

Since $P_2 > 1.5\ P_1$, the condition of tube rupture is to be considered.

Flow through one opening of the tube $= 236 d_0^2 C_0 \sqrt{\frac{(\Delta P)}{\rho}}$

Where: d_o = 0.62 inch

C_o = 0.7

ΔP = (518 - 1.25 x 100) = (518 - 125) = 393 psi

ρ = 62.4 lb/cuft

Flow through one opening = 236 (.62) (0.7) $\sqrt{393/62.4}$ = 159.4 GPM

Since the calculated flow is greater than the normal flow of 88 GPM, the latter is to be used.

$$A = \frac{GPM\sqrt{G}}{23.56K_pK_wK_v\sqrt{P_1 - P_b}}$$

$$K_p = 1 \;,\; K_w = K_v = 1 \;,\; P_1 - P_b \;=\; 100 \times 1.25 - 0 \;=\; 125$$

$$A = \frac{88}{23.56 \times \sqrt{125}} = 0.33 \;\; in^2$$

Recommended size = 1½G2½

Sizing Relief Device for Two-Phase Flow During Runaway Reaction

The sizing of relief device for two-phase flow during runaway reaction requires three steps.

Estimation of the Rate of Heat Release Per Unit Mass, q

$$q = \frac{1}{2}C_v\left[\left(\frac{dT}{dt}\right)_s + \left(\frac{dT}{dt}\right)_m\right] \tag{16-12}$$

Where:

q = rate of heat release per unit mass, J/kg·s

C_v = liquid heat capacity per unit mass at constant volume, J/kg·K

$\left(\frac{dT}{dt}\right)_s$ = rate of temperature rise at the set pressure, K / s

$\left(\frac{dT}{dt}\right)_m$ = rate of temperature rise at the relieving pressure, K / s

(Note: The last two parameters are determined experimentally using the vent sizing package).

(Note: The last two parameters are determined experimentally using the vent sizing package).

Estimation of the Required Mass Flux, G_T

$$G_T = \frac{0.9\psi \Delta H_v}{V_{fg}} \sqrt{\frac{g_c}{C_p T_s}} \qquad (16\text{-}13)$$

Where:

G_T = mass flux, kg/m²·s

ψ = dimensionless factor dependent on length/diameter ratio of the pipe to relieve the pressure through the relief device. This is shown in Figure 8.

ΔH_v = Latent heat of vaporization, J/kg

g_c = Newton's law conversion factor, 1 kg·m/s²·N

v_{fg} = V_g - V_f, m³/kg

v_g = Vapor specific volume, m³/kg

v_f = Liquid specific volume, m³/kg

C_p = Specific heat capacity of liquid, J/kg·K (Note: 1 J = N·m)

T_s = Temperature corresponding to set pressure, K

The Required Vent Area is Then Computed by the Leung Equation:

$$A = \frac{m_0 q}{G_T \left[\sqrt{\frac{V}{m_0} \times \frac{\Delta H_v}{V_{fg}}} + \sqrt{C_v \Delta T} \right]^2} \qquad (16\text{-}14)$$

A = Vent area, m²

m_o = Mass within protected vessel before relief, kg

V = Volume of the vessel, m³

ΔT = Temperature corresponding to relieving pressure - temperature corresponding to set pressure, K

Example 16-11. A 3,650 gallon stirred tank herbicide reactor undergoes run-away reaction on loss of cooling medium. The reactor is to be protected by a rupture disc with a set pressure of 100 psig and a relief pressure of 121 psig. The equivalent length to diameter ratio of the discharge pipe is 175. Determine the diameter of the rupture disc. The following information is available.

Data

Volume (V): 3,650 gallon = 13.82 m^3
Reaction mass (m_o): 9,951 kg
Set temperature (T_s): 421.7 K

Data from VSP

Maximum temperature (T_m): 430.34 K
$(dT/dt)_s$: 0.876 K/s
$(dT/dt)_m$: 1.893 K/s

Physical Property Data

	100 psig	121 psig
V_f (liquid) m^3/kg	0.001	0.0012
V_g(vapor) m^3/kg	0.07454	0.0654
C_p (liquid) kJ/kg K	1.8634	1.9152
ΔH_v kJ/kg	392.82	380.80

Solution.

Step 1: Estimate rate of heat release per unit mass.

$$q = (1/2)\, C_v \left[(dT/dt)_s + (dT/dt)_m\right]$$

Assuming $C_v = C_p$

$$\begin{aligned} q &= (1/2)(1.8634)[0.876 + 1.893] \\ &= 2.5798 \text{ kJ/kg·s} \\ &= 2579.8 \text{ J/kg·s} \end{aligned}$$

Step 2: Estimate the required mass flux

$$G_T = 0.9\psi \frac{\Delta H_v}{V_{fg}} \sqrt{\frac{g_c}{C_p T_s}}$$

Assume: L/D of discharge pipe = 175, ψ = 0.7 from Figure 8.

$$G_T = 0.9\,(0.7) \frac{392820}{(0.07454\ -\ 0.001\ )} \sqrt{\frac{1}{1863.4 \times 421.7}}$$

$$= 3796.257\ \ \mathrm{kg/m^2 \bullet s}$$

<u>Step 3</u>: Estimation of vent area and hence relief diameter.

$$A = \frac{m_0 q}{G_T \left[\sqrt{\frac{V}{m_0} \times \frac{\Delta H_v}{V_{fg}}} + \sqrt{C_v \Delta T} \right]^2}$$

$$V_{fg} = V_g - V_f = 0.07454 - 0.001 = 0.07354\ \mathrm{m^3}/kg$$

$$\Delta T = T_m - T_s = 430.34 - 421.7 = 8.64K$$

$$A = \frac{(9951)(2579.8)}{3{,}796.257 \left[\sqrt{\frac{13.82}{9{,}951} \times \frac{392820}{0.07354}} + \sqrt{1863.4 \times 8.64} \right]^2}$$

$$= 0.149\ \mathrm{m^2}$$

Allowing 20%,

$$d = (4 \text{ x } .149 \text{ x } 1.2/3.14)^{0.5} = 0.4773\text{m}$$

Venting Capacities of Atmospheric Tanks

In addition to the emergency relief requirements, the flammable liquid storage tanks need to be equipped with a vent to allow the tank to breathe during filling and emptying operations. Another purpose of venting the tank is to normalize internal tank pressures caused by expansion and contraction of the liquid in the tank due to temperature changes (thermal changes). API-RP2000 provides

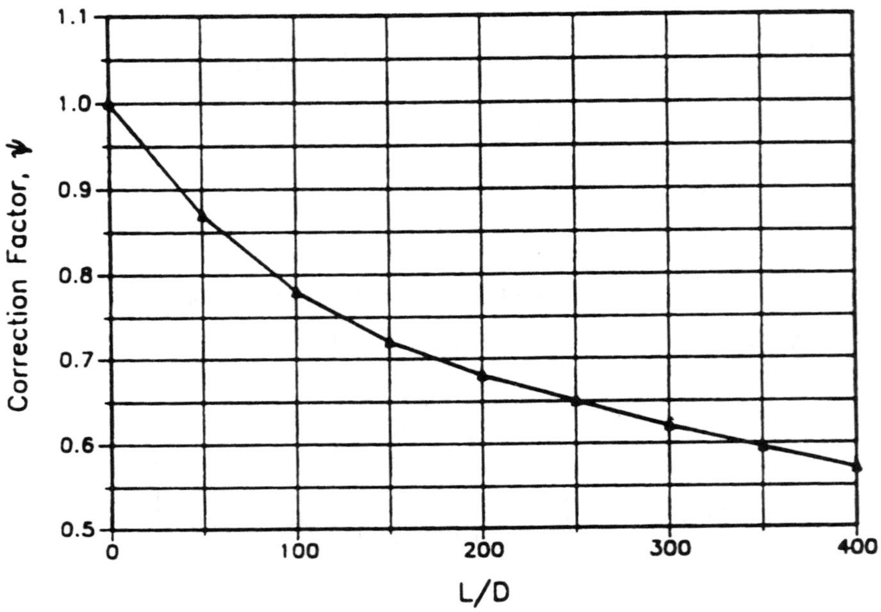

Figure 16-8. Correction Factor, ψ, For Two-Phase Flashing Flow Corresponding to Equivalent Length To Diameter Ratio of Pipe. Crowl/Louvar, CHEMICAL PROCESS SAFETY: Fundamentals with Applications, (c) 1989,p278.Capacity Correction Factor, K_V, Due to Viscosity. Reprinted, courtesy of Prentice Hall, Upper Saddle River, NJ.

guidelines for calculating the venting capacities to be provided in case of both thermal and liquid movement situations. For this purpose, the flammable liquids are classified into two groups viz. (1) liquids with flash points below 100°F and (2) liquids having flash points above 100°F. API guideline provides the requirements of venting for normal breathing which includes both thermal venting as well as venting required during emptying or filling of the vessel.

Thermal Venting

The venting capacities required during breathing-in and breathing-out due to temperature changes (thermal venting) are given in Table 10.

Venting Requirements Due to Filling and Emptying

The venting capacities required in each case are as follows:
Liquids with flash points below 100°F.

(1) Out-breathing (pressure) capacity- 1 ft^3/h of air for each 3.5 gallons per hour of tank filling rate.

(2) In-breathing (vacuum) capacity - 1 ft^3/h of air for each 7.5 gallons per hour of tank emptying rate.

Liquids with flash points above 100°F.

(1) Out-breathing (pressure) capacity - 1 ft^3/h of air for each 7 gallons per hour of the tank filling rate

(2) In-breathing (vacuum) capacity - 1 ft^3/h for each 7.5 gallons per hour of the emptying rate.

Normal breathing rate is the sum of the thermal venting rate as given in Table 10 plus that calculated due to either filling or emptying the tank.

TABLE 10

THERMAL VENTING CAPACITIES (ft^3/h of air) AS A FUNCTION OF TANK CAPACITY

Tank Capacity		**Vacuum**	**Pressure**	
gallons	42 gal. barrels	all stocks	flash point below 100°F	flash point above 100°F
2,500	60	75	75	-
5,000	119	125	125	60
10,000	238	250	250	137
15,000	357	350	350	200
20,000	476	480	480	270
25,000	595	600	600	350
30,000	714	730	730	425
35,000	833	850	850	500
42,000	1,000	1,000	1,000	600
84,000	2,000	2,000	2,000	1,200
126,000	3,000	3,000	3,000	1,800

168,000	4,000	4,000	4,000	2,400
210,000	5,000	5,000	5,000	3,000
420,000	10,000	10,000	10,000	6,000
630,000	15,000	15,000	15,000	9,000
840,000	20,000	20,000	20,000	12,000
1,050,000	25,000	24,000	24,000	15,000

* Do not interpolate figures. Take larger size.

Example 16-12. Calculate the normal venting capacity for a 65,000 gallon vertical cone-roofed tank holding naphthalene (flash point 175°F) with 150 gpm filling rate and 400 gpm pump-out rate.

Solution.

<u>Pressure Venting</u>

Required venting for normal breathing $= \dfrac{150 \text{x} 60}{7} + 1200 = 2486\ \text{ft}^3/\text{h}^*$

<u>Vacuum Venting</u>

Requirements for normal breathing $= \dfrac{400 \text{x} 60}{7.5} + 2000 = 5200\ \text{ft}^3/\text{h}^*$

* Note that the thermal venting value is not obtained by interpolation from Table 10, but the value for the next larger size tank is used.

When evaporation losses are not a factor, vents without pressure and vacuum relief valves are used. These vents are often referred to as non-conservation or free venting type vents. Pipe sizes should be selected to provide required pressure and vacuum relief under operating and emergency conditions. If it is necessary or desirable to reduce evaporation losses of the stored liquid, vents with pressure and vacuum relief valves are provided. These are called conservation vents. Flame arresters should be provided on storage tanks under the following conditions:

(1) On tanks containing liquids with flash points below 110°F.

(2) Tanks containing liquids with flash points above 110°F if exposed to combustible occupancy, construction or other tanks which contain liquids having flash points below 110°F.

(3) Tanks where the bulk of the contents can be heated to the flash point under normal operating conditions.

REFERENCES

1. Daniel A. Crowl and Joseph F. Louvar, *Chemical Process Safety: Fundamentals with Applications*, Prentice Hall, Englewood Cliffs, N.J.

2. *Sizing, Selection, and Installation of Pressure-Relieving Devices in Refineries, Part I* - Sizing and Selection. API Recommended Practice 520. Fifth edition, July 1990, American Petroleum Institute, Washington D.C.

3. *Guide For Pressure-Relieving and Depressuring Systems.* API Recommended Practice 521, third edition, 1991, American Petroleum Institute Washington D.C.

4. R. A. Crozier, Jr. *Sizing Relief Valves for Fire Emergencies*, Chemical Engineering, October 28, 1985.

5. *Venting Atmospheric and Low-Pressure Storage tanks*, (Non-Refrigerated and Refrigerated), API-Standard 2000, Second Edition, American Petroleum Institute, Washington D.C.

6. J.C. Leung, *Simplified Vent Sizing Equations for Emergency Relief Requirements in Reactors and Storage Vessels*, AIChE Journal, Vol 32, No 10, p1622, 1986.

Chapter

17

Biochemical Engineering

In last several years, principles and techniques of chemical engineering and biochemistry have been applied in the design, operation, and analysis of processes for the production of a variety of materials such as antibiotics and industrial organic chemicals using living microorganisms as catalysts. This has given rise to a special field known as biochemical engineering. It involves the application of the principles of biochemistry, cell biology, and chemical engineering for successful design of biochemical process plants. This chapter covers some basic aspects of biochemical engineering. More details can be found in the references[1,2,3] provided at the end of this chapter.

MICROBIOLOGY AND STRUCTURE OF CELLS:

Microbiology is the study of living organisms which cannot be seen by a naked eye. Generally most microorganisms have a diameter of 1 μm or less and yet have their own most sophisticated mechanisms including information and control systems. The cell theory first proposed by Schleiden and Schwann states that all living systems consist of cells and their products. The various types of cells are classified into two groups based on their cellular structures : procaryotes and eucaryotes.

PROCARYOTES : These are single-cell organisms and do not contain a nucleus enclosed within a membrane. Their shapes may be spherical, rod-like or spiral. Their dimensions may range from 0.5 to 3 μm. Individual cell structure is as follows : The cell is surrounded by a rigid wall which gives it its structural strength. Inside this wall is the thinner cell membrane called a plasma membrane which controls the rate of mass transfer between the interior of the cell and its surroundings.

Nuclear zone within the cell is the control center for the cell operations. There are some grainlike dark spots called ribosomes in the cell which are the sites of biochemical reactions. The fluid occupying much of the remaining space in the cell is called cytoplasm. The interior of the cell has also regions which appear like bubbles. These are termed storage granules. Inspite of having common structural and biochemical features, these cells do possess diverse properties. For example, some are capable of photosynthesis whereas the others are not.

EUCARYOTES: These are also single cell organisms which, unlike procaryotes, possess a nucleus enclosed within a membrane. They are considerably larger (1,000 to 10,000 times) in size than the procaryotes. Their internal structure is more complex. The interior region of the cell is divided into various compartments, each of which has a special structure and functions in the cell activity.

Eucaryotic cell is surrounded by a wall which may differ in thickness from cell to cell. Next to it is plasma membrane as found in procaryotes. A membrane system termed endoplasmic reticulum starts from the cell membrane and terminates at the nucleus membrane of the cell. Ribosomes, the sites of biochemical reactions, are situated on the surface of the endoplasmic reticulum. The nucleus is also surrounded by various other unit membranes and controls not only the catalytic activity at the ribosomes but also the nature of the reactions that can take place. There are several other unit membrane regions around the nucleus termed organelles. They consist of mitochondria which utilize oxygen in energy generation. In phototropic cells, the organelle chloroplast uses light as the energy source. The other organelles in the cell, viz. golgi complex, lysosomes and vacuoles serve in general to isolate chemical reactions or compounds from the cytoplasm.

CLASSIFICATION OF MICROORGANISMS : All unicellular organisms and all other organisms which contain multiple cells of the same type are termed as protists. On the other hand, the plants and animals are characterized by possesion of a diversity of cells. Protists exhibit different characteristics in regard to their energy and nutritional needs, method of reproduction, growth and product release rates as well as capability, and means of motion. Based on these differences, biologists have classified them into procaryotes and eucaryotes. These groups are further divided into subgroups as follows:

Procaryotes: Bacteria and Blue-green algae
Eucaryotes: Fungi, Algae and Protozoa

The fungi are further classified into the subgroups of molds and yeasts.

BACTERIA: Bacteria are single cell organisms which exist in three basic shapes: spiral, spherical, and rodlike. Most of them cannot use light energy but are capable of motion and reproduce by division into two daughter cells. They may be gram-positive or gram-negative. This classification depends upon their response to the staining test conducted with the dye crystal violet and iodine solution.

A major property of bacteria is their ability to form endospores which is a dormant form of the cell capable of resisting heat, radiation, and harmful chemicals. They revert to normal active vegetative state when returned to an environment favorable to cell functioning. Spore-forming bacteria are either aerobic or anaerobic. The former multiply in oxygen environment while the latter grow in oxygen-free surroundings. The property of endospore formation is of commercial importance since it allows to isolate and store a desired bacterial species.

YEASTS : These are an important subgroup of fungi. They are incapable of extracting energy from sunlight and live in areas of low relative humidity. The dimensions of these single cells range from 5 to 30 μm in length and 1 to 5 μm in width. Yeasts multiply both by fission and by sexual fusion. They are commercially important because of their extensive utilization in the production of beer, wine and industrial ethyl alcohol and glycerol. They are also used in baking and as protein supplement in animal feed.

MOLDS: These are also fungi with vegetative structure called mycelium consisting of branchlike enclosed tubes. Inside the tubes is the moving cytoplasm containing many nuclei. The long thin filaments of cells in the mycellium are called hyphae. Molds do not contain chlorophyll, are generally non-motile and require oxygen for normal function. They reproduce either by fission or sexually through formation of spores. Most important molds, aspergillus and penicilium have been used for the commercial production of antibiotics, organic acids, and biological catalysts. Actinomycetes group of microorganisms are classified as bacteria. They resemble fungi because they also have a structure consisting of hyphae.

ALGAE AND PROTOZOA: These are relatively large eucaryotes with sophisticated and organized structure. Most algae require energy in the form of light and are motile. They are used as foodstuffs.
Protozoa cannot use the energy of sunlight. They play an important part in biological waste treatment.

VIRUSES: Viruses are the smallest microorganisms. They are parasites of plants and animals. They contain no water and have no metabolic activity of their own. They multiply by directing the host cells to synthesize a new virus which results in the damage to and death of the host cells. Bacteriophages are viruses that are parasitic on bacteria.

ANIMAL AND PLANT CELLS: Biologists have also developed methods to produce useful chemicals employing animal and plant cells. Growth of the cells can be achieved by attachment on to a solid surface or in some cases by culture growth in a liquid medium.

GROWTH REQUIREMENTS OF CELLS

Growth of micro-organisms consumes energy which is supplied either through oxidation or by light. The cells require nitrogen, carbon, and other minerals for their growth. These are made available by various chemicals of life such as fatty acids, lipids, sugars, polysaccharides and monosaccharides such as glucose and cellulose. DNA (dioxyribonucleic acid) and RNA (ribonucleic acid) are important biopolymers. DNA stores information for the synthesis of RNA, which in turn coordinates the synthesis of other protein molecules. Details of the chemistry and functions of these and other chemicals such as aminoacids and polypeptides are available elsewhere.[4] Of more importance from the engineering point of view is the catalysis of chemical reactions within the cell and the application of chemical kinetic principles to model these biochemical catalytic reactions. This aspect is discussed a little later in this chapter.

CELL METABOLISM:

Functions of a living cell are varied. There are several reactions taking place within the confines of a cell and therefore it can be compared to a chemical reactor. Although a living organism is involved in the reaction processes within the cell, the familiar laws of material and energy balances as well as thermodynamic principles hold true in the case of the biological systems also. The processes that take place in the cell overall have been given the name "cell metabolism". The processes occurring in the cell are very complex and yet there is some order to them. It has been found by the microbiologists that the metabolic reactions in a cell are organized into sequences called metabolic pathways. There is a connection between branching pathways which join the reaction sequences and pathways which feed on themselves.

The activities within the cell can take place either in presence or absence of oxygen. These two categories of cell function are called aerobic and anaerobic respectively as mentioned earlier. A third class of cells grows in either environment, of which yeast is an example. These are called facultative anaerobics. Microbes use two kinds of energy: light and chemical. Those that utilize light as the source of energy are called phototrophs and those employing chemical energy are known as chemotrophs. The chemotrophs are further divided into two groups: Lithotrophs which oxidize inorganic chemicals and organotrophs utilize organics for the production of energy. Cell metabolism requires the use of elements such as carbon, nitrogen, sulfur and phosphorus. Also needed are hydrogen and oxygen. Typical sources of these elements are listed in Table 17-1.

The cell extracts its metabolic energy from the free energy (G) released in a chemical reaction. The non-utilized energy is lost in the form of entropy and is given by the equation:

$$G = H - T S \tag{17-1}$$

Table 17-1
Sources of elements for cell growth

Element	Source
Carbon	CO_2 , sugars, proteins, fats
Nitrogen	Proteins, NH_3 , NO_3^-
Sulfur	Proteins, SO^{-2}_4
Phosphorus	PO_4^{-3}

and the free energy change of the chemical reaction is given by:

$$\Delta G = -RT \ln K_e \tag{17-2}$$

where R is the gas constant, T is temperature, K, and K_e is the equilibrium constant (cf equation 13-4). For the reaction to proceed ΔG should be negative. Cells generate and use energy in quanta by removal of hydrogen or electrons from a substrate through the formation of organic compounds with high energy bonds. These compounds contain phosphate or sulfur as indicated in Figure 17-1.

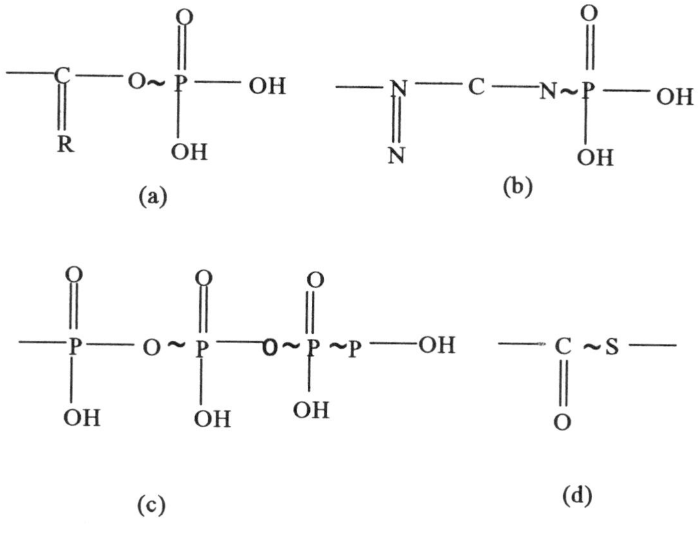

Figure 17-1: High energy bonds (~) in phosphate and sulfur compounds

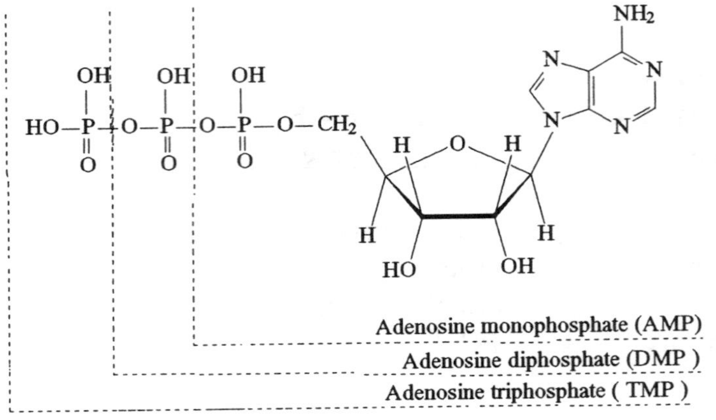

Figure 17-2 : Structures of Adenosine phosphates

ATP (adenosine triphosphate) which has the structure shown in Figure 17-2 is one of the most important high energy compounds in the metabolism of the cell. P-O-P bonds in this compound are in a high energy state in an aqueous medium. This energy is released in the following manner:

$$ATP + H_2O \rightarrow ADP + P \qquad \Delta G = -7.3 \text{ kcal/mol}$$

$$ADP + H_2O \rightarrow AMP + P \qquad \Delta G = -7.3 \text{ kcal/mol}$$

$$AMP + H_2O \rightarrow A + P \qquad \Delta G = -3.4 \text{ kcal/mol}$$

ATP is generated from ADP (adenosine diphosphate) by phosphorylation, the energy for which is provided by fermentation, photosynthesis or respiration.The oxidative phosphorylation occurs during photosynthesis and respiration. In the substrate level of phosphorylation, the intermediates of higher energy level than ATP are formed. These in turn transfer their phosphate groups to ADP to generate ATP. Thus, for example, phosphoenolpyruvate which has the structure:

CH_2=CCOOH
|
OPO_3H

undergoes the following transformation:

$$\begin{array}{ll} H_2C = \underset{\displaystyle COOH}{\underset{|}{C}}\text{-}OPO_3H_2 + ADP \rightarrow & H_3C - \underset{\displaystyle COOH}{\underset{|}{C}}=O + ATP \end{array}$$

FERMENTATION: THIS process involves biochemical breakdown (catabolism) of organic compounds which serve as either electron donors or acceptors. Examples of fermentation are the decompositions of glucose to produce ethyl alcohol and acetic acid. Lactic acid, butanol and acetone have also been obtained as products of fermentation.

RESPIRATION : This is a mechanism of ATP generation in which the electron donors are either organic or inorganic compounds (such as H_2S) with usually oxygen as an electron acceptor. When oxygen is the electron acceptor, the process is called to be aerobic. It is termed anaerobic if the electron acceptor is other than oxygen e.g. nitrate or sulfate. In respiration, hydrogen atoms are transferred from the fuel to oxygen, with simultaneous release of energy.

PHOTOSYNTHESIS : This complex process uses light energy from the sun, provides the cells with organic molecules for its reactions and releases oxygen into the atmosphere. It is the reverse of respiration. It plays an important role in closing the carbon and oxygen cycles by reducing the carbon oxidized by respiration process. Overall photosynthetic reaction is represented by the following equation

$$6CO_2 + 6H_2O + light \rightarrow C_6H_{12}O_6 + 6O_2$$

Examples of photosynthetic reactions are:

$$2H_2S + CO \xrightarrow{light} CH_2O + O_2$$

CH_2O represents the carbohydrate formed in the reaction. Also,

$$2CH_3CHOHCH_3 + CO_2 \xrightarrow{light} CH_2O + 2CH_3COCH_3 + H_2O$$

The light is absorbed by chlorophylls and ATP is generated by photophosphorylation. Bacteria do not use water as an H donor but utilize organic substrates, reduced sulfur compounds or molecular hydrogen. Glucose breakdown in various ways is central to a cell's metabolism. The breakdown occurs in a sequence of reactions called a biochemical pathway. There are three pathways for the breakdown of glucose in cells. These are:

- Embden-Meyerhoff (EM pathway)
- Entner-Douderoff (ED pathway)
- Hexose monophosphate (HMP pathway)

The cells get energy and carbon skeletons for their growth and reproduction by a series of chemical reactions catalyzed by enzymes through such pathways. The simplest of these, the Embden-Meyerhoff or EM pathway is shown in Figure 17-3. In this pathway, the following sequence of reactions takes place:

(1) ATP reacts with D-glucose in presence of Mg^+ to give ADP D-glucose 6-phosphate with ΔG = - 4.0 kcal/mol. Hexokinose is the enzyme used by most cells. This enzyme acts also on hexoses and their derivatives. The enzyme glucokinase is used by some cells but phosphorylates only D-glucose and unlike hexokinase is not inhibited by glucose 6-phosphate. However, both require divalent Mg^+ or Mn^+ which combine with ATP to form the true substrate.

(2) Next regeneration of D-glucose occurs by the action of enzyme glucose 6-phosphatase. In this reaction step, D-glucose 6-phosphate combines with water to form D-glucose and phosphate with Δ G = - 3.3 kcal/mol.

(3) In the third step, glucose phosphate isomerase isomerises glucose 6-phosphate to fructose 6-phosphate with ΔG =+0.4 kcal/mol.

(4) In this step, the enzyme 6-phospho-fructokinase catalyzes the reaction of ATP with D-fructose - 6-phosphate to produce ADP and D-fructose 1,6-diphosphate accompanied by ΔG = -3.4 kcal/mol.

(5) Next breakdown of fructose 1.6-diphosphate to dihydroxyacetone phosphate and glyceral- dehyde 3-phosphate takes place by the catalytic action of the enzyme fructose diphosphate aldolase with Δ G = +5.73 kcal/mol. The reaction is reversible and results in the formation of dihydroxyacetone phosphate and

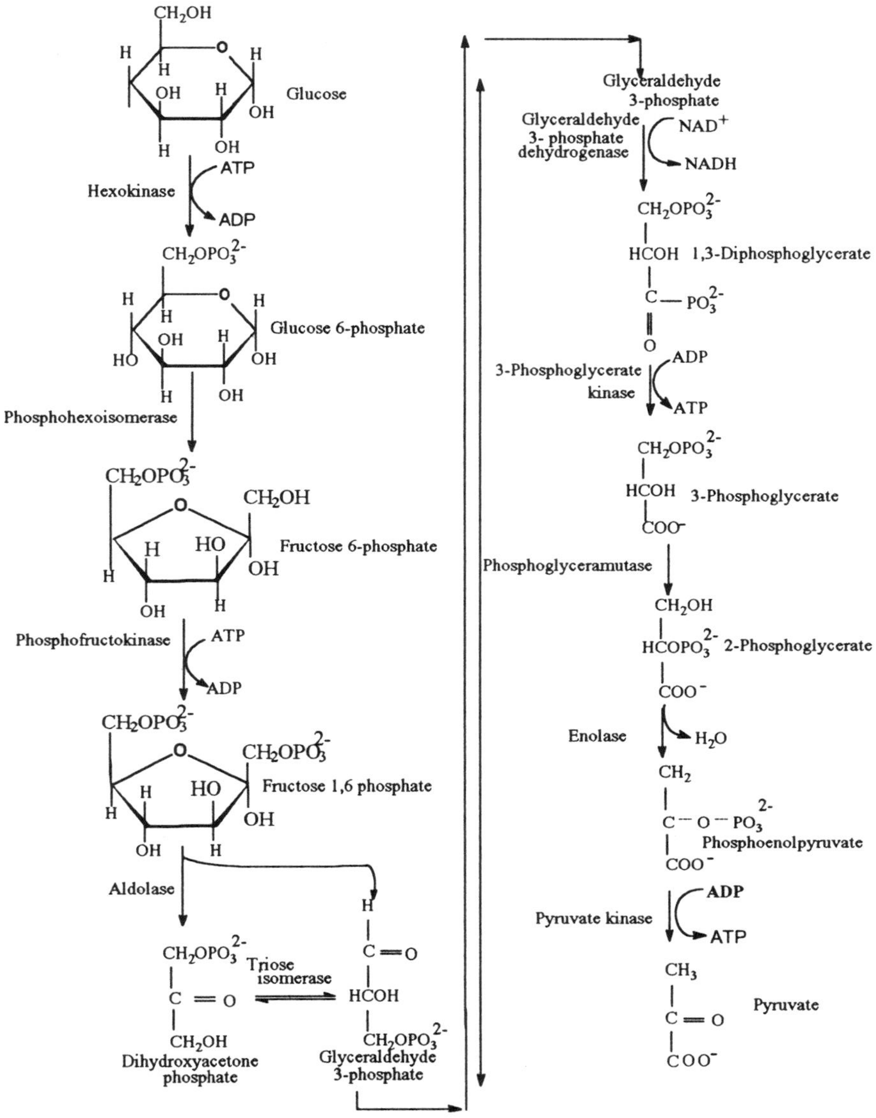

Figure 17-3 : Embden-meyerhoff (EM) pathway

D-glyceraldehyde 3-phosphate. Interconversion of the above triose phosphates is effected through catalysis by the enzymetriose-phosphate isomerase with $\Delta G = +1.83$ kcal/mol. This interconversion completes the first stage of glycolysis.

(6) The second stage of glycolysis consists of oxidoreduction as well as phosphorylation by which ATP is generated from ADP. The enzyme glyceraldehydephosphate dehydrogenase catalyzes the reaction :

$$\text{D-3-phosphoglyceraldehyde} + NAD^+ + P \rightarrow \text{D-3-phosphoglyceroyl phosphate} + NADH + H$$

with $\Delta G = +1.5$ kcal/mol. NAD^+ and NADH are the oxidized and reduced forms of nicotinamide adenine dinucleotide respectively.

(7) Next the enzyme phosphoglycerate kinase catalyzes the reaction between 3-phosphoglyceroyl phosphate and ADP to produce ATP and 3-phosphoglycerate with a $\Delta G = -4.5$ kcal/mol.

(8) In this step, 3-phosphoglycerate is converted to 2-phosphoglycerate. This reaction is reversible and the free energy of reaction ΔG is $+ 1.06$ kcal/mol.

(9) 2-phosphoglycerate is converted into phosphoenolpyruvate and water by the catalytic action of the enzyme enolase with a $\Delta G = + 0.44$ kcal/mol.

(10) In the next step, phosphoenolpyruvate and ADP react by the catalytic action of the enzyme pyruvate kinase with $\Delta G = -7.5$ kcal/mol.

(11) Finally the reduction of pyruvate to lactate takes place with the transfer of electrons originally donated by 3-phosphoglyceraldehyde and carried by NADH. The reaction is catalyzed by lactate dehydrogenase and can be represented as follows:

$$\text{pyruvate} + NADH + H^+ \rightarrow \text{lactate} + NAD^+ \qquad \Delta G = -6.0 \text{ kcal/mol}$$

Lactate, the end product of glycolysis under anaerobic conditions, diffuses through the cell membrane as a waste. The overall reaction of this pathway is given by :

$$\text{glucose} + 2P_i + 2\,ADP \rightarrow 2 \text{ lactate} + 2ATP + 2H_2O$$

Thus glucose is converted to two lactate molecules, two ATP molecules and two water molecules. For additional details of this pathway and others such as ED and hexose monophosphate pathways, reference[4] should be consulted.

ENZYME AND CHEMICAL REACTIONS IN THE CELL:

Enzymes are globular proteins which catalyze chemical reactions in the cell without undergoing a permanent change in themselves. They influence the rate of reaction but do not affect the reaction equilibria. Unlike chemical catalysts, they are generally specific in catalyzing only one reaction in conjunction with certain substrates. Enzymes also often need assistance from other nonprotein compounds termed cofactors, which combine with the inactive protein (apoenzyme) to yield an active complex. Cofactors can be either metal ions such as ca, Mn, Zn, Na or coenzymes. Similar to synthetic catalysts, the enzymetic ones can loose activity. Enzyme activity is measured in terms of a turnover number which is defined as the number of substrate molecules per catalyst site per unit time. The enzyme catalysts lose activity with increase in temperature and are useful only at or very near the temperatures at which they are found.

A simple Mathematical kinetic model for the enzyme catalyzed reactions was proposed by Henri and later by Michaelis and Menton.[5] They assumed that the enzyme E and the substrate S combine first to give a complex ES which then decomposes irreversibly into a product P and the enzyme E . Thus :

$$E + S \underset{k_{-1}}{\overset{k_1}{\Leftrightarrow}} (ES) \qquad (17\text{-}3)$$

$$(ES) \overset{k_2}{\rightarrow} E + P \qquad (17\text{-}4)$$

If s and p are the molar concentrations of the substrate and the product per unit volume at time t, respectively, we can write the rate equation :

$$-\frac{ds}{dt} = k_1 se - k_{-1}(es) \qquad (17\text{-}5)$$

and

$$\frac{d(es)}{dt} = k_1 se - (k_{-1} + k_2)(es) \qquad (17\text{-}6)$$

with initial conditions : s(0) = s and (es) (0) = 0 at t = 0. e and (es) are molar concentrations of the enzyme and the intermediate complex (ES) respectively. These are a pair of differential equations which cannot be solved analytically. To simplify the analysis a quasi-stationary state is assumed whereby

$$\frac{d(es)}{dt} = 0$$

This is approximately true since e/s is very small because the enzyme (catalyst) concentration e is very small compared to s, the substrate concentration at the start of the reaction. With this approximation and assuming that e + (es) = e_0 , Michaelis and Menton eliminated e and (es) from the above two equations and obtained the following expression:

$$r_p = \frac{dp}{dt} = -\frac{ds}{dt} = \frac{k_2 e_0 s}{\left[\frac{(k_{-1} + k_2)}{k_1}\right] + s} \qquad (17\text{-}7)$$

where r_p is the rate of reaction. From Equation 17-7, it can be deduced that the maximum rate of reaction is given by :

$$r_{max} = k_2 e_0 \qquad (17\text{-}8a)$$

Also, K_m , called the Michaelis-Menton or Saturation Constant is defined by

$$K_m = \frac{(k_{-1} + k_2)}{k_1} \qquad (17\text{-}8b)$$

If we substitute the above expressions for r_{max} and K_m in Equation 17-7, the following relation is obtained :

$$-r_p = \frac{-r_{max} s}{K_m + s} \qquad (17\text{-}9)$$

with initial condition $s(0) = 0$.

$$K_m = s \text{ when } r = r_{max}/2 \tag{17-9a}$$

The solution of the equation gives a relation between time t and substrate concentration s at any given time t and is as follows:

$$r_{max}\, t = s_0 - s + K_m \ln \frac{s_0}{s} \tag{17-10}$$

The assumption of the quasi-stationary state is found to be fairly applicable in many other situations of enzyme catalyzed reactions since the values of Michaelis-Menten constant, K_m, are very small, of the order of 10^{-2} to 10^{-5}. The linear dependence predicted by Michaelis-Menten equation is usually observed initially at low enzyme concentrations but deviations occur when the enzyme concentrations become larger. Similarity between Michaelis-Menten kinetics and Langmuir and Hinshelwood kinetics described in chapter 13 should be noted. Kinetic parameters for several enzyme reactions are given by Laidler.[6]

INTERPRETATION OF KINETIC DATA :

The kinetic parameters r_{max} and K_m are evaluated from experimental data by graphical means. For this purpose, the equation 17-9 is rewritten in the following forms which allow to obtain linear plots:

$$\frac{1}{r} = \frac{1}{r_{max}} + \frac{K_m}{r_{max}} \frac{1}{s} \tag{17-11a}$$

and

$$\frac{s}{r} = \frac{K_m}{r_{max}} + \frac{s}{r_{max}} \tag{17-11b}$$

Also,

$$r = r_{max} - K_m \tag{17-11c}$$

The linear plots of equations 17-11a and 17-11c are known as Lineweaver-Burk and Eadie-Hofstee plots respectively after their authors. In the former, 1/r is plotted against 1/s (Figure 17-4a) while Eadie-Hofstee plot graphs r against r/s (Figure 17-4c). From such plots the values of r_{max} and K_m are evaluated.

Limitation of Michaelis-Menten equation is that it does not include in its analysis the possibility of a back reaction for the final product, assumes the presence of only one enzyme substrate complex and no interaction between the enzyme and the product. Nevertheless, it is an important tool for the analysis of the enzyme catalyzed reactions. A more complete mechanism can be represented by the following series of reversible reactions:

$$E + S \underset{k_{-1}}{\overset{k_1}{\Leftrightarrow}} E.S \underset{k_{-2}}{\overset{k_2}{\Leftrightarrow}} E.P \underset{k_{-3}}{\overset{k_3}{\Leftrightarrow}} E + P$$

with four constants to evaluate experimentally viz. the Michaelis-Menten constants for the forward and reverse reactions and the maximum rates in each direction.

Example 17-1: Initial rates of an enzyme catalyzed reaction for various substrate concentrations are listed in Table 17-2a. Calculate the values of the parameters K_m and r_{max} using Lineweaver-Burk, Eadie-Hofstee and s/r vs s plots of the data.

Solution: The values of 1/s , 1/r and s/r are given in Table 17-2a.

The Lineweaver-Burk and Eadie-Hofstee plots were prepared and are shown in Figures 17-4a and 17-4c respectively. The plot of s/r vs s is shown in Figure 17-4b.

It is necessary to employ caution in the determination of the kinetic parameters K_m and r_{max}. The Lineweaver - Burk plot separates the depedent and the independent variables clearly. The large r the values near r_{max} that are more accurately known will tend to cluster near the origin while less accurately measured values will fall away from the origin and will greatly influence the value of the slope K_m / r_{max} . The plot of s/r vs s tends to spread out the data points for larger values of r, which allows the determination of the slope

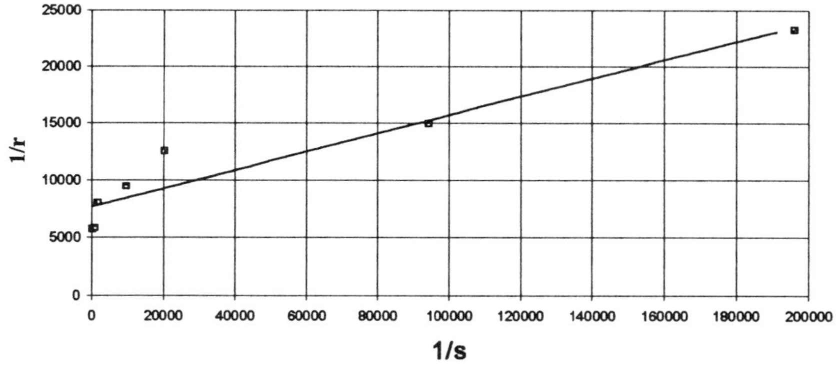

Figure 17-4a: Lineweaver - Burk plot of 1/r vs 1/s for Example 17-1

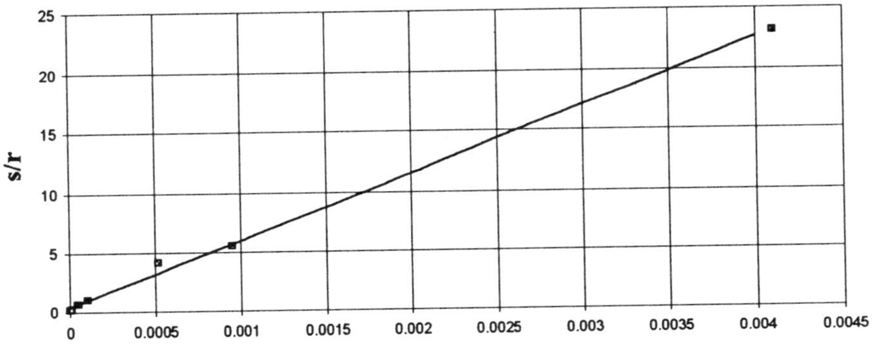

Figure 17-4b: Plot of s/r vs s for example 17-1

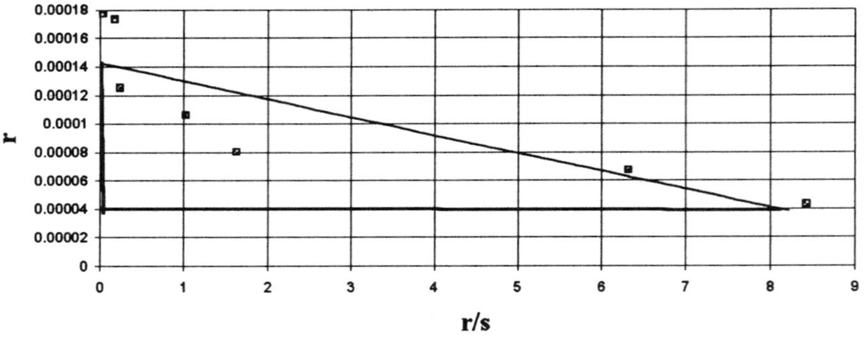

Figure 17-4c : Eadie-Hofstee plot of r vs r/s for example 17-1

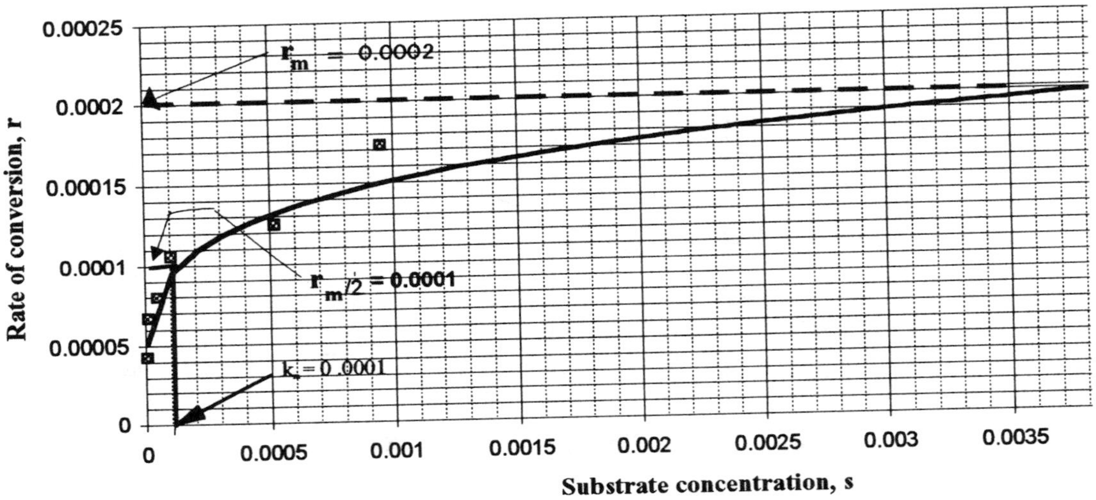

Figure 17-4d : Rate of conversion vs Substrate concentration

$1/r_{max}$ relatively more accurately. In Eadie-Hofstee plot, the measured variable r appears in both x and y coordinates and is subject to larger error. Therefore, it is better to determine r_{max} by estimating r_{max} from the intercept of the Lineweaver-burk plot or from the slope of the s /r vs s plot.
The plot of r vs s is shown in Figure 17-4d. From this plot, r_{max} is estimated at 0.0002. Using the equation 17-9a, the value of K_m is read from the graph at $r_{max}/2 = 0.0001$ and found to be 0.0001.

OTHER ENZYME CATALYZED REACTIONS: Kinetic expressions for several other situations of enzyme catalyzed reactions have been derived by various authors. Such reactions are :

- Reversible reactions
- Two substrate reactions
- Reactions involving cofactor activation
- Substrate activation and inhibition
- Deactivation of enzymes
- Multiple substrates reacting on a single enzyme.
- Inhibitor Effect in enzyme catalyzed reactions.

Many of the derivations are of complex nature. Also, relaxation methods developed by Eigen and coworkers[7] are widely used to determine the kinetic constants involved in a mathematical model of the reaction. The relaxation method involves the application of an external perturbation such as temperature, pressure, or an electric field to a reaction mixture and monitoring the response of the reaction continuously. The detailed treatment of enzyme and transient kinetics is outside the scope of this book. Reference[5] discusses these kinetics quite exhaustively.
The enzyme catalytic activity is affected by other system factors such as pH, temperature, fluid forces (hydrostatic pressure, interfacial tension), chemicals, and irradiation. For example, the temperature effect follows the Arrhenius relation up to a temperature of 40 to 50^0C. However, at higher temperatures, the enzyme activity decreases due to deactivation of the enzyme.

IMMOBILIZATION OF ENZYMES:
If an enzyme is used in dissolved form in solution, some portion of it will leave with the product withdrawn from the solution. Besides the loss of the catalyst and subsequent need of its removal from the product, new amount of the catalyst enzyme must be supplied to the reactor in order to continue the reaction

at a reasonable rate. To avoid the loss of enzyme with the product, they are immobilized i.e. they are confined in a given space so that they can be used continuously. These enzymes retain their activity longer than the free enzymes in solution and can be isolated in such a way as to increase a desired enzyme activity in multistep and simultaneous reactions.

Several methods are in use for enzyme immobilization. A chemical method involves covalent bonding and cross linkage with a water-insoluble matrix. Physical methods entrap the enzyme in a porous hollow fibre, in spun fibres, within an insoluble gel matrix or within a microcapsule. Other methods use physical adsorption on activated carbon, silica gel, and other solids. Ionic bonding involves cation and anion exchange with materials such as Amberlite, Dowex 50 or Amberlite IR-45. Membrane ultrafiltration can also be used in continuous fashion to retain enzymes of larger molecular size while allowing the product of smaller molecular size to pass through the membrane.

INDUSTRIAL APPLICATIONS OF ENZYME CATALYSIS: Enzymes in both the natural and immobilized forms have been extensively used in the production of a large number of products. These include antibiotics, steroids, organic acids and alcohols, and vitamins. For example, d-glucose is converted to d-fructose using the enzyme glucose isomerase commercially. The Tanabe-Seiyaku process uses immobilized aminoacylase to commercially manufacture l-amino acid in a column reactor. A list of some industrially important enzymes is given by Baily[1].

BATCH AND CONTINUOUS MICROBIAL GROWTH:

BATCH CULTURE: For practical applications, batch or continuous cultivation of microbial cultures is important rather than the growth of the single cells. In batch cultivation, after a liquid medium is seeded with an inoculum of living cells, nothing is added or removed from the medium. This results in overall growth of the cells as shown in Figure 17-5. Initially the increase in the number of cells is very small. This stage of the batch culture is called the lag phase. After this, the cells grow rapidly. This phase of rapid growth is called exponential growth. Then for a short period, the growth rate is maximum and almost constant. This phase is termed as maximum stationary phase because the cell population achieves its maximum size during this period. Exhaustion of the

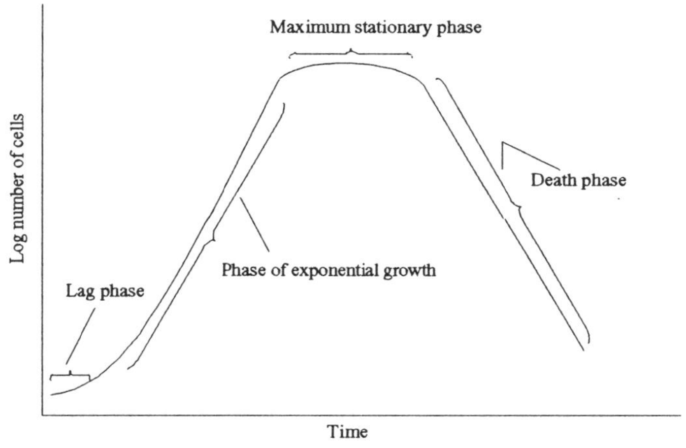

Figure 17-5 : Typical curve for the growth of cells in a batch cultivation

nutrient and a high concentration of toxic metabolites take place. Eventually due to unavailability of the nutrient, the cells begin a death phase which is also exponential. After the lag phase, the increase in cells can be represented by the relation:

$$\frac{dn}{dt} = \mu\, n \qquad (17\text{-}12)$$

with initial condition, $n = n_0$ at $t = t_{lag}$. The rate of the increase in the cells is proportional to the number of the cells present at a given time. The solution of this equation yields the following relation between n and t :

$$\ln \frac{n}{n_0} = \mu\left(t - t_{lag}\right) \qquad (17\text{-}13)$$

where t_{lag} is the time at end of the lag phase. The rate of growth can also be expressed in terms of concentration of the cells in the solution. If t_d is the time required to double the population, it can be deduced from the above equation that

$$t_d = \frac{\ln 2}{\mu} \qquad (17\text{-}14)$$

Thus during exponential growth, only a single parameter μ or t_d is required to characterize the cell population. μ is the specific growth rate and has the unit of time^{-1}. It represents the rate of growth per unit amount of biomass. Nutrients are not necessarily supplied in great excess because that can inhibit or even poison the cell growth. It is common practice to limit the total growth by supplying limited amount of one nutrient.

A functional relationship between specific growth rate μ and the concentration of the essential nutrient compound was proposed by Monod[8] and is as follows:

$$\mu = \frac{\mu_{max} C_i}{K + C_i} \qquad (17\text{-}15)$$

where μ = specific growth rate, in h^{-1}
μ_{max} = maximum specific growth rate, in h^{-1}
K = substrate utilization constant
C_i = substrate concentration in appropriate units.

K is equal to the substrate concentration when μ is equal to half μ_{max} and is a measure of the affinity of the micro-organism for its substrate. This relation is of the same form as the Langmuir adsorption isotherm and the standard rate equation for enzyme catalyzed reaction with only one substrate. Other forms of relationships have also been proposed to describe specific growth of cells in more complex situations. Effect of toxins has also been mathematically treated[1].

CONTINUOUS CULTURE: In the continuos cultivation of culture in an ideal CSTR (continuous stirred tank reactor), medium is added and product is withdrawn from the reactor continuously. For a steady state operation of such biochemical reactor, a material balance on a component of the system can be written in the same manner as for a continuous chemical reactor as follows:

$$F(C_{if} - C_i) + V_R r_{fi} = 0 \tag{17-16}$$

where V_R - total volume of the culture in the reactor ,
F - volumetric flow rate of feed and effluent liquid streams
C_{if} - molar concentration of component i in the feed stream
C_i - concentration of component i in the reaction mixture and in the effluent stream.

A rearrangement of the equation 17-16, yields the following relation:

$$r_{f_i} = \frac{F}{V_R}\left(c_i - c_{if}\right) = D\left(c_i - c_{if}\right) \tag{17-17}$$

The parameter D is termed the dilution rate and is given by $D=F/V_R$. Actually it is the reciprocal of the mean residence time (V_R/F). The parameter D can be interpreted as the number of tank liquid volumes passing through the vessel per unit time.

Frequently the cell mass formed by the cell growth is proportional to the mass of substrate consumed. On this basis, the yield factor $Y_{x/s}$ is defined as the ratio of the cell mass formed to the amount of substrate consumed, whereby:

$$Y_{x/s} = \frac{\text{mass of cells formed}}{\text{substrate consumed}} \tag{17-18}$$

With the definition of the yield factor, Equation 17-17 can be rewritten for the material balance on the substrate in terms of substrate concentrations and the specific growth rate, and x, the cell mass concentration as follows:

$$D(s_f - s) - \frac{1}{Y_{x/s}}\mu x = 0 \tag{17-19}$$

A functional relationship between specific growth rate and the concentration of the nutrient was proposed by Monod [8, 9] and is given by Equation 17-20 below:

$$\mu = \frac{\mu_{max} s}{K_s + s} \tag{17-20}$$

where μ = specific growth rate, h^{-1}

μ_{max} = maximum specific growth rate, h^{-1}

s = substrate concentration, mass/unit volume

K_s = substrate utilization constant , which equals the limiting substrate concentration at which μ equals one-half of μ_{max}.

If the expression for μ from Equation 17-20 is substituted in Equation 17-19, the following relation results:

$$Dx_f + \left(\frac{\mu_{max} s}{K_s + s} - D \right) x = 0 \tag{17-21}$$

A similar material balance on the cell mass yields the following equation:

$$D(s_f - s) - \frac{\mu_{max} sx}{Y_{x/s}(s + K_s)} = 0 \tag{17-22}$$

Equations 17-21 and 22 together constitute the Monod "chemostat model " for bioreactors. Usually the feed to the bioreactor is sterile (i.e. $x_f = 0$) . For this case, the following relation is obtained

$$x_{sf} = Y_{x/s}\left(s_f - \frac{DK_s}{\mu_{max} - D} \right) \tag{17-23}$$

and

$$s_{sf} = \frac{DK_s}{\mu_{max} - D} \tag{17-24}$$

Where x_{sf} = cell concentration in the reactor and effluent
s_{sf} = substrate concentration in the reactor and in the effluent.

The subscript sf denotes sterile feed i.e. $x_f = 0$.

From Equation 17-23, it is apparent that if the dilution rate D → 0, the effluent cell concentration is given by

$$X_{sf} = Y_{x/s}\, s_f \qquad (17\text{-}25)$$

On the other hand, when D increases, x_{sf} decreases. When D approaches μ_{max},, x_{sf} diminishes more rapidly. When dilution rate becomes just greater than μ_{max} , the steady state solution for the cell concentration can be only x = 0 . This condition allows to deduce a relationship between D_{max} and μ_{max} as follows:

$$D_{max} = \frac{\mu_{max} s_f}{K_s + s_f} \qquad (17\text{-}26)$$

The rate of cell production is the product of dilution rate D and the cell concentration x or Dx. Using the value of x given by Equation 17-23 and differentiating Dx with respect to D and equating

$$\frac{d(Dx)}{dD} = 0$$

the maximum dilution rate at which the cell production is maximum can be obtained as

$$D_{max} = \mu_{max}\left(1 - \sqrt{\frac{K_s}{K_s + s_f}}\right) \qquad (17\text{-}27)$$

Product formation in the fermentation processes is frequently analysed with the use of another product yield factor Y_{px} which is defined as follows:

$$Y_{px} = \frac{\text{mass of product formed}}{\text{increase in cell mass}} \qquad (17\text{-}28)$$

Y_{px} is constant for some fermentations. These are called Type I fermentations. For others, Y_{px} varies. Using this definition of the product yield factor, the steady state mass balance for cell production can be written as

$$D(p_f - p) + Y_{px}\mu\ x = 0 \qquad (17\text{-}29)$$

where, p_f = product concentration in feed
and p = product concentration in the reactor effluent.

The design analysis of the batch and continuous biological stirred tank and plug flow reactors by the immobilized enzyme catalysis involves the application of principles of mass and heat transfer, agitation and many other unit operations in a similar manner as is done in the case of chemical reactors. Exhaustive treatment of the biological reactors will be found in a text by Baily and Ollis[1]. Unlike chemical reactors , however, maintaining a sterile environment in the biological reactors is very important . Because of specific applications, it is necessary to prevent contamination from unwanted species so that a desired species is free to grow and produce desired product with maximum efficiency. Sterilization therefore plays an important role in biological process reactors. Sterilization can be achieved by several methods. More widely used are: heating, chemical addition, and filtration.

Fermentation and biological waste treatment are two most important applications of biochemical processing. Industrial fermentations are carried out in both the batch and continuous manner. Many of these processes use only one species of microorganism. In some processes, although multiple organisms are present, the activity of a single species may predominate. Many fermentations are initially conducted under aerobic conditions and later finished anaerobically. A typical aerobic fermentation process involves the following operations:

- Medium preparation
- Medium sterilization
- Inoculum preparation
- Air filtration
- Cell and product recovery.

MEDIUM FORMULATION : Medium formulation involves a consideration of a number of factors. Most important is the material balance. Depending upon the biomass to be produced and a given cellular stoichiometry and efficiency, sufficient amount of nutrients need to be provided. Theoretical elemental requirement of the process can be determined from the typical organic and inorganic components of microorganisms[3]. The efficiencies with which elements are incorporated into cellular material are to be obtained experimentally on laboratory or pilot plant scale. Preparation of inoculum requires a careful multiplication of few cells to produce a dense suspension of 1 to 20% by volume of the production scale fermentator. Stock culture is usually maintained in dormant state by lyophilization (freeze drying) of liquid cell suspensions or the spore formation when possible. The stock culture is then transferred to one or more seed vessels where the microorganism is allowed to grow further before transferring to the large fermentator. Fermentator usually consists of a vertical vessel equipped with an agitator and baffles, a sparger for introduction of air required for aeration, a feed pipe for aseptic addition of inoculum and heat exchanger coils. Necessary instrumentation and controls are also provided to maintain the required production conditions to maximize the amount of the desired product. Various other designs such as horizontal vessels, air lift fermentators, pulsating column, packed columns, vessels with external pump circulation, and submerged type reactors with energy input by compression etc. are in use. Selection of a particular design is dictated by various factors such as scale of production, nature of the particular fermentation process, power requirement, heat removal, and avoidance of cell damage.[1]

The fermentation in the production vessel may require anywhere between less than a day or sometimes more than two weeks. Typical duration is 4 - 5 days in case of processes for the manufacture of antibiotics. It is very essential to maintain aseptic conditions in the production vessel to maximize product formation. Additional information on fermentation processes may be found in the references.[1,2,6]

The products of bioreaction broths are complex mixtures of biomass, dissolved molecules, and electrolytes. Separation, concentration, and isolation of these components is rendered more difficult because the desired products are usually in dilute solution, have similar physical and chemical properties, and many of them are sensitive to shear stresses and high temperature. Depending upon the nature of the product broth formed, the process of isolation, concentration and purification may involve the use of several of the chemical engineering unit operations such as filtration, ion exchange, evaporation, crystallization, centrifugation, sedimentation, solvent extraction, drying, and membrane separation processes. Most of these are discussed in previous chapters. For a thorough treatment of the applications of unit operations in biochemical field, references [1,2,6,11,12,13] should be consulted.

MEMBRANE SEPARATIONS:

The membrane separation processes have found extensive application in the biochemical product separation and purification. These use semipermeable membranes for separation or concentration of the desired component in a solution. Three processes viz reverse osmosis (RO), ultrafiltration (UF), and crossflow (or micro) filtration (CFF) are shown in Figure 17-6. They depend upon the principle of osmotic reversion and/or physical size difference of the molecular species to be separated and use only hydraulic pressure difference as the driving force. The separation is achieved without change of phase or interphase mass transfer at modest temperatures. Distinction between these unit operations is not exactly defined but one useful definition is based on the smallest molecules or particles which are retained by various membranes. Each process is defined by the pore size of the membrane as follows:

Reverse osmosis (RO)	-	0.0001 to .001 mm (1 to 10 Å)
Ultrafiltration (UF)	-	0.001 to 0.02 mm (10 to 200 Å)
Microfiltration (MF)	-	0.02 to 10 mm (200 to 10,000 Å)

A variety of membranes are in use. These are prepared by different processes using various materials such as cellulose acetate, cellulose nitrate. Polymers such as nylon, polyvinyl chloride, copolymers of PVC, acrylonitrile and PTFE and polybisphenol-A-carbonate etc. are also used. Types of membranes in use are: track-etch Nucleopore, Loeb-Sourirajan anisotropic membrane with asymetric structure, hollow-fibre membranes, composite hollow-fibre membranes and dynamic membranes. Membranes can also be made of ceramic and sintered metal. An extensive review of membrane production and their properties is given by Porter[12].

REVERSE OSMOSIS : This process uses membranes that are permeable to small molecules like water but not to larger molecules. It depends upon the fact that the applied pressure on the solution side of the membrane must exceed the osmotic pressure of the solution to reverse the flow of the solvent towards the pure solvent side of the membrane. Passage of solvent from the solution to the pure solvent side under hydraulic pressure results in the concentration of the

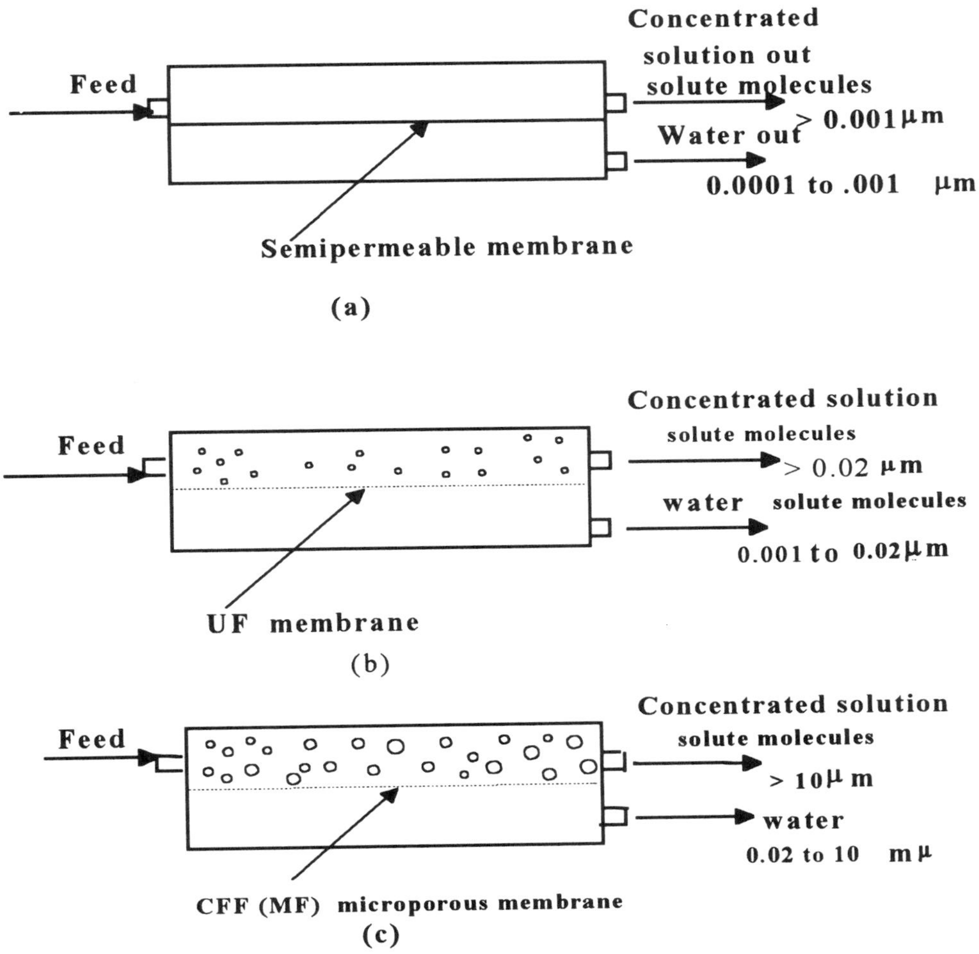

Figure 17-6 : Membrane processes
(a) Reverse osmosis
(b) Ultrafiltration
(c) Crossflow or MF filtration

solution on the solution side. Although the mechanism by which reverse osmosis occurs across a semipermeable membrane is not fully known, experts in the field believe that the membrane must exhibit a preference for the solvent (water) as agaist the dissolved solute at the surface/solution interface. Permeation of the solvent molecules occurs by a two step process of preferential sorption of the solvent molecules at the surface of the membrane, followed by the diffusion through the microscopic capillaries. Solute molecules of sufficient size are prevented from permeation basically by a sieving action by the surface of the membrane. Solutes having comparable molecular size with that of the solvent, may pass through the membrane.

The passage of water through a semipermeable membrane may be described by the equation

$$Q_w = \frac{K_w A}{d}(\Delta P - \Delta\pi) \tag{17-30}$$

where Q_w = rate of solvent (usually water) permeation, l/h
K_w = permeability coefficient of the membrane. It is a function of the type of the membrane, temperature, and pH of the solution.
A = membrane area, m^2
d = membrane thickness, m
ΔP = Hydraulic pressure differential across the membrane, Bar
$\Delta\pi$ = Osmotic pressure differential across the membrane, Bar

Solvent (water) passage per unit surface area of the membrane is termed permeate flux or simply flux . Thus using equation 17-30, the flux is given by

$$F = \frac{Q_w}{A} = \frac{K_w}{d}(\Delta P - \Delta\pi) \quad 1/m^2\text{-}h \tag{17-30a}$$

Actual fluxes gradually tend to be lower than predicted because of a phenomenon known as concentration polarization. This means that when solvent passes through the membrane, the solute left behind concentrates in the solution at the membrane surface. This concentration increases the osmotic pressure of the solution at the interface which in turn decreases the driving force. The localised solute concentration at the interface may be so high that the osmotic pressure may be very close to the hydraulic pressure which will result in a very low and even zero flux.

In reverse osmosis, solute molecules whose dimentions are close to the pore size of the membrane will plug some of the pores. This will reduce the permeability. Asymetric microporous membranes resist plugging because they have pores of conical shape with their diameters increasing in the direction of solvent flow with distance. Periodic reversal of solvent flow is useful to dislodge the cake at the surface which would allow to maintain reasonable flux rates for a longer period of time.

Although reverse osmosis is most often used for the purification and desalination of water, it is possible to use it to fractionate mixtures difficult to separate by other means. It is also useful in the concentration and desalting of dyestuffs and production of thermally sensitive and chemically unstable products such as drugs, biologicals, and foods. Several industrial applications of the reverse osmosis have been summarized by Norman[13].

ULTRAFILTRATION: This is also a pressure driven membrane process which separates solution components on the basis of molecular size and shape. Solvent and smaller-size solute molecules which pass through the membrane are collected as permeate. Larger solute molecules that are retained by the membrane are recovered as a concentrated and purified retenate. Water transport throgh the membrane pores is by viscous flow. As mentioned earlier, ultrafiltration involves solutes of molecular dimensions ranging between (0.001 to 0.02 mm, 10 to 200 À). The molecular weights of the solutes are greater than 300 but less than 300,000. Since the osmotic pressures of the solution are very low because of high molecular weights of the solutes, lower operating pressures of the order of 5 to 100 psi are required. The lower operating pressures reduce the energy requirements of pumping and the equipment cost cosiderably as compared to reverse osmosis which requires operating pressures of the order of 300 to 1000 psi. Ultrafiltration is used to concentrate, purify, and fractionate macromolecular solutes in the feed liquid.

CROSSFLOW (MICRO) FILTRATION: This is similar to ultrafiltration but the membrane pores are of larger size than those in UF membranes. Macrosolutes are passed throgh the membrane but larger molecules of colloids and cells are retained. Flow of solvent and solute through the membrane is convective. Microfiltration is mainly used to concentrate cells and to recover enzymes and /or proteins from fermentation broths.

MEMBRANE PROCESS EQUIPMENT:

In order to achieve reasonable capacities from the plant, membranes are packed in compact equipment modules. The designer of membrane modules need to consider several aspects such as physical support for the membranes, packing density to effectively manage the fluid flow, minimum system volume for product recovery, cleaning, ease of maintenance, and replacement. The different types of designs that have been developed are : (a) flat plate, (b) spiral wound, (c) pleated cartridge, (d) shell and tube, and (e) hollow fibres enclosed in cylindrical shell. For an exhaustive account of the operation and design of these modules, references[11,12] should be consulted.

ACTIVATED SLUDGE PROCESS:

Biological treatment of wastes, both domestic and industrial, is very important. Both of these wastes contain, besides water, a complex mixture of insoluble solids and dissolved substances. These contaminants have to be reduced or chemically transformed to acceptable low limits before the sewage waste or the industrial waste water are disposed off because of environmental considerations. Various types of unit operations and processes are employed to treat the wastes before disposal[9].

In primary treatment, easily separable components, the wet solids, collectively called the sludge are removed. In the secondary treatment stage, the suspended as well as soluble components are removed. Most commonly used method at this stage is biological oxidation. In certain cases a tertiary treatment is needed to remove some or all of the remaining contaminants.

Unit operations such as adsoption, deep bed filtration, reverse osmosis and electrodialysis are commonly used in the tertiary treatment.

The strength of sewage and waste water is measured in terms of biochemical oxygen or chemical oxygen demand. The biochemical oxygen demand (BOD) of sewage or waste water is defined as the milligrams of dissolved oxygen which is consumed by one liter of the waste for a specified incubation length of time at 20 ^{0}C. The length of incubation time is often shown as subscript e.g in BOD_5 , the subscript 5 refers to a period of 5 days over which the oxygen consumption of the waste sample is measured. The ultimate BOD is the amount of dissolved oxygen consumed in an incubation that is continued until biological oxidation of carbonaceous material stops. The chemical oxygen demand (COD) is defined as the number of milligrams of oxygen which a liter of a sample will absorb from a hot acidic solution of potassium dichromate. COD value is greater than BOD because more components of waste water can be oxidized by potassium dichromate than by biological treatment. Both are used as indicators of the degree of damage that will be caused by the waste water to the environment.

BOD of a waste sample is determined usually at 20^0 C with incubation period of 5 days but other longer time periods (e.g. 7 days) and temperatures can also be used. The incubation is carried out by addition of dilution water to the sample. The dilution water may be unseeded or seeded with additional bacterial culture. The dissolved oxygen is measured before and after incubation and BOD is calculated from the amount of oxygen consumed by the sample.
Since biochemical oxidation is a slow process, theoretically it takes an infinite time for the oxidation to go to completion. The kinetics of the BOD reaction rate are, for practical reasons, formulated assuming first order reaction kinetics and can be represented by the equation

$$\frac{dL_t}{dt} = -kL_t \qquad (17\text{-}31)$$

The solution of this equation is given by

$$L_t = Le^{-kt} \qquad (17\text{-}31a)$$

where L or BOD_L = BOD remaining at time t = 0 (i.e. the total or ultimate first-stage BOD initially present)
L_t = The amount of the first-stage BOD remaining in the sample at time t
and k = reaction rate constant.

The amount of BOD that has been utilized at any time, equals

$$y_t = L - L_t = L(1- e^{-5t}) \qquad (17\text{-}31b)$$

Thus the n-day BOD equals

$$y_n = L\text{-}L_n = L(1\text{-}e^{-nk}) \qquad (17\text{-}31c)$$

Example 17-2 : Determine 3-day and 1-day BODs and ultimate first-stage BOD for a waste water whose 5-day, 20°C is 225 mg/l. The reaction rate constant k (base e) is .25 d^{-1}.

Solution:

1. Determination of ultimate BOD:

$$y_5 = L - L_5 = L(1-e^{-kt})$$

Substitution of values given in the example yields

$$225 = L(1-e^{-5(0.25)}) = L(1-0.2865)$$

which gives L = ultimate BOD = 225/(1-.2865) = 315.4 mg/l

2. Determination of 3- day BOD:

$$L_3 = Le^{-0.25(3)} = 315.4(.4724) = 149 \text{ mg/l}$$

$$y_3 = L - L_3 = 315.4 - 149 = 166.4 \text{ mg/l}$$

Also $y_3 = L(1 - e^{-3x0.25}) = 315.4 \times (1 - e^{-0.75}) = 166.4$

3. Determination of 1-day BOD:

$$L_1 = Le^{-.25(1)} = 315.4(0.7788) = 245.6 \text{ mg/l}$$

$$y_1 = L - L_1 = 315.4 - 245.6 = 69.8 \text{ mg/l}$$

The temperature dependence of the BOD reaction rate constant is given by the equation

$$k_T = k_{20}\theta^{(T-20)} \qquad (17\text{-}31d)$$

and is derived from the van't Hoff-Arrhenius relationship. The value of θ has been found to be 1.056 in the temperature range between 20 and 30°C and 1.135 in the temperature between 4 and 20°C.

The activated sludge process is schematically shown in Figure 17- 7. The process consists of a continuous-flow biological reactor which is aerated. The reactor is fed with a clarified effluent stream from a primary sedimentation tank in which suspended solids are removed. The overflow from the reactor vessel is continuously fed to a secondary sedimentation tank where the sludge settles and is removed from the bottom. The clarified and biologically treated effluent leaves the vessel at the top. A portion of the sludge collected from the secondary sedimentation tank is recycled to the biological reactor which provides a continuous inoculation. This recycling increases the mean residence time of the microorganisms in the reactor allowing them to adapt to the available nutrients. The residence time of the sludge must also be long enough in order to oxidize the adsorbed organics. Two other variations of the activated sludge process are : contact stabilization and step-feed process. In the former, the recycle sludge is subjected to an additional aeration time before it is mixed with the waste to be treated in the reactor. In the step-feed process, the effluent stream is divided and fed to the aeration tank at two different points. There are several other variants of the activated sludge process [7]. The mathematical modeling of the activated sludge is based on the assumption of the sludge as a single species whose growth follows simple Monod kinetics with an additional endogenous

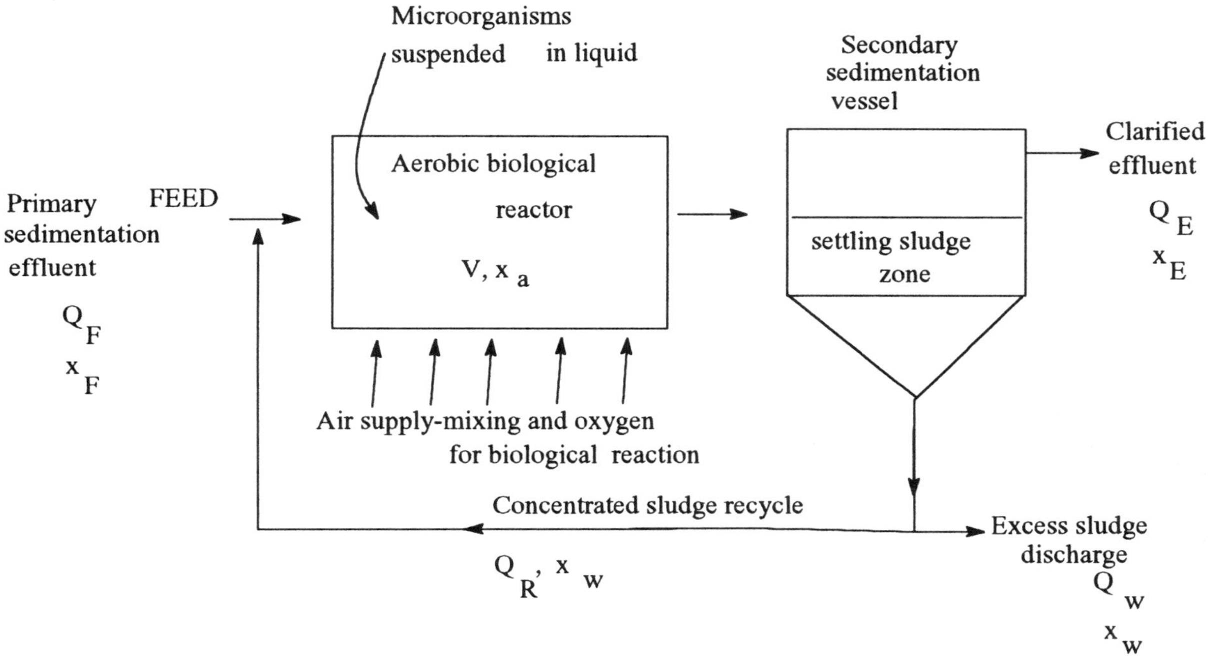

Figure 17-7 : Activated sludge process flow diagram.

metabolism decay term. The substrate limiting concentration is usually expressed in terms of BOD. The detail treatment of activated sludge process and a list of Monod parameters for the utilization of various substrates in aerobic mixed cultures will be found elsewhere.[9] In an alternate process, a trickling filter is widely used to remove organic matter and for nitrification. The filter consists of a vessel loosely filled with a suitable packing material to which microorganisms are attached. The waste water percolates or trickles through the packing. Packing media are rocks of various sizes or plastic packings such as pall rings. Depth of rock packing varies between 3 to 8 feet (.3 to 2.5 m) while that for plastic media, it varies between 30 to 40 feet (9 to 12 m). The major difference between activated sludge process and the trickling filter process is the means of oxygen transfer. In activated sludge process, forced aeration is used whereas in trickle bed filter, the air circulates by natural convection. Design of this process is discussed in detail by Tchobanoglous[10].

Design of the activated sludge process involves consideration of many factors such as loading criteria, type of reactor, sludge production, requirements and transfer, nutrient needs, solids separation and most importantly the effluent characteristics that must be obtained to meet environmental standards. A simplified approach is used to describe the process mathematically. As an example, for the simple configuration shown in Figure 17-7 assuming x_F, the active solids concentration in the feed, Q_F, is equal to zero, a material balance for the active solids under steady state conditions is given by:

$$Q_E x_E + Q_W x_W = V x_a \left(\frac{\mu_{max} s_a}{s_a + K_s} - K_e \right) \tag{17-32}$$

where Q_E = Clarified effluent rate, m^3/h
Q_F = Feed rate, m^3/h
Q_W = Waste sludge rate, m^3/h
K_e = decay constant, h^{-1}
K_S = Monod parameter, mass/unit volume
s_a = Concentration of active solids in the reactor

In the above derivation, it is also assumed that the density ρ of the streams is constant. A mean residence time θ_s for the activated sludge, also called the sludge age is defined as follows:

$$\theta_s = \frac{\text{active solids retention in the reactor}}{\text{active solids effluent rate}}$$

$$= \frac{V x_a}{Q_E x_e + Q_w x_w} \tag{17-33}$$

The sludge age for most sewage treatment plants varies between 6 to 15 days. Combining the definition of θ_s with equation 17-32, the following expression results:

$$\frac{1}{\theta_s} = \mu_{max} \frac{s_a}{s_a + K_s} - K_e \tag{17-34}$$

If the kinetic constants are known for a given waste, it will be possible to calculate the required solids retention time to achive a design- specified BOD level for that waste. Typical Monod kinetic constants for the activated sludge process are given by Tchobanoglous.[10]

Assuming $Y_{x/s}$, the yield factor, to be constant, a steady state material balance over the aeration basin or the main bioreactor gives the following relation:

$$Vx_a = \frac{Y_{x/s} Q_F \theta_s (s_0 - s_a)}{1 + K_e \theta_s} \tag{17-35}$$

Another assumption in the derivation of Equation 17-34 is that the substrate is not separated in the clarifier i.e. $s_a = s_e = s_w$ where s_w is the substrate concentration in the waste sludge.

If x_w is known, an active solids balance on the bioreactor alone results in the relation:

$$\theta_s Q_F x_F + \theta_s Q_R x_w + Vx_a = (Q_F + Q_R) x_a \theta_s \tag{17-36}$$

Where Q_R = recycle sludge flow rate, m^3/h
V = volume of aeration basin, m^3

Since $x_F = 0$ usually, Equation 17-36 on simplification yields:

$$V = \theta_s \left[Q_F + Q_R - Q_r \frac{x_w}{x_a} \right] \tag{17-37}$$

$$= Q_F \theta_s \left[1 + \alpha - \alpha \frac{x_w}{x_a} \right] \tag{17-37a}$$

wherein α is the ratio of recycle sludge to the feed.

Both empirical and theoretical parameters have been used for the design and control of the activated sludge process.[10] Two parameters commonly used are (1) food-to-microorganism ratio (F/M) and (2) the mean cell residence time θ_C.

The food-to-microorganism ratio is defined as:

$$\frac{F}{M} = \frac{S_0}{\theta x} \tag{17-38}$$

where F/M = food-to-microorganism ratio, d^{-1}
F = mass of substrate consumed /day by microorganisms
M = mass of microorganisms
S_0 = influent BOD or COD concentration, g/m^3
θ = hydraulic retention time of the aeration tank
= V/Q , day (d)
V = aeration tank volume, m^3
Q = influent waste water flow rate, m^3/d
x = concentration of volatile suspended solids in the aeration tank, g/m^3

The organic material portion of the suspended solids in the waste that acts as the food to the microorganisms is called volatile suspended solids and are referred to as vss.

The specific utilization rate of the substrate can be related to the food-to-micro-organism ratio by

$$U = \frac{(F/M)E}{100} \tag{17-39}$$

where E = process efficiency, %

$$= \frac{[s_0 - s]}{s_0}(100) \tag{17-39a}$$

Substitutiting F/M from Equation 17-37 and the value of the efficiency from Equation 17-39a into Equation 17-39 yields

$$U = \frac{s_0 - s}{\theta x} \tag{17-40}$$

The mean cell residence time can be defined either in terms of volatile suspended solids in aeration tank or total mass of volatile suspended solids in aeration tank, settling tank and the sludge return line as follows :

$$\theta_c = \frac{Vx}{Q_e x_e + Q_w x_w} \tag{17-41}$$

where θ_c is the mean cell residence time based on the aeration tank volume. Based on total system volume, the mean residence time can be given by

$$\theta_{ct} = \frac{x_t}{Q_e x_e + Q_w x_w} \tag{17-42}$$

where θ_{ct} = mean cell residence time based on total system volume.
x_t = total mass of volatile suspended solids in the system including aeration tank, settling tank and sludge return line.

The design is usually based on θ_c since almost all the substrate conversion occurs in the aeration tank. The relationship between θ_c , F/M and U is

$$\frac{1}{\theta_c} = Y\frac{F}{M}\frac{E}{100} - K_d = YU - K_d \tag{17-43}$$

The quantity of the sludge that will be produced per day affects the design of sludge handling and disposal facilities. An estimate of daily sludge waste can be made using the following equation :

$$W_x = Y_{obs}Q(s_0 - s)(\text{ g/kg})^{-1}\text{x}10^{-3} \tag{17-44}$$

where

W_x = net waste sludge produced each day, measured in terms of volatile suspended solids
Y_{obs} = observed yield, g/g

The observed yield can be calculated by using the following equation

$$Y_{obs} = \frac{Y}{1+K_d} \tag{17-45}$$

Depending upon where the sludge waste is withdrawn (aeration tank or recycle line), the sludge waste calculation can be simplified further. If the wasting is done from the aeration tank and if θ_c is used, the pumping rate may be estimated by

$$\theta_c = \frac{Vx_a}{Q_{wa}x_a + Q_e x_e} \tag{17-46}$$

where Q_{wa} is waste sludge flow rate from the aeration tank. In the case when the concentration of solids in the effluent from the settling tank is very low, $Q_e x_e \approx 0$ and then

$$Q_{wa} \approx \frac{V}{\theta_c} \tag{17-47}$$

On the other hand, if the sludge is wasted from the recycle return line and if it is assumed that $Q_e x_e \approx 0$, the sludge wastage rate can be calculated from the following equation

$$Q_{ww} = \frac{Vx}{\theta_c x_w} \tag{17-48}$$

where Q_{ww} is the sludge waste flow rate from the recycle return line. and x_w is the solids concentration in the return line.

If the food-to-microorganism control is used, the wasting flow rate from the return line can be calculated by the following equation

$$W_x = Q_{ww}x_w(g/Kg)^{-1} \times 10^{-3} \tag{17-49}$$

It is possible to treat wastes containing fermentable organic compounds biologically under anaerobic condition. Although it has a number of applications, a major one is the treatment of excess sludge solids produced in sewage treatment processes. In this process, the first step involves solubilization of solid sludge material by enzymes synthesized by various bacteria. The next step involves the action of acid producing bacteria termed acid formers, and best takes place in the range of pH from 4. to 6.5. The major product of this step is acetic acid. Propionic and butyric acids are also formed. Acetic acid is the most important substrate on which the methane bacteria act to give out methane and carbon dioxide. Thus anaerobic process produces a fuel which can be used to reduce energy costs. As a consequence of anaerobic digestion, the sludge is in a better condition for further treatment. It is less putrefactive and is easier to dewater. After dewatering and drying, the sludge can be used as a fertilizer, dumped on land or incinerated. The kinetic coefficients required for the design of activated sludge processes are experimentally obtained using either batch reactor operated with or without recycle of the sludge. For a continuous flow, completely mixed stirred tank reactor without sludge recycle, it is possible to derive from the mass balances for the substrate and the microorganisms the following equations:

$$\frac{1}{\theta} = \frac{\mu_{max} s}{K_s + s} - K_d \tag{17-50}$$

and

$$s_0 - s - \theta \frac{k\,x\,s}{K_s + s} = 0 \tag{17-51}$$

If Equation 17-50 is solved for $s/(K_s+s)$ and resulting expression is substituted in Equation 17-51 and then simplified using the relation $k = \mu_{max}/Y$, the effluent steady state cell concentration is obtained as

$$x = \frac{Y(s_0 - s)}{1 + k_d\theta} \tag{17-52}$$

where k is the maximum rate of substrate utilization per unit mass of microorganisms.

In a similar manner, the effluent substrate concentration s is found to be given by

$$s = \frac{K_s(1+k_d\theta)}{\theta(Yk-k_d)-1} \qquad (17\text{-}53)$$

Example 17-3: Determine the kinetic coefficients k, K_s, μ , Y, and k_d from the data in Table 17-3, obtained using a small complete- mix bioreactor with sludge recycle.

Table 17-3 a

Data for Example 17-3

Run No.	s_0 mg/L	s mg/L	q day	x mg/L	q_c day
1	400	10.00	0.17	3950.00	3.10
2	400	14.30	0.17	2865.00	2.10
3	400	21.00	0.17	2100.00	1.60
4	400	49.50	0.17	1050.00	0.80
5	400	101.60	0.17	660.00	0.60

Solution: The rate of substrate utilization is given by

$$r_{su} = -\frac{kxs}{K_s+s} = -\frac{s_0-s}{\theta}$$

from which a linearized form is obtained by dividing by x and taking inverse as follows

$$\frac{x\theta}{s_0-s} = \frac{K_s}{k}\frac{1}{s} + \frac{1}{k}$$

By plotting $x\theta/(s_0\text{-}s)$ versus $1/s$, or by linear regression analysis, the values of K_s and k can be determined. Also we have the following relation

$$\frac{1}{\theta_c} = Y\frac{s_0 - s}{x\theta} - k_d$$

which also gives a straight line on plotting $1/\theta_c$ versus $(s_0\text{-}s)/x\theta$. The slope and intercept of the line can be obtained by plotting or by linear regression analysis. The following calculations are set up for the determination of the kinetic coefficients:

Table 17-3b

Calculations for Example 17-3b

s_0 - s mg/L	$x\theta$ mg/(d.L)	$x\theta/(s_0$ - s) d	1/s $(mg/L)^{-1}$	$1/\theta_c$ d^{-1}	$(s_0$ - s$)/x\theta$ d^{-1}
390.00	659.65	1.69	0.10	0.32	0.59
385.70	478.46	1.24	0.07	0.48	0.81
379.00	350.70	0.93	0.05	0.63	1.08
350.50	175.35	0.50	0.02	1.25	2.00
298.40	110.22	0.37	0.01	1.67	2.71

The two plots are shown in Figures 17-8a and b respectively. The intercept of the line in Figure 17-8a is 1/k. Therefore,

Y intercept = 1/k = 0.215512

and k = 1/0.215512 = 4.64 d^{-1}

Slope of line = K_s/k = 14.73915

Then K_s = 14.73915 x k = 14.74915 x 4.64 = 68.39 mg/L

From the plot of figure 17-8b, intercept of the line = - k_d = - 0.04953

from which k_d = 0.04953 d^{-1}

Slope of line = Y = .638633

and μ = kY = 4.64 (.638633) = 2.963 d^{-1}

Example 17-4:

A waste water is to be treated aerobically in a continuous-flow batch reactor without recycle. Calculate: (1) minimum θ_c . (2) If actual cell residence time, θ_c , used for design is 3 days, determine the following :(a) the effluent substrate concentration (b) the specific utilization rate (c) the food-to-micro-organism ratio and the concentration of the micro-organisms in the reactor. Initial concentration of the substrate in waste water is 250 mg/L and concentration of microorganism can be taken as negligible. the system kinetic coefficients determined experimentally are:

K_s = 60 mg/L   k = 5 d^{-1} k_d = .05 d^{-1} Y = .6 .

Solution: The various quantities can be calculated by substituting the various constants in expressions derived from mathematical treatment of the steady state operation of a batch reactor without recycle.[10]

(1) Minimum θ_c is given by

$$\frac{1}{\theta_c} = Y\frac{ks_0}{K_s + s_0} - k_d$$

$$= (0.6)\frac{5(250)}{60+250} - 0.05$$

$$= 2.37 \ d^{-1}$$

therefore, minimum θ_c = 1/2.37 = .422 d

(2) a. Substrate concentration

$$s = \frac{K_s(1+\theta k_d)}{\theta_c(Yk - k_d) - 1} = \frac{60(1+3\times 0.05)}{3(.6\times 5 - 0.5) - 1}$$

$$= 8.79 \ \mathrm{mg/L}$$

b. Effluent concentration

$$x = \frac{Y(s_0 - s)}{1 + k_d\theta_c} = \frac{0.6(250 - 8.79)}{1+.05(3)} = 125.9 \ \mathrm{mg/L}$$

c. specific utilization rate

$$U = \frac{s_0 - s}{\theta_c x} = \frac{250 - 8.79}{3(125.9)} = 0.64 \text{ d}^{-1}$$

d. Food-to-micro-organism ratio

$$= \frac{s_0}{x\theta_c} = \frac{250}{125.9(3)} = 0.662$$

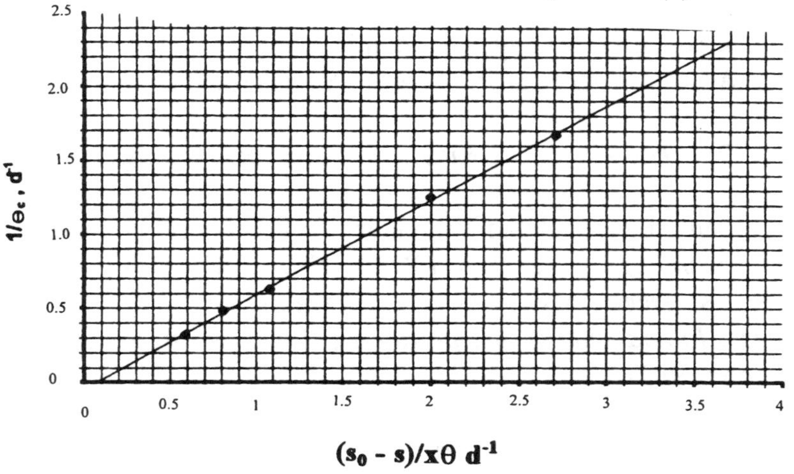

Figure 17-8b. Plot of data in Example 17-3 to determine the kinetic coefficients Y and k_d .

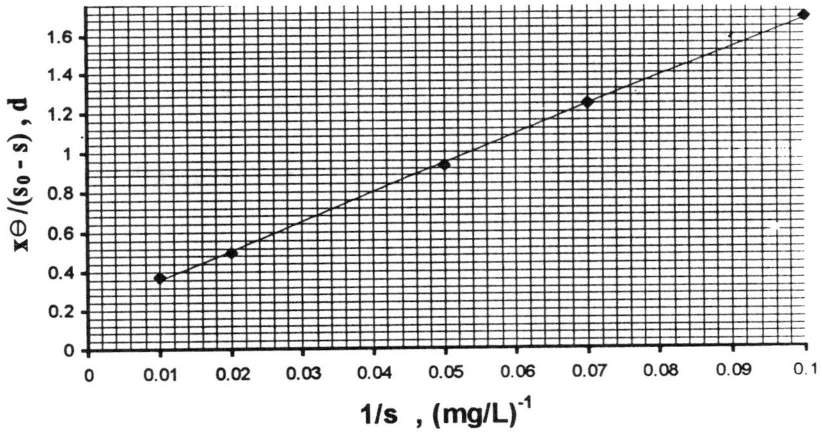

Figure 17-8a - Plot of xΘ/(s_0 - s) versus 1/s for Example 17-3

REFERENCES:

(1) Bailey James E., and Davis F. Ollis, *Biochemical Engineering Fundamentals*; McGraw Hill Book CO., 2nd Ed. (1986).

(2) Vogel H. C., *Fermentation and Biochemical Engineering Hand book*, Noyes Publications, Park Ridge, N.J., (1983).

(3) Aiba Shuichi, Humphrey A. E., and Nancy F. Millis; *Biochemical Engineering*, Academic Press, New York (1965).

(4) Lehninger A. L., *Biochemistry,* Worth Publishers Inc; new York (1975).

(5) Laidler K.T., *The Chemical Kinetics of Enzyme Action*, Oxford (1956).

(6) Stanburry Peter F. and Allan Whitaker, *Principles of Fermenta - tion Technology*; Pergamon Press, New York, (1984).

(7) Jenkins R., K. Deeny and T. Eckoff, *The Activated Sludge Process: Fundamentals of Operation* ; Butterworth Publishers, Boston (1983).

(8) Monod J., *Reserches sur les Croissances des Cultures Bacteri - ennes*, Hermann and Cie, Paris (1942).

(9) Eckenfelder, J., *Industrial Water Pollution control*, McGraw Hill Co., New York, (1966)

(10) G.Tchobanoglous, *Wastewater Engineering*, Metcalf & Eddy Inc., McGraw-Hill Book Company, (1979)

(11) JUAN A. ASENJO, *Separation Processes in Biotechnology*, Marcel Dekker, New York, (1990)

(12) M.C.Porter, *Handbook of Separation Technics*, McGraw-Hill Book Co, New York, (1988)

(13) Norman N. LI, *Perry's Chemical Engineers Handbook*, McGraw-Hill Book Co, New York, (1984)

Index

About the Authors

Dilip K. Das, B.S., M.S.,P.E., is pricipal process engineer for Ciba-Geigy where he serves as an interdivisional consultant with responsibilities including process design, simulation, systems engineering, technical support of production, cost control and project management. He has worked for Rhône-Poulenc, Inc., as senior project manager of agro-chemical projects, and for C F Braun Company as senior engineer. He has also worked for Stauffer Chemical Company as senior process engineer and for C. F. Braun Company as senior engineer. He received an M.S. in Chemical Engineering from the University of Washington and a B. Ch.E. (honors) in chemical engineering from Jadavpur University, Calcutta, where he received a gold medal for excellence. Mr. Das has several publications and a US patent to his credit.

Rajaram K. Prabhudesai, Ph.D., P.E., is senior process engineer with Stauffer Chemical Company, where he is responsible for process design, systems engineering, process economics, and plant start-up. Previously he worked for AMF as principal research engineer and for Coca Cola Company as principal chemical engineer. He has written numerous papers for professional journals and contributed to Perry's *Chemical Engineering Handbook* and Scliweitzer's *Handbook of Separation Techniques for Chemical Engineer* (both McGraw-Hill). Dr. Prabhudesai received his M.S. from the University of Bombay and his Ph.D. in chemical engineering from the Universitv of Oklahoma.